Sustainable Civil Infrastructures

About this Series

Sustainable Infrastructure impacts our well-being and day-to-day lives. The infrastructures we are building today will shape our lives tomorrow. The complex and diverse nature of the impacts due to weather extremes on transportation and civil infrastructures can be seen in our roadways, bridges, and buildings. Extreme summer temperatures, droughts, flash floods, and rising numbers of freeze-thaw cycles pose challenges for civil infrastructure and can endanger public safety. We constantly hear how civil infrastructures need constant attention, preservation, and upgrading. Such improvements and developments would obviously benefit from our desired book series that provide sustainable engineering materials and designs. The economic impact is huge and much research has been conducted worldwide. The future holds many opportunities, not only for researchers in a given country, but also for the worldwide field engineers who apply and implement these technologies. We believe that no approach can succeed if it does not unite the efforts of various engineering disciplines from all over the world under one umbrella to offer a beacon of modern solutions to the global infrastructure. Experts from the various engineering disciplines around the globe will participate in this series, including: Geotechnical, Geological, Geoscience, Petroleum, Structural, Transportation, Bridge, Infrastructure, Energy, Architectural, Chemical and Materials, and other related Engineering disciplines.

More information about this series at http://www.springer.com/series/15140

Murad Abu-Farsakh · Khalid Alshibli
Anand Puppala
Editors

Advances in Analysis and Design of Deep Foundations

Proceedings of the 1st GeoMEast International Congress and Exhibition, Egypt 2017 on Sustainable Civil Infrastructures

Springer

Editors
Murad Abu-Farsakh
Louisiana State University (LSU)
Louisiana, LA
USA

Anand Puppala
University of Texas at Arlington
Arlington, TX
USA

Khalid Alshibli
University of Tennessee, Knoxville
Tennessee, TN
USA

ISSN 2366-3405 ISSN 2366-3413 (electronic)
Sustainable Civil Infrastructures
ISBN 978-3-319-61641-4 ISBN 978-3-319-61642-1 (eBook)
DOI 10.1007/978-3-319-61642-1

Library of Congress Control Number: 2017946465

Printed on acid-free paper

This Springer imprint is published by Springer Nature
The registered company is Springer International Publishing AG
The registered company address is: Gewerbestrasse 11, 6330 Cham, Switzerland

Preface

Toward building sustainable and longer civil infrastructures, the engineering community around the globe continues undertaking research and development to improve existing design, modeling, and analytical capability. Such initiatives are also the core mission of the Soil-Structure Interaction Group in Egypt (SSIGE) to contribute to the ongoing research toward sustainable infrastructure. This conference series "GeoMEast International Congress and Exhibition" is one of these initiatives.

Ancient peoples built their structures to withstand the test of time. If we think in the same way, our current projects will be a heritage for future generations. In this context, an urgent need has quickly motivated the SSIGE and its friends around the globe to start a new congress series that can bring together researchers and practitioners to pursue "Sustainable Civil Infrastructures." The GeoMEast 2017 is a unique forum in the Middle East and Africa that transfers from the innovation in research into the practical wisdom to serve directly the practitioners of the industry.

More than eight hundred abstracts were received for the first edition of this conference series "GeoMEast 2017" in response to the Call for Papers. The abstracts were reviewed by the Organizing and Scientific Committees. All papers were reviewed following the same procedure and at the same technical standards of practice of the TRB, ASCE, ICE, ISSMGE, IGS, IAEG, DFI, ISAP, ISCP, ITA, ISHMII, PDCA, IUGS, ICC, and other professional organizations who have supported the technical program of the GeoMEast 2017. All papers received a minimum of two full reviews coordinated by various track chairs and supervised by the volumes editors through the Editorial Manager of the SUCI "Sustainable Civil Infrastructure" book series. As a result, 15 volumes have been formed of the final +320 accepted papers. The authors of the accepted papers have addressed all the comments of the reviewers to the satisfaction of the track chairs, the volumes editors, and the proceedings editor. It is hoped that readers of this proceedings of the GeoMEast 2017 will be stimulated and inspired by the wide range of papers written by a distinguished group of national and international authors.

Publication of this quality of technical papers would not have been possible without the dedication and professionalism of the anonymous papers reviewers. The names of these reviewers appear in the acknowledgment that follows. For any additional reviewers whose names were inadvertently missed, we offer our sincere apologies.

We are thankful to Dr. Hany Farouk Shehata, Dr. Nabil Khelifi, Dr. Khalid M. ElZahaby, Dr. Mohamed F. Shehata, and to all the distinguished volumes editors of the proceedings of the GeoMEast 2017. Appreciation is extended to the authors and track chairs for their significant contributions. Thanks are also extended to Springer for their coordination and enthusiastic support to this conference. The editors acknowledge the assistance of Ms. Janet Sterritt-Brunner, Mr. Arulmurugan Venkatasalam in the final production of the 15 edited volumes "Proceedings of GeoMEast 2017".

Contents

Evaluation of Ultimate Pile Compression Capacity from Static Pile Load Test Results

Kedar C. Birid(✉)

Toyo Engineering India Pvt. Ltd., Mumbai, India
kedar.birid@toyo-eng.com

Abstract. The in-situ static pile load testing is often carried out on the test pile by applying 1.5 times to 2.5 times the design pile capacity. Due to practical and time constraints, it is not possible to always load and test the pile up to the failure. The load-settlement behavior of the pile during pile load testing under this loading often does not reach the ultimate pile capacity. Hence, in order to utilize the maximum available pile capacity in the design of pile foundations, an extrapolation of load-settlement data are required to evaluate the ultimate pile capacity. Various methods were proposed in the past by researchers such as Chin, Decourt, Davisson, De Beer, Brinch Hansen etc. to evaluate extrapolated ultimate pile capacity. These methods have been adopted in this paper to estimate ultimate pile capacity using load-settlement data of 23 nos. static pile load tests performed on driven piles and drilled shafts. The ultimate capacities evaluated from different methods have been discussed and compared with each other and with ultimate pile capacities for piles tested up to the failure. Based on this comparison, it has been observed that when the test load is high and close to the ultimate load limit, the accuracy in estimated ultimate load can be achieved by all the methods. However, study on piles tested under very less or partial load revealed the overestimation of ultimate pile load except for the Davisson method. Each method estimated different values of ultimate load under different test loads and no specific method can be recommended based on accuracy to evaluate the ultimate pile capacity.

1 Introduction

For projects involving pile foundations, it is usually necessary to confirm the actual ultimate compression capacity of the pile with respect to the theoretical ultimate pile capacity. Often this is confirmed by performing a static load test on the test pile. The ultimate pile compression capacity can roughly be defined as the load for which rapid pile movement occurs under sustained or slight increase of the applied load or when the pile plunges. However, often distinct plunging ultimate load is not obtained during the test. Therefore, the pile ultimate capacity or failure load must be determined by some criterion using load-settlement data recorded in the test. Various researchers in past suggested different methods for evaluation of pile ultimate capacity.

As per Fellenius (2001), the concept of an ultimate load, a failure load or capacity, is a fallacy, and a design based on the ultimate load is a quasi-concept and of uncertain relevance for the pile assessment. An old definition of capacity has been the load for

M. Abu-Farsakh et al. (eds.), *Advances in Analysis and Design of Deep Foundations*,
Sustainable Civil Infrastructures, DOI 10.1007/978-3-319-61642-1_1

which the pile head movement exceeds a certain value, usually 10% of the diameter of the pile, or a given distance, often 1.5 inches. However, these definitions do not consider the elastic shortening of the pile, which can be substantial for long piles, while being negligible for short piles. In reality, a movement limit relates only to a movement allowed by the superstructure to be supported by the pile, not to the capacity of the pile in the static loading test. As such, the 10% or any other ratio to the pile diameter is meaningless from both the points-of-view of the pile-soil behavior as well as the structure. Similarly, 1.5-inch maximum movement criterion can be just right for the structure, but it has nothing to do with the pile-soil behavior. The maximum deflection at the design load depends on the sensitivity of the structure to the movement, the desired rigidity of the foundation, and local experience. However, it remains necessary to arrive at some acceptable value of ultimate capacity, based on some method, for the design purpose. As a result, practitioners and academicians have introduced various pile capacity interpretation methods. Some methods were established to decouple the effect of pile shaft stiffness from soil resistance, some that separate side shear from end bearing, and others to try to better understand the pile-soil behavior. Few of these methods are Davisson offset limit, the De Beer yield limit, Hansen ultimate load, the Chin-Kondner extrapolation, Decourt methods and Mazurkiewicz's graphical method. The 2003 International Building Code permitted the use of the original Davisson, Brinch-Hansen 90%, Chin-Kondner, and other methods approved by the building official. Pile Driving Contractors Association proposed an amendment to the 2006 International Building Code, and Chin-Kondner was replaced by the Butler-Hoy criterion. Commentary on the code change stated that extrapolation methods should be avoided (Perko 2009). Some of these methods have been considered in this paper to evaluate ultimate pile capacities using load settlement data of static pile load test conducted on different types of piles of various projects.

2 Methods to Evaluate Ultimate Pile Capacity

2.1 The Davisson Offset Limit Load

The ultimate load, as proposed by Davisson (1972), is the load corresponding to the movement that exceeds the elastic compression of the pile by a value of 0.15 inches (4 mm), plus soil quake, a factor equal to the diameter of the pile divided by 120 as shown in Fig. 1. Soil quake is the deformation (or pile movement) required to mobilize the strength of the soil below the pile tip (NeSmith and Siegel 2009). This method is probably the best known and widely used in North America and other regions worldwide because it provides the lowest estimate of axial compression capacity from the actual load-settlement curve without any requirement of extrapolation. The method is based on the assumption that capacity is reached at a certain small toe movement and tries to estimate that movement by compensating for the stiffness (length and diameter) of the pile. It is primarily intended for test results from driven piles tested according to quick methods. However, Davisson's method requires the pile to be loaded near failure to be applicable.

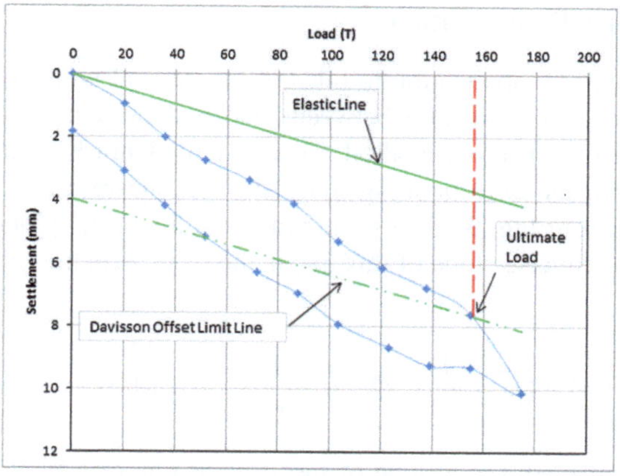

Fig. 1. Davisson's offset limit load method

2.2 The Hansen 80% Criterion (Fellenius 2001)

Hansen (1963) proposed a definition for pile capacity as the load that gives four times the movement of the pile head as obtained for 80% of that load. This '80%-criterion' can be estimated directly from the load movement curve, but is more accurately determined in a plot of the square root of each movement value divided by its load value and plotted against the movement as shown in Fig. 2. Following simple relations can be derived for computing the capacity or ultimate resistance, Qu, according to the Hansen 80%-criterion for the Ultimate Load:

$$Q_u = \frac{1}{2\sqrt{C_1 C_2}} \qquad Q_u = \frac{1}{2\sqrt{0.0006 * 0.0335}} = 111.52T$$

Where Qu = capacity or ultimate load, C_1 = slope of the straight line (see Fig. 2), C_2 = y-intercept of the straight line (see Fig. 2).

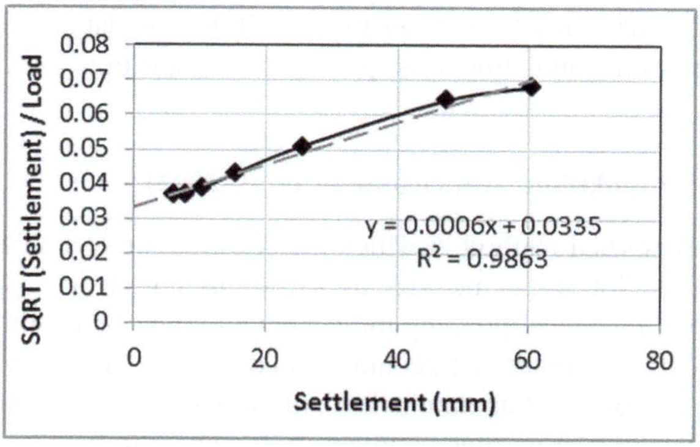

Fig. 2. Hansen's 80% criteria

2.3 Chin-Kondner Extrapolation

Chin (1970) proposed an application to piles of the general work by Kondner (1963). Chin assumes that the relationship between load and settlement is hyperbolic. The method is similar to the Hansen method. To apply the Chin-Kondner method, divide each settlement with its corresponding load and plot the resulting value against the settlement. As shown in Fig. 3, after some initial variation, the plotted values will fall on straight line. The inverse slope of this line is the Chin-Kondner Extrapolation of the ultimate load.

$$Q_u = \frac{1}{C_1} \quad Q_u = \frac{1}{0.0082} = 121.95T$$

Where Qu = applied load, C_1 = slope of the straight line (see Fig. 3).

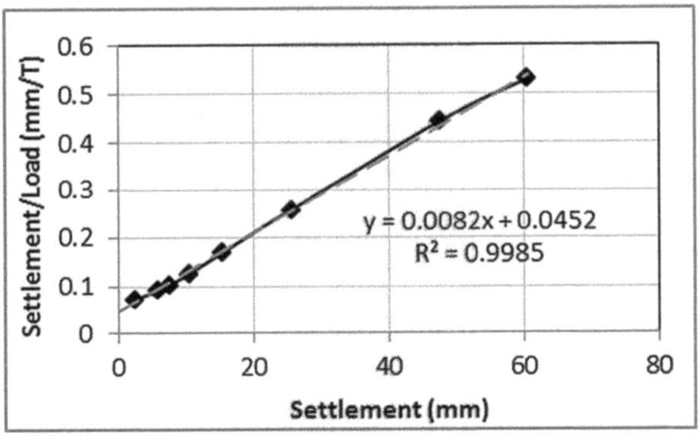

Fig. 3. Chin-Kondner method

Normally, the correct straight line does not start to materialize until the test load has passed the Davisson Offset Limit. As an approximate rule, the Chin-Kondner Extrapolation load is about 20% to 40% greater than the Davisson limit. When this is not a case, it is advisable to take a closer look at all the test data. The Chin method is applicable on both quick and slow tests, provided constant time increments are used.

2.4 Decourt Extrapolation (Abdelrahman et al. 2003)

Decourt (1999) proposes a method in which the construction is similar to that used in the Chin-Kondner and Hansen methods as shown in Fig. 4. To apply the method, divide each load with its corresponding movement and plot the resulting value against the applied load. The Decourt extrapolation load limit is the value of load at the intersection. The Decourt extrapolation load limit is equal to the ratio between the y-intercept and the slope of the line as given in the equation below.

$$Q_u = \frac{C_2}{C_1} \quad Q_u = \frac{24.796}{0.2061} \quad Q_u = 120.31T$$

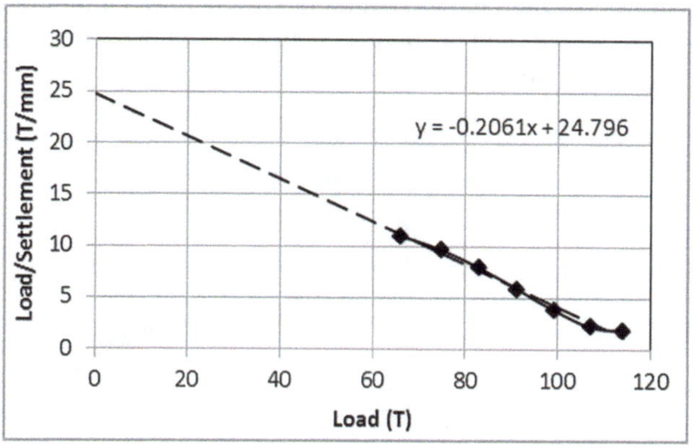

Fig. 4. Decourt method

2.5 Mazurkiewicz's Method (Abdelrahman et al. 2003)

This method is based on the assumption that the load–settlement curve is approximately parabolic. Series of equal pile head settlement lines are arbitrarily chosen using equal intervals and the corresponding loads are marked on the abscissa as shown in Fig. 5. For the marked loads on the load axis, a 45-degree line is drawn to intersect the next vertical line running through the next load point. These intersections fall approximately on a single straight line and the intersection of this line with the load axis defines the ultimate failure load.

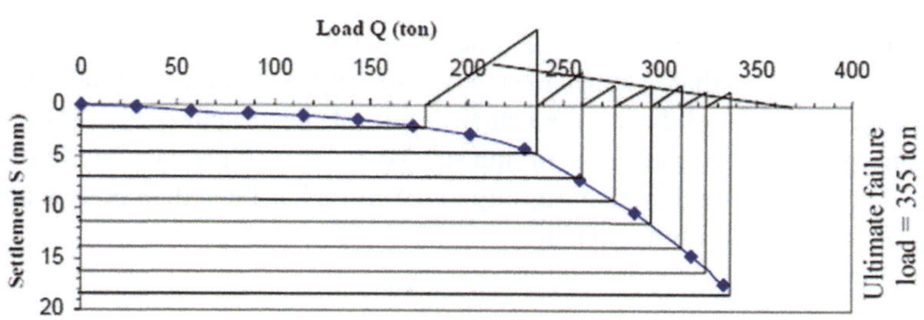

Fig. 5. Mazurkiewicz's method

2.6 De Beer Yield Load (Fellenius 2001)

De Beer (1968) made use of the logarithmic linearity by plotting the load-movement data in a double-logarithmic diagram as shown in Fig. 6. If the ultimate load is reached in the test, two line approximations will appear; one before and one after the ultimate load (provided the number of points allow the linear trend to develop). The slopes are meaningless, but the intersection of the lines is useful as it indicates where a change occurs in the response of the piles to the applied load. De Beer called the intersection the Yield Load. All previously mentioned methods determine a failure load except for De Beer's. Therefore, one should distinguish between the failure load and the limit load to adopt the proper factor of safety.

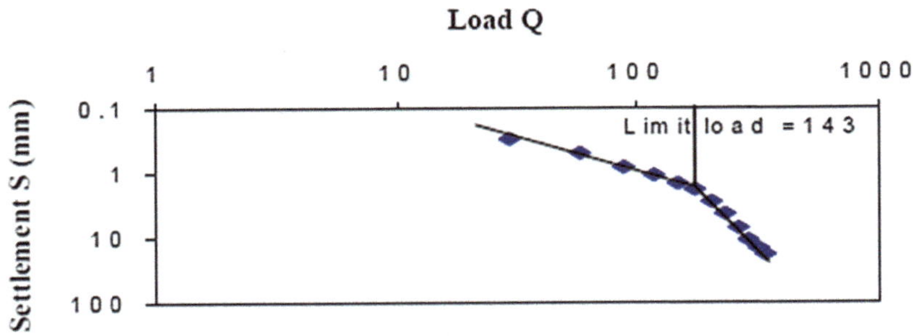

Fig. 6. De Beer method

3 Static Pile Load Test Data

The pile load-settlement data from different projects in different countries such as India, Egypt, Nigeria, Singapore and USA are analysed using different extrapolation methods explained in the previous section. The static load testing has been carried out on drilled and driven piles with different diameters and lengths. Table 1 summarises the load and settlement results all the test piles considered in this study.

The pile design capacity mentioned in Table 1 has been calculated based on the theory of plasticity with factor of safety of 2.5 or 3.0 depending on the probable accuracy of the Geotechnical report as per the author's experience. The test load has been applied in the range of 0.86 to 5.5 times the calculated design load. The reason for applying test load less than 2 times or as low as 0.86 times the design load is that the pile capacities considered during the actual design of the structure were less based on the local experience and loadings from the superstructure with respect to the calculated capacity. The test load with higher ratio up to 5.5 times has been deliberately applied to test the pile near to its ultimate value.

4 Results of Ultimate Pile Capacities

The estimated values of ultimate pile capacities from various extrapolation methods are summarised in Table 2.

It can be observed from Table 2 that Davisson's method is applicable just to four pile load tests, i.e. 17.4%, of all analysed cases, while Hansen's, Chin's,

Table 1. Load and settlement values of pile foundations used in the analysis

Sr. no.	Region	Type of pile	Pile diameter, mm	Length, m	Design load, T	Test load, T	Total settlement, mm	Net settlement, mm
1	Dahej, Gujarat, India	Driven cast in situ	600	15.00	54	105.30	3.51	0.91
2	Dahej, Gujarat, India	Driven cast in situ	600	15.00	54	79.88	2.97	1.04
3	Dahej, Gujarat, India	Driven cast in situ	600	15.00	93	79.88	2.66	0.77
4	Dahej, Gujarat, India	Drilled	500	13.80	32	174.85	10.11	1.81
5	Dahej, Gujarat, India	Drilled	500	18.00	32	123.19	8.44	1.56
6	Dahej, Gujarat, India	Drilled	500	18.00	32	79.46	5.73	2.05
7	Dahej, Gujarat, India	Drilled	500	12.00	34	63.58	2.09	0.49
8	Dahej, Gujarat, India	Drilled	500	18.00	56	123.19	6.04	2.53
9	Dahej, Gujarat, India	Drilled	600	21.36	54	107.30	5.94	1.95
10	Dahej, Gujarat, India	Drilled	500	13.23	35	77.94	7.04	1.99
11	Dahej, Gujarat, India	Drilled	500	13.53	38	63.58	5.58	2.45
12	Kakinada, A.P., India	Drilled	600	30.00	98	242.19	4.25	3.08
13	Kakinada, A.P., India	Drilled	600	30.00	98	180.00	3.23	2.16
14	Alexandria, Egypt	CFA	600	18.00	170	255.00	1.39	0.40
15	Alexandria, Egypt	CFA	600	18.00	170	340.00	2.57	0.73
16	Port Harcourt, Nigeria	Driven steel pile	508	28.00	100	193.00	37.16	20.73
17	Port Harcourt, Nigeria	Driven steel pile	508	53.00	95	252.60	30.65	5.62
18	Port Harcourt, Nigeria	PHC	400	28.00	78	114.10	60.50	50.00
19	Port Harcourt, Nigeria	PHC	400	28.00	78	112.50	40.70	33.64
20	Port Harcourt, Nigeria	PHC	400	28.00	78	73.43	50.00	50.00
21	Pulau Bukom in Singapore	Drilled	300	26.00	70	133.00	14.38	3.50
22	Port Comfort, Texas, USA	ACIP	600	21.33	70	360.00	5.38	1.83
23	Port Comfort, Texas, USA	ACIP	600	21.33	110	360.00	8.75	4.28

CFA: Continuous Flight Auger Pile
ACIP: Auger Cast in Place Pile
PHC: Pretensioned Hollow Concrete Pile (Driven)

Mazurkiewicz's, Decourt's and De Beer's methods were applicable to 17, 23, 23, 23 and 3 tests, representing 74%, 100%,100%, 100% and 13% respectively of all analysed cases. Consequently, the only three methods which are applicable to all pile load tests are Chin's, Mazurkiewicz's, and Decourt's methods.

5 Ulimate Capacities from Partial Load Data for Failed Piles

Out of five piles tested in the Nigeria project, four piles were loaded to failure during the testing. Often the piles are tested up to 1.5 to 2.0 times the design load, during which there are no signs of ultimate failure load. Hence, the test results of four failed piles are used to estimate the ultimate load using above various methods and the values are compared with the actual load at failure. For failed piles, the ultimate load can simply be found out from plunging curve as shown in Fig. 7a. However, if the test loading is done only up to 75T, the curve would be obtained as shown in Fig. 7b which

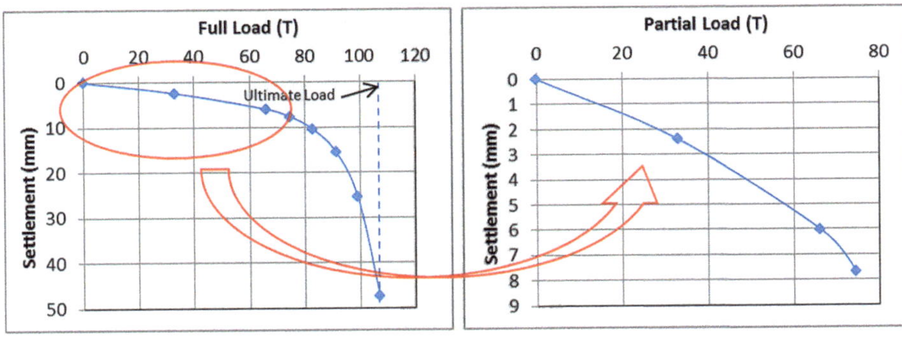

Fig. 7. Full load and partial load test curves

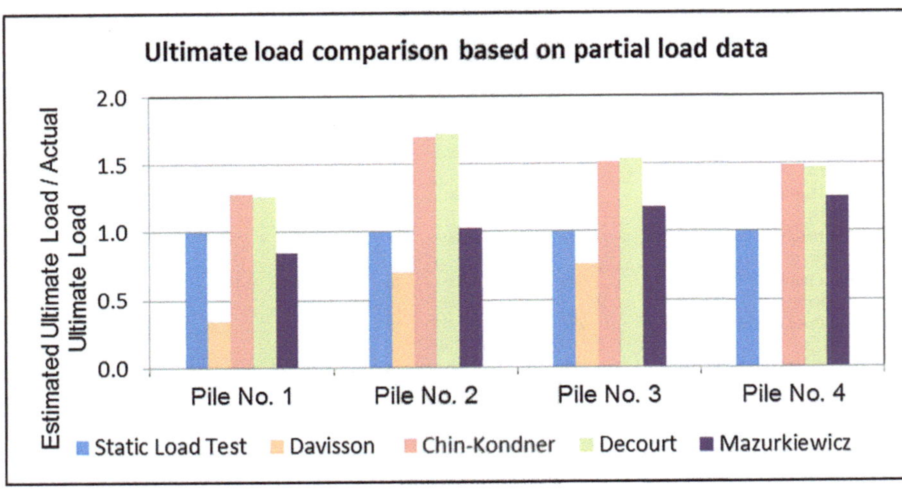

Fig. 8. Ultimate load evaluated from partial load test data for failed piles

does not show plunging failure. Similar load-settlement curves are drawn only with partial loads for 4 piles which have failed and the ultimate load are estimated using various methods. The ultimate load is then compared with the actual failure load to assess the accuracy of different methods for ultimate load estimation. The test results are shown in Table 3 and in Fig. 8.

It can be observed from Table 3 that the estimated ultimate load value based on partial test results is 13%–48% higher than those estimated based on full test data. The capacities estimated from Chin-Kondner and the Decourt method does not differ by a large extent. The capacities from Davisson's method are most conservative and are lower by 24% to 66% with respect to actual capacities.

Among all the methods, estimation by Mazurkiewicz's method is more accurate with variation ranging between −15% to +26%.

Table 2. Ultimate pile capacities from different methods

Sr. no.	Region	Type of pile	Methods of pile ultimate capacity, T						
			Static load test	Davisson	Chin-Kondner	Hansen 80%	Decourt	Mazurkiewicz	De Beer
1	Dahej, Gujarat, India	Driven cast in situ	NA	NA	158.73	154.30	145.79	160	–
2	Dahej, Gujarat, India	Driven cast in situ	NA	NA	153.84	146.17	138.15	140	–
3	Dahej, Gujarat, India	Driven cast in situ	NA	NA	144.93	114.28	122.35	100	–
4	Dahej, Gujarat, India	Drilled	NA	155	357.14	274.41	386.05	250	–
5	Dahej, Gujarat, India	Drilled	NA	120	227.27	197.79	286.99	210	–
6	Dahej, Gujarat, India	Drilled	NA	NA	133.33	92.45	150.21	98	–
7	Dahej, Gujarat, India	Drilled	NA	NA	98.03	102.57	113.01	80	–
8	Dahej, Gujarat, India	Drilled	NA	NA	181.81	172.00	193.02	155	–
9	Dahej, Gujarat, India	Drilled	NA	NA	169.49	116.69	164.67	120	–
10	Dahej, Gujarat, India	Drilled	NA	54	344.82	–	366.93	94	–
11	Dahej, Gujarat, India	Drilled	NA	63	113.63	67.96	130.01	76	–
12	Kakinada, A.P., India	Drilled	NA	NA	294.11	246.03	282.55	280	–
13	Kakinada, A.P., India	Drilled	NA	NA	285.71	190.34	301.81	200	–
14	Alexandria, Egypt	CFA	NA	NA	1000.00	–	1059.68	900	–
15	Alexandria, Egypt	CFA	NA	NA	1111.11	–	1070.09	1000	–
16	Port Harcourt, Nigeria	Driven steel pile	NA	NA	256.41	233.12	261.69	205	–
17	Port Harcourt, Nigeria	Driven steel pile	NA	NA	588.23	–	612.06	400	–

(*continued*)

Table 2. *(continued)*

Sr. no.	Region	Type of pile	Methods of pile ultimate capacity, T						
			Static load test	Davisson	Chin-Kondner	Hansen 80%	Decourt	Mazurkiewicz	De Beer
18	Port Harcourt, Nigeria	PHC	105	NA	121.95	111.52	120.31	110	85
19	Port Harcourt, Nigeria	PHC	112	NA	120.48	117.65	113.21	120	120
20	Port Harcourt, Nigeria	PHC	73	NA	74.62	81.71	88.50	76	74
21	Pulau Bukom, Singapore	Drilled	NA	NA	416.67	159.39	497.19	360	–
22	Port Comfort, Texas, USA	ACIP	NA	NA	1111.11	–	1158.82	650	–
23	Port Comfort, Texas, USA	ACIP	NA	NA	2500.00	–	2730.92	500	–

Hence, it is suggested to test the pile to maximum possible load and near to ultimate load, instead of evaluating ultimate load from test with very less test load, as such test does not indicate any signs of failure and tends to overestimate the pile ultimate load.

6 Discussion and Choice of Evaluation Method

Davisson's assumption that an offset of 3.8 mm + D (inches)/120 from the elastic line represents the movement necessary to mobilize toe resistance. The soil quake proposed by Davisson is specifically for driven piles and is not appropriate where soil resistance beneath the pile toe has not been fully mobilized at the beginning of load testing. The Davisson study evaluated piles installed by driving where a compressed soil plug forms during placement. In contrast, cast-in-place piles and other types of drilled shafts do not compress the soil beneath the pile toe during installation. Thus, a greater downward movement of the pile toe would be required to mobilize the end resistance for cast-in-place piles if all other conditions are equal.

Analysis by Zheng et al. (2007) confirmed this based on the results of load tests performed on the displacement cast-in-place piles (NeSmith and Siegel 2009).

It is difficult to make a rational choice of the best capacity criterion to use, because the preferred criterion depends heavily on the extent of test load, one's past experience and conception of what constitutes the ultimate resistance of a pile.

The Davisson Offset Limit is very sensitive to errors in the measurements of load and movement and requires well maintained equipment and accurate measurements. This method offers the benefit of allowing the engineer, when proof testing a pile for a certain allowable load, to determine in advance the maximum allowable movement for this load with consideration of the length and size of the pile. However, this method has failed to provide the ultimate capacity in this case study as the pile settlement has hardly approached near yield or ultimate load during the testing. The Davisson offset line is thus unable to intercept the load-settlement curve which is required to evaluate ultimate load.

The Brinch-Hansen 80%-criterion usually gives a Qu-value, which is close to what one subjectively accepts as the true ultimate resistance, determined from the results of the static loading test. The value is smaller than the Chin-Kondner value. These two methods are always obtained by extrapolation.

Mazurkiewicz and Hansen ultimate load methods indicate the most conservative results, less than the values obtained using Davisson, Chin, and Decourt methods. It is simple in its construction, more reliable, especially for piles loaded near failure. However, Hansen method has also failed to provide ultimate load in some cases. Mazurkiewicz ultimate load values are on average 51% lesser than Chin's ultimate load.

Chin's method is affected by the limit of loading as the pile is loaded near failure, the greater predicted value of ultimate load. It has also been found that the Chin–Kondner extrapolation ultimate load is 80% to 500% greater than the Davisson ultimate load. Chin–Kondner extrapolation ultimate load is also 22% greater than Hansen's ultimate Load on average for the 15 pile load tests. A review of the load test data available to the author indicates that at pile-head deflections of about 5% of the pile

Table 3. Pile ultimate load based on full load test and partial load test data

Method of analysis	Pile no. 1			Pile no. 2			Pile no. 3			Pile no. 4		
	From full load, Tons	From partial load, Tons	Actual/ estimated	From full load, Tons	From partial load, Tons	Actual/ estimated	From full load, Tons	From partial load, Tons	Actual/ estimated	From full load, Tons	From partial load, Tons	Actual/ estimated
Failure load from load test	195.0	–	–	105.0	–	–	112.0	–	–	73.0	–	–
Davisson	68.0	67.0	0.34	75.0	74.0	0.70	85.0	85.0	0.76	72.0	–	–
Chin-Kondner	256.4	250.0	1.28	121.9	178.6	1.70	120.5	169.5	1.51	74.6	108.0	1.49
Hansen 80%	233.1	EG	–	111.5	EG	–	117.7	EG	–	81.7	EG	–
Decourt	261.7	247.0	1.27	120.3	181.1	1.73	113.2	172.4	1.54	88.5	107.8	1.48
Mazurkiewicz	205.0	165.0	0.85	110.0	107.0	1.02	120.0	132.0	1.18	76.0	92.0	1.26

EG: Erratic graph

diameter, sufficient mobilization of the pile toe has occurred to allow for a reasonable extrapolation according to Chin's method.

The Decourt method has an advantage that a plot prepared, while the static loading test is in progress, allows the user to 'eyeball' the projected capacity directly once a straight-line plot starts to develop. Extrapolation, by this method, shows ultimate load 4% lesser than Chin's method.

The De Beer method requires the pile to be loaded near failure, otherwise the plotted values of the load settlement fall on approximately one straight line and the limit load is not defined (Abdelrahman et al. 2003).

It is a sound engineering rule never to interpret the results from a static loading test to obtain an ultimate load larger than the maximum load applied to the pile in the test. For this reason, the allowable load cannot and must not be determined by dividing the limit loads according to Chin-Kondner and Decourt methods with a factor of safety.

7 Conclusions

The analysis results into an observation that irrespective of the type of method, more the pile loaded near the ultimate load, more accurate estimation of the ultimate load can be made. If the test load is very less compared to the ultimate load, variation in ultimate load can be obtained by different methods. Hence, no conclusion can be reached about the suitability of method for ultimate load evaluation.

There is a considerable variation in the methods of "Failure Load" interpretation used in the industry.

For small and non-complex projects, such level of sophistication or lack thereof, is acceptable if the uncertainty is covered by a judiciously large factor of safety. For larger projects, however, this approach is costly. For these, the test pile should be instrumented and the test data evaluated carefully to work out the various influencing factors.

Combining an instrumented static loading test with dynamic testing, which can be performed on many piles at a relatively small cost, can extend the application of the more detailed results of the instrumented static test.

Design of pile foundation shall not be completely based on the capacity value and more emphasized on the settlement of the pile under sustained load.

Pile capacity determined from a predefined maximum deflection can depend mainly on the structural properties and elasticity of the pile, and can have less to do with the pile-soil behavior. Additionally, load test equipment utilized in pile load testing is limited in the amount of load that can be applied to the pile and often cannot reach the full ultimate capacity of the pile.

Only three methods applicable to all tests used in the study are Chin's, Mazurkiewicz's, and Decourt's methods. Davisson and De Beer methods need the pile to be loaded to failure to be applicable and thus cannot rely upon for non-failed piles.

As per England (1994) and England and Fleming (1994), all pile testing methods for determining bearing capacity, from a continuous rate of penetration test to wave analysis system, appear to introduce complications related to inability of soils to reach a stable state in terms of effective stress during the load period. Hence, no specific method of failure load estimation is workable under all the circumstances.

References

Abdelrahman, G.E., Shaarawi, E.M., Abouzaid, K.S.: Interpretation of axial pile load test results for continuous flight auger piles. In: Proceedings of the 9th Arab Structural Engineering Conference, 2003, Abu Dhabi, UAE, pp. 791–801 (2003)

Chin, F.K.: Estimation of ultimate load of piles not carried to failure. In: Proceedings, 2nd Southeast Asia Conference on Soil Engineering, pp. 81–92 (1970)

Davisson, M.T.: High capacity piles. In: Proceedings of the Soil Mechanics Lecture Series on Innovations in Foundation Construction, ASCE, Illinois Section, Chicago, pp. 81–112 (1972)

De Beer, E.E.: Proefondervindlijke bijdrage tot de studie van het grensdraag vermogen van zand onder funderingen op staal. Tijdshift der Openbar Verken van Belgie, No. 6, 1967 and No. 4, 5, and 6 (1968)

Decourt, L.: Behavior of foundations under working load conditions. In: Proceedings of the 11th Pan-American Conference on Soil Mechanics and Geotechnical Engineering, Foz DoIguassu, Brazil, August 1999, vol. 4, pp. 453–488 (1999)

England, M.: New techniques for reliable pile installation and pile behavior design and analysis. In: Transportation Research Record, Issue Number: 1447, pp. 39–48. Transportation Research Board Publisher (1994)

England, M., Fleming, W.G.K.: Review of foundation testing methods and procedures. Proc. Inst. Civ. Eng. Geotech. Eng. **107**, 135–142 (1994)

Fellenius, B.H.: What capacity value to choose from the results a static loading test. We have determined the capacity, then what? Two articles reprinted from Deep Foundation Institute, Fulcrum, Winter 2001, pp. 19–22 and Fall 2001, pp. 23–26 (2001)

Hansen, J.B.: Discussion on hyperbolic stress-strain response. Cohesive soils. J. Soil Mech. Found. Eng. **89**(SM4), 241–242 (1963). (American Society of Civil Engineers, ASCE)

Kondner, R.: Hyperbolic stress-strain response of cohesive soils. J. SMFD **89**(SM1), 115–143 (1963). (ASCE)

NeSmith, W., Siegel, T.: Shortcomings of the Davisson offset limit applied to axial compressive load tests on cast-in-place piles. In: Proceedings of Contemporary Topics in Deep Foundations, International Foundation Congress and Equipment Expo 2009, Orlando, Florida, United States, March 15–19, 2009, pp. 568–574 (2009)

Perko, H.A.: Helical piles, pp. 205–214. Wiley, London (2009)

Zheng, W., Hart, T.P., Roldan, R.A.: Load test analysis on augered pressure grouted displacement piles. In: Proceedings of the 32nd Annual Conference on Deep Foundations, Colorado Springs, Deep Foundations Institute, pp. 25–36 (2007)

Behavior of Bored Piles in Two Soil Layers, Sand Overlaying Compressible Clay (Case Study)

Adel H. Hammam(✉) and A.E. Abdel Salam

Housing and Building National Research Center, Cairo, Egypt
adelhamam@yahoo.com

Abstract. Piles foundations have the function of transferring loads from the superstructure through weak compressible strata onto stiffer or more compact soils or onto rock. This is soil profile is considered the ideal profile for pile foundations. While for soil profile including strong soil strata overlaying compressible soils, the estimation of piles behavior will be more difficult. Forty six residential buildings consisted of twelve stories were constructed at the north of Nile delta in Egypt. Bored piles with lengths equal to or more than 27.0 m were chosen to support the raft foundations of these buildings. Soil investigations concluded that the soil consists of dense sand overlaying compressible clay that extended down to 40.0 m, while the ground water was found near ground surface. Soil properties were determined through drilling four boreholes at the site with 60.0 m depth and disturbed and undisturbed samples were collected. Six cone penetration tests (CPTU) with 25.0 m depth were achieved to estimate soil properties and also to estimate bearing capacity of the piles. Five preliminary pile load tests were carried out on different types of piles. Four rotary drilling bored piles and one contentious flight auger were tested having lengths ranged between 27.0 m and 37.0 m, while its diameter ranged between 0.60 m and 0.7 m. The static load tests continued to load equal to 200 t which is equal to 250% the allowable pile load. Some tests completed within 48 h and some within 7 days to check the time-load-effect. In the light of these measurements, comparisons have been achieved between the bearing capacities of piles estimated by the results of CPTU and those measured by field tests. Moreover, the ultimate bearing capacities of piles estimated by Egyptian and Canadian Codes have been compared with those measured by field tests. Good agreement was noticed between the bearing capacity of piles estimated by methods depended directly on CPTU. While the indirect methods depended on soil parameters derived from CPTU produced lower values than the measured. It was also noticed that the estimated bearing capacities of piles based on soil properties derived from lab tests did not soundly match with the measured values.

1 Introduction

Pile foundations are considered more expensive relatively to shallow foundations, so it should be only selected when shallow foundations cannot satisfy an acceptable factor of safety against bearing failure in the foundation soils or acceptable settlements during

M. Abu-Farsakh et al. (eds.), *Advances in Analysis and Design of Deep Foundations*,
Sustainable Civil Infrastructures, DOI 10.1007/978-3-319-61642-1_2

the life of the structure. These two independent design criteria can be arise due several reasons among them are nature and magnitude of structural loads, settlement-sensitive structures and type of soil. Hence, the main goal of piles is to achieve the design criteria by transferring structural loads to the deep, stable and strong soil strata. Sometime the case can be more complex when the stronger soil layers with relatively small thickness exist above weak soil that extended to great depths. In this case, the methodology of using pile foundations includes some uncertainties and the behavior of piles cannot be certainly expected. The uncertainties of pile behavior arise from non-homogeneity of soil and alteration of soil properties after piles installation. Moreover, most of the rules or approaches for estimating the bearing capacity of piles took in consideration the general case that weak soil is overlaying strong soil. As pile behavior depends mainly on soil properties and method of pile instillation, great efforts have been achieved to determine soil properties throughout lab and field tests. Cone penetration test (CPTU) has been widely used for several decades because it is the most effective in-situ test method for obtaining practically continuous soil properties reliably. Data from the CPT can be used directly in foundation design or in the estimation of soil parameters. Un-drained shear strength (s_u) and angle of shear resistance (Φ) are the most important quantities for geotechnical design in clay and sand respectively. It is knowledge that north of Nile delta in Egypt contains soft clays. Soft clay exists sometimes near ground surface overlying stiff clay or sand and sometimes it exists at relatively deep depths underlying sand. At Port-Said area, dense sand with thickness 8.0 m to 10.0 m is existed overlying compressible clay which extended down to about 40 m while the ground water was found near ground surface. On this soil profile, forty six residential buildings consisted of twelve stories were constructed at Port-Said area, north of Nile delta in Egypt. Soil stratification at the site failed to support shallow foundations with acceptable and allowable settlements. So, bored piles with lengths ranged between 27.0 m and 34.0 m were chosen to support the raft foundations of these buildings. Several preliminary pile load tests were carried out up to load equal to 200 t. Some tests completed within 48 h and some within 7 days to check the time-load-effect. Six CPTU with depth 25.0 m and four boreholes with depth 60.0 m were carried out at the site.

This paper shows comparison between the bearing capacities of piles estimated by the results of CPTU and those measured by field tests. Moreover, the ultimate bearing capacities of piles estimated by several methods included Egyptian and Canadian Codes have been compared with the measured values. The differences between the behaviors of bored pile and CFA pile were also discussed.

2 Cone Penetration Tests, CPTU

The cone penetration test has been recognized as one of the most widely used in situ tests. Cone penetration testing has gained rapid popularity in the past twenty years. The cone penetration test consists of advancing a cylindrical rod with a conical tip into the soil and measuring the forces required to push this rod. The friction cone penetrometer measures two forces during penetration. These forces are: the tip resistance (q_c), which is the soil resistance to advance the cone tip and the sleeve friction (f_s), which is the sleeve friction developed between the soil and the sleeve of the cone penetrometer. The friction ratio (R_f) is defined as the ratio between the sleeve friction and tip resistance and is expressed in percent. The penetrometer is capable of registering pore water pressure (u_2) induced during advancement of the penetrometer tip using an electronic pressure transducer. These measurements are measured by electrical methods, at a minimum of every 50 mm of penetration. The resistance parameters are used to classify soil strata and to estimate strength and deformation characteristics of soils.

Six CPTU were carried out at the site down to depths 25.0 m below ground surface. The standard CPTU (according to ASTM D 5778 with cross-section area of 10 cm^2) was performed by pushing the cone into the ground at a rate of 20 mm/s. Data of the tip resistance q_c, sleeve friction f_s, and pore water pressure u_2, are collected every 12 mm penetration using electric data acquisition equipment and a portable computer. Figures 1 and 2 shows the measured and derived parameters respectively for CPTU # 3 for example.

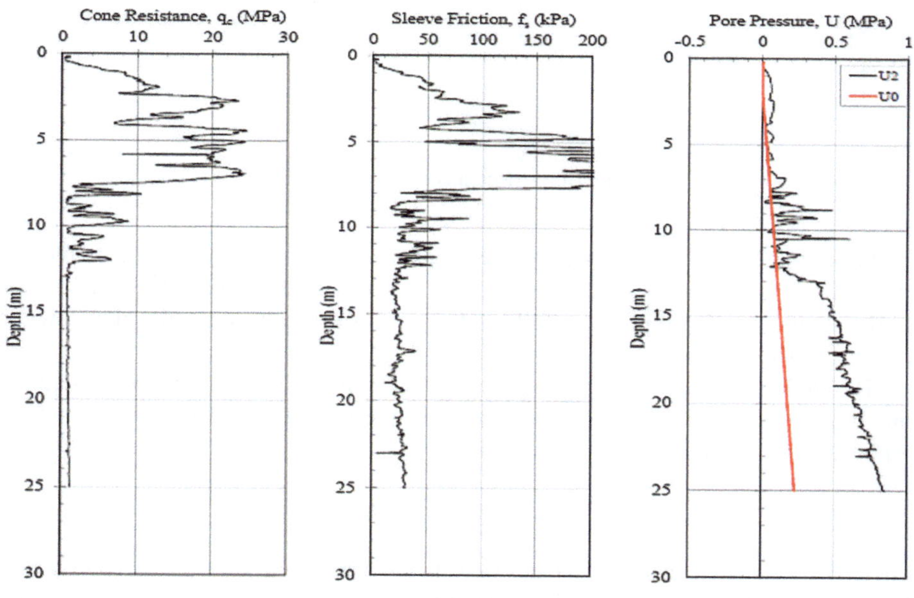

Fig. 1. Measured parameters for CPTU # 3

2.1 Soil Classification and Soil Parameters

There are several methods for soil classification, each method took in consideration certain parameters.

Methods of Robertson et al. (1986) and Jefferies and Davies (1993) are considered the famous ones that depends on the three measured parameters, q_c, f_s and u_2 besides total σ_v, effective σ'_v overburden pressures and in situ pore water pressure u_o. Accordingly, soil in the site could be classified as follows:

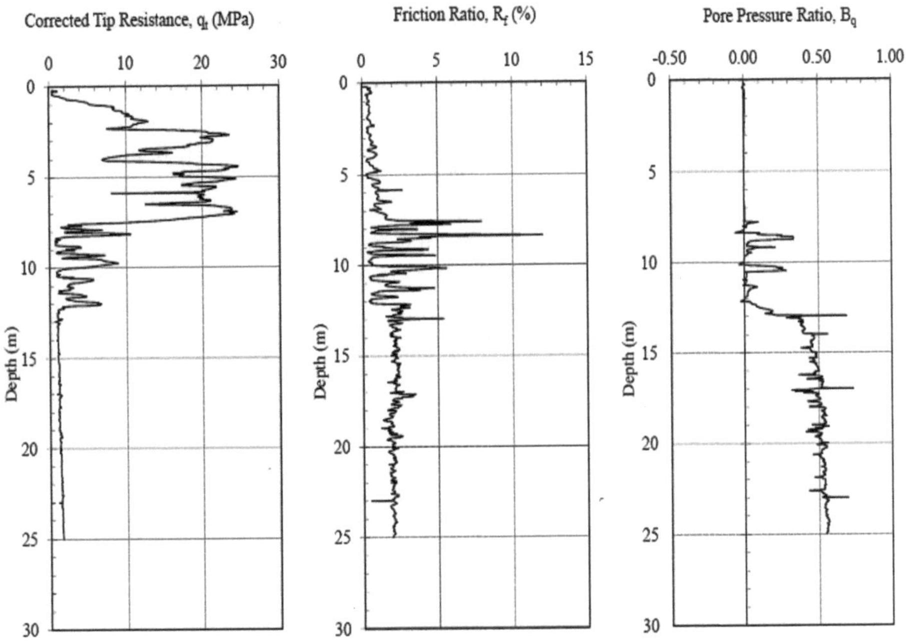

Fig. 2. Derived parameters for CPTU # 3

From 0 to 2.0 m top soil layers of intercalated silty clay and sand mixtures followed by sand layers down to 8.0 m. From 8.0 m down to 12.0 m intercalated silty sand, sandy silt and clayey silt mixtures followed by firm (soft to medium) silty clay down to 25.0 m, as shown in Fig. 3.

The undrained shear strength ($s_u = c_u$) of the soil is commonly used for stability and bearing capacity analyses. The classical approach to evaluating s_u from CPT readings is via the corrected cone resistance:

$$S_u = (q_t - \sigma_v)/N_k \quad (1)$$

where q_t = corrected cone resistance, σ_v = total overburden stress N_k = cone factor. Knowledge of the cone factor N_k is essential for reliable estimation of s_u, and numerous attempts have been made by researchers to develop accurate N_k values by empirical approaches (Lunne and Kleven 1981; Aas et al. 1986; Lunne et al. 1986; Stark and Juhrend 1989). In general, most of the researchers agreed that N_k values ranged between 15 and 20.

The friction angle Φ of sand could be also estimated through different methods (Meyerhof 1976; Robertson and Campanella 1983; Kulhawy and Mayne 1990). The results of (s_u and Φ) are shown in Fig. 3.

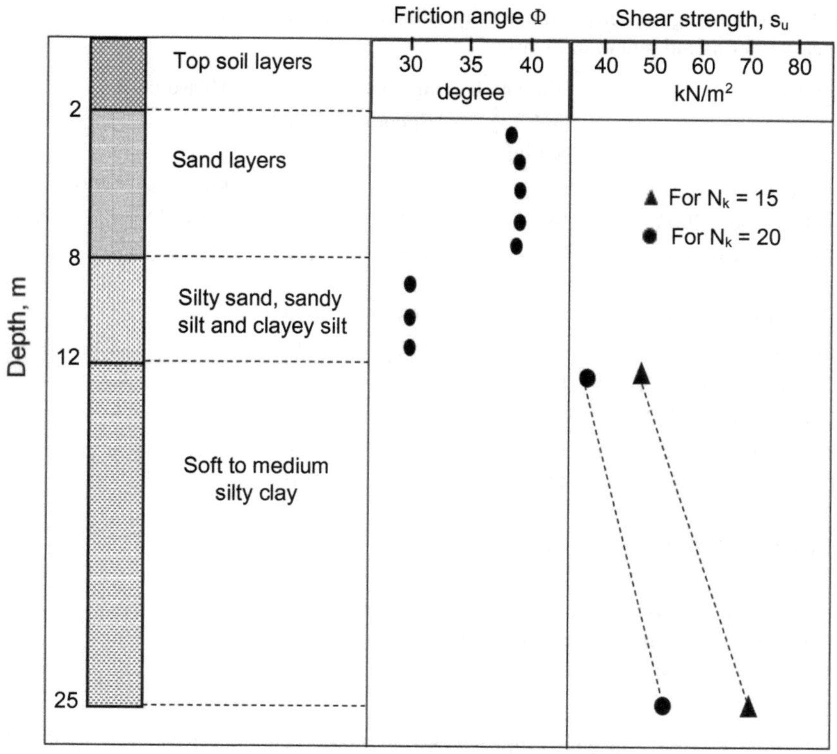

Fig. 3. Shows the derived soil classification and soil parameters from CPTU

3 Estimation of Pile Bearing Capacity

Two methods were used to estimate pile bearing capacity, static formula and cone penetration test. Static formula used soil parameters estimated from laboratory or derived from CPT data. Two approaches for application of cone data to pile design have been evolved, direct and indirect methods. Direct methods more or less equal the cone resistance with the pile end resistance. Some methods use the cone sleeve friction in determining the pile shaft capacity. Indirect methods employ soil parameters estimated from cone data as based for estimating pile bearing capacity. Although the site under study with its complicated profile is located in Egypt, the Egyptian code for foundations did not give direct approach depending on CPT to estimate pile bearing capacity for similar soils. Hence, the Canadian code was used for this estimation. Tables 1 and 2 show the estimated values of pile bearing capacities for different methods.

Table 1. Pile bearing capacity estimated by different methods (rotary drilling bored piles with diameter = 0.60 m and length = 27 m)

The method	Ultimate bearing capacity (t)			Measured (t)
	End bearing	Friction	Total	
*Egyptian Code, 2001	18	78	96	Load test = 200 Ultimate load ≈ 223 t to 244 t
*Canadian Code, 2006	14	134	157	
*de Ruiter and Beringen (1979)	14	138	152	
**Egyptian Code, 2001	11	63	74	
**Canadian Code, 2006	9	105	114	
***Canadian Code, 2006	15	215	230	
***Schmertmann (1978)	42	227	269	
***Bustamente (1982)	15	215	230	
***Tumay and Fakhroo (1981)	42	163	205	
***Eslami and Fellenius (1997)	16	238	254	

Notes: * Depends on soil properties derived from CPT
** Depend on soil properties derived from laboratory tests
*** Depend on q_c and f_c

Table 2. Pile bearing capacity estimated by different methods (CFA pile with diameter = 0.70 m and length = 27 m)

The method	Ultimate bearing capacity (t)			Measured (t)
	End bearing	Friction	Total	
*Egyptian Code, 2001	24	91	115	Load test = 200 Ultimate load ≈ 335
*Canadian Code, 2006	19	156	175	
*de Ruiter and Beringen (1979)	19	161	180	
**Egyptian Code, 2001	15	73	88	
**Canadian Code, 2006	12	122	134	
***Canadian Code, 2006	20	250	270	
***Schmertmann (1978)	57	265	322	
***Bustamente (1982)	20	250	270	
***Tumay and Fakhroo (1981)	57	190	247	
***Eslami and Fellenius (1997)	22	278	300	

Notes: * Depends on soil properties derived from CPT
** Depend on soil properties derived from laboratory tests
*** Depend on q_c and f_c

4 Pile Load Tests

Six preliminary pile load tests were carried out before starting the project to verify the calculated ultimate and allowable bearing capacities of piles. Two bored piles with length 34 m were excluded because they did not match with depth of CPTU. Three bored piles had diameter 0.60 m and length of 27.0 m while one CFA pile had diameter 0.70 m and depth of 27.0. The piles were loaded up to 200 t which equal to 250% allowable load and the tests completed within different durations ranged between 48 h and 7 days to check the time-load-effect, Fig. 4 shows load-settlement curves. Two different methods were used for estimating the ultimate loads, Modified Chin Method and Brinck Hansen Method as recommended by Egyptian code. The expected ultimate loads for the piles were reported in Tables 1 and 2.

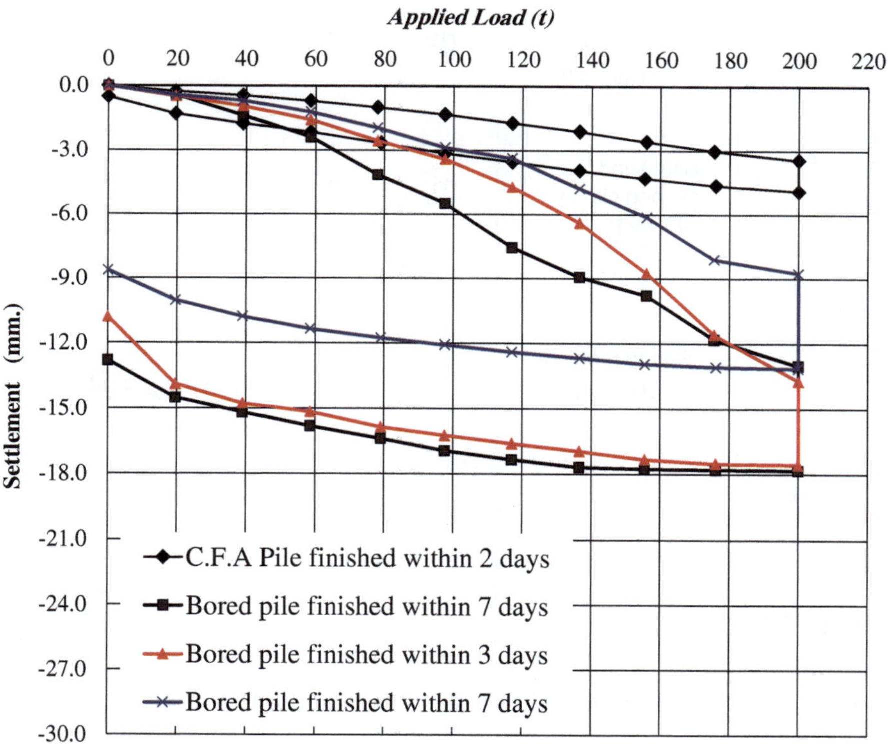

Fig. 4. Shows load-settlement curves for piles with length of 27.0 m

5 Discussion of Tests Results

5.1 Soil Parameters

Soil parameters (Φ and c_u) are considered the best important parameters for estimating the bearing capacity of foundations. The accuracy the determination of soil parameters is the accuracy of foundations bearing capacity. These parameters are usually determined from different lab tests that depend mainly on the accuracy of extracted undisturbed samples. Although well-controlled undisturbed samples were achieved during drilling works in our case study, the disturbed behavior dominated which could be noticed on the shape of *e – log б′* curves for consolidation test. The values of c_u for silty clay layer were highly affected by the samples disturbance. It can be noticed that values of (Φ and c_u) determined from lab tests are obviously lower than those derived from CPT, see Fig. 5.

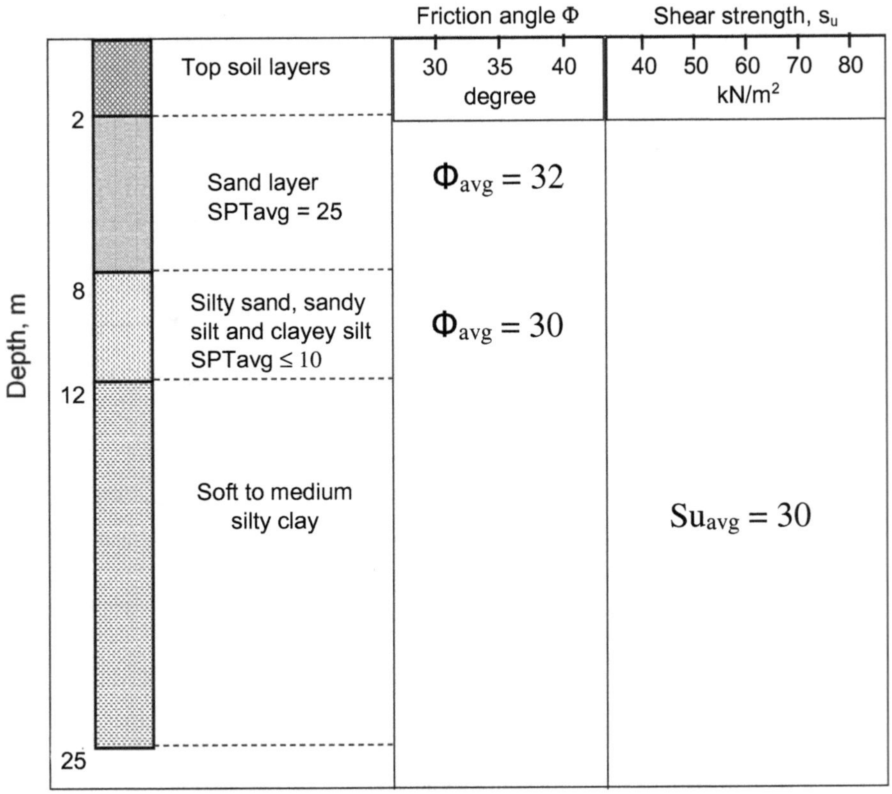

Fig. 5. Shows soil parameters determined from lab tests

5.2 The Ratio c_u/σ'_v

Due the difficulties of determining of c_u in lab especially for soft clay, several empirical relationships have been proposed between c_u and the effective overburden pressure σ'_v (Bowles 1984; Jamiolkowski et al. 1985; Bazaraa et al. 1986). Most of the correlations agreed that the ratio c_u/σ'_v range between 0.22 and 0.30 for most soft clays. These correlations depended on estimating undrained shear strength by different lab tests. For our case study, this ratio ranged between 0.2 and 0.25. According to the data of CPTU, the ratio c_u/σ'_v ranged between 0.3 and 0.4 with an average value 0.37. This improvement of c_u could be attributed to two factors, the first is to the limited disturbance of soil during cone tests and the second is to the existence of dense sand overlaying soft clay.

5.3 Pile Bearing Capacity

Although the ultimate bearing capacities estimated by methods depending on soil parameters derived from CPT are obviously lower than those measured from pile load tests, but they are still better than those estimated by methods depending on soil properties derived from laboratory tests, see Tables 1 and 2. On the other hand, the methods that depended directly on q_c and f_c produced good agreement with the measured ultimate bearing capacities. The Egyptian code gave the lowest values although its approach is similar to that in the Canadian code. The reason is that Egyptian code put some precautions for bored pile installation led to decrease the values of soil parameters Φ and c_u. It could be also noticed that CFA pile produced relatively higher ultimate bearing capacity than those of bored piles as shown in Table 2 and Fig. 4.

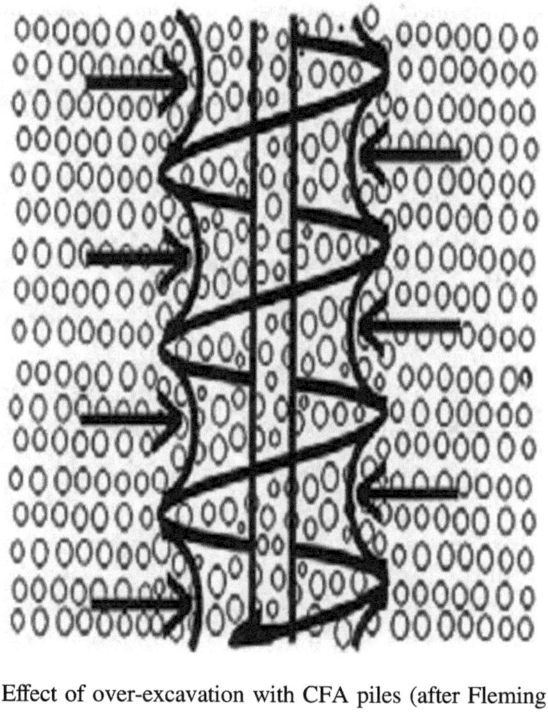

Fig. 6. Effect of over-excavation with CFA piles (after Fleming 1995)

5.4 CFA Methods in Sand Overlying Firm Clay

The key component of the CFA pile system, contributing to the speed and economy of these piles, is that the pile is drilled in one continuous operation using a continuous flight auger, thus reducing the time required to drill the hole. While advancing the auger to the required depth, it is essential that the auger flights be filled with soil so that the stability of the hole is maintained. If the auger turns too rapidly (over-rotation), with respect to the rate of penetration into the ground, then the continuous auger conveys soil to the surface. This action can result in a reduction of the horizontal stress necessary to maintain stability of the hole, lateral movement of soil towards the hole, ground subsidence at the surface and reduced confinement of nearby installed piles, see Fig. 6 (Brown 2005).

Brown et al. (2007) reported that the installation of CFA piles can be problematic in the very soft soils. In these soils, the installation of CFA piles can present problems concerning ground stability due to soft-ground conditions, which can produce necks or structural defects in the pile. Also if hard soil is overlain by soft soil, the installation of CFA piles is typically difficult. The problem occurs when the hard stratum is encountered and the rate of penetration is slowed with respect to rotation because of the difficult of drilling. Accordingly, decompression of the ground inside the hole, loosening of the in-situ soil around the hole and ground subsidence adjacent to the pile are expected, see Fig. 7.

In our case study the situation is conversely, CFA have balanced auger rotation and penetration rates in the upper dense sand. As soon as the auger reaches the underlying

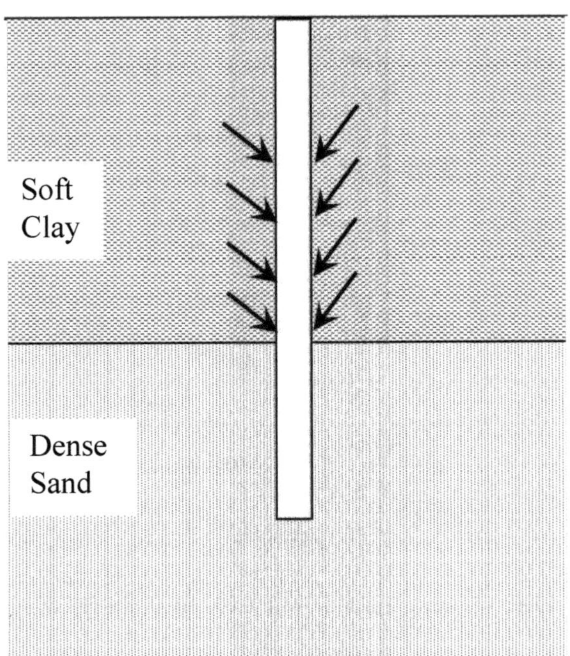

Fig. 7. Difficult conditions for CFA piles (hard soil overlain by Soft clay)

compressible clay its penetration rate will increase with respect to the rotation (over-penetration). This over-penetration is considered the best situation for CFA and the defects that can be occur in soft soils overlying hard soil are limited.

Whatever, it can attribute the better behavior and the higher ultimate bearing capacity of CFA pile than those of bored piles to the contentious drilling of CFA and placing the concrete by pumping under pressure.

5.5 Time-Load-Effect

The assessment of creep resistance of pile is particularly relevant in the design of raft foundations supported on piles in poor ground when there is no underlying bearing stratum. King et al. (2000) studied the creep failure of CFA pile in soft clay through 12 pile load tests. All the piles were instrumented and loaded up to failure and the durations of loading for some tests continued for over 300 h. They found that the creep

Table 3. Results of field pile load tests

Test # 1			Test # 2		Test # 3		Test # 4	
Load (t)	Time (h)	Set. (mm)	Time (h)	Set. (mm)	Time (h)	Set. (mm)	Time (h)	Set. (mm)
19.5	1	0.27	1	0.44	1	0.47	1	0.43
39	1	0.46	1	1.40	1	0.95	1	0.69
58.5	1	0.71	1	2.39	1	1.59	1	1.22
78	3	1.02	3	4.14	3	2.56	3	1.96
97.5	3	1.34	3	5.50	3	3.39	3	2.68
117	3	1.73	3	7.54	3	4.71	3	3.35
136.5	3	2.11	3	8.93	3	6.39	3	4.78
156	3	2.59	3	9.79	3	8.74	3	6.13
175.5	3	3.01	3	11.81	3	11.58	3	8.08
200	0	3.44	0	13.0	0	13.68	0	8.75
200	12	4.91	144	17.81	48	17.55	144	13.12
175.5	0.25	4.63	0.25	17.78	0.25	17.5	0.25	13.06
156	0.25	4.31	0.25	17.75	0.25	17.3	0.25	12.93
136.5	0.25	3.94	0.25	17.67	0.25	16.93	0.25	12.65
117	0.25	3.53	0.25	17.33	0.25	16.56	0.25	12.38
97.5	0.25	3.14	0.25	16.93	0.25	16.23	0.25	12.07
78	0.25	2.67	0.25	16.38	0.25	15.84	0.25	11.76
58.5	0.25	2.16	0.25	15.81	0.25	15.15	0.25	11.35
39	0.25	1.78	0.25	15.18	0.25	14.76	0.25	10.79
19.5	0.25	1.32	0.25	14.51	0.25	13.86	0.25	10.04
0	0	0.99	0	13.18	0	11.49	0	8.80
0	12	0.52	3	12.81	4	10.81	4	8.63

Test # 1: CFA and completed within 2 days between 13 and 15 Nov. 2013
Test # 2: Rotary drilling and completed within 7 days between 3 and 10 Nov. 2013
Test # 3: Rotary drilling and completed within 3 days between 10 and 13 Nov. 2013
Test # 4: Rotary drilling and completed within 7 days between 3 and 10 Nov. 2013

loads were ranged between 70% and 50% of peak loads depending mainly on pile diameter, i.e. the high ratio belong to the small diameters.

In our case study, Table 3 shows that the test load of 200 t was left about 144 h for two tests and 48 h for one test. Creep behavior, i.e. increasing settlement at constant or peak load, could be obviously noticed for each pile test but unfortunately the creep load was not measured. The creep behavior is considered a phenomenon related generally to the peak or failure load. So, creep behavior could not be noticed for loads much less than the failure which could explain the unnoticeable creep behavior for CFA.

6 Conclusions

This paper presented the behavior of bored piles in sand overlaying compressible clay. Several approaches were used to estimate the ultimate bearing capacity of piles depending on soil parameters and on CPTU. Soil parameters were estimated from lab tests or derived from CPTU data. The important points can be concluded as follows.

1. The soil profiles in which strong soil stratum overlying compressible clay are not the ideal profile for pile foundations whether displacement, bored or CFA piles. Hence the analysis of pile behavior in these soil profiles should be distinguished.
2. The behavior of pile depends mainly on the accuracy of determining soil parameters that are estimating from laboratory tests or field tests. Still there are difficulties for extracting undisturbed samples from soft clays especially at deeper depths so that the values of shear strength can be very low relatively to that in filed.
3. Cone penetration testing (CPT) is a fast and reliable field means of conducting site investigations for exploring soils and soft ground with limit disturbance. The values of soil parameters that have been derived from the data of CPT, were obviously higher than those estimated from lab tests. So that the estimated bearing capacities of piles depending on lab soil parameters were obviously less than those depending on soil parameters derived from CPT.
4. The ratio of c_u/σ'_v derived from CPT was ranged between 0.3 and 0.4 with an average value 0.37. These values are considered higher than that customary for soft clays. This improvement of shear strength could be attributed to the limited disturbance of soil during cone tests. The unique soil profile, dense sand overlaying compressible clay, may also participate in this improving.
5. The direct approaches for estimating ultimate bearing capacities of piles from CPT produced good agreement with the measured ultimate bearing capacities.
6. Egyptian code approach produced the lowest and illogical values of ultimate bearing capacities because of the precautions for bored pile installation that led to decrease the values of soil parameters Φ and c_u.
7. The precautions for CFA piles in soft clays should be discussed with the soil profile. It was found that the behavior of CFA piles in sand overlying compressible clay was better than those of bored piles.
8. In soil profile when there is no underlying bearing stratum, the creep resistance of pile should be assessed.

References

Aas, G., Lacasse, S., Lunne, T., Hoeg, K.: Use of in situ tests for foundation design on clay. In: Proceedings of In situ 1986, Use of In Situ Tests in Geotechnical Engineering, pp. 1–30 (1986). ASCE GSP 6, Blacksburg, Virginia

Bazaraa, A., Sherif, M.M., Mashhour, M.: Some geotechnical properties of port said silty clay. Egypt. Soc. SMFE **1**, 27–34 (1986)

Bowles, J.E.: Foundation Analysis and Design. McGraw-Hill Book Company, Singapore (1988)

Brown, D.A.: Practical considerations in the selection and use of contentious flight auger and drilling displacement piles. Geotechnical Special Publication No. 132, ASCE, 2005, pp. 1–11 (2005)

Brown, D.A., et al.: Design and construction of continuous flight auger (CFA) piles. Geotechnical Engineering Circular No. 8, Report No. FHWA-HIF-07-03 (2007)

de Ruiter, J., Beringen, F.L.: Pile foundations for large North Sea structures. Mar. Geotech. **3**(3), 267–314 (1979)

Eslami, A., Fellenius, B.H.: Pile capacity by direct CPT and CPTU methods applied to 102 case histories. Can. Geotech. J. **34**, 886–904 (1997)

Fleming, W.K.: The understanding of continuous flight auger piling, its monitoring and control. Proc. Inst. Civ. Eng. Geotech. Eng. **113**(3), 157–165 (1995)

Jamiolkowski, M., Ladd, C.C., Germaine, J.T., Lancellotta, R.: New developments in field and laboratory testing of soils. In: Theme Lecture, 11th ICSMFE, San Francisco (1985)

Jefferies, M.G., Davies, M.P.: Use of CPTu to estimate equivalent SPT N60. Am. Soc. Test. Mater. ASTM Geotech. Test. J. **16**(4), 458–468 (1993)

King, G.J.W., et al.: The influence of rate of loading on the behavior of continuous-flight-auger bored piles in soft clay. Geotech. Geol. Eng. **18**, 139–153 (2000)

Lunne, T., Eidsmoen, D., Howland, J.D.: Laboratory and field evaluation of cone penetrometer. American society of civil engineers. In: Proceedings of In-Situ 1986, 23–25 June, pp. 714–729 (1986). ASCE SPT 6, Blacksburg

Lunne, T., Kleven, A.: Role of CPT in North Sea foundation engineering. In: Symposium on Cone Penetration Engineering Division, October 1981, pp. 49–75 (1981)

Meyerhof, G.G.: Bearing capacity and settlement of pile foundations. J. Geotech. Eng. ASCE 102(3), 197–228 (1976)

Robertson, P.K., Campanella, R.G.: Interpretation of cone penetrometer tests, Part I sand. Can. Geotech. J. **20**(4), 718–733 (1983)

Robertson, P.K., Campanella, R.G., Gillespie, D., Greig, J.: Use of piezometer cone data. In: In-Situ 1986 Use of In-situ testing in Geotechnical Engineering, pp. 1263–1280. Specialty Publication (1986). GSP 6, ASCE, Reston, VA

Schmertmann, J.H.: Guidelines for cone penetration test. Performance and Design. U.S. Department of Transportation, Report No. FHWA-TS-78-209, Washington, DC, p. 145 (1978). Bustamente (1982)

Stark, T.D., Juhrend, J.E.: Undrained shear strength from cone penetration tests. In: Proceedings of the 12th International Conference on Soil Mechanics and Foundation Engineering, Rio de Janeiro, vol. 1, pp. 327–330 (1989)

Tumay, M.T., Fakhroo, M.: Friction pile capacity prediction in cohesive soils using electric quasi-static penetration tests. Interim Research Report No. 1, Louisiana Department of Transportation and Development, Research and Development Section, Baton Rouge, LA, 275 (1982)

Laboratory Study of Plug Length Development and Bearing Capacity of Pipe Pile Models Embedded Within Partially Saturated Cohesionless Soils

Mahmood R. Mahmood(✉), Karim H. Al-Helo, and Ali M. AL-harbawee

Building and Construction Engineering Department, University of Technology, Baghdad, Iraq
mahmoudal_qaissy@yahoo.com, karim.alhelo@yahoo.com, star_of_sky91@yahoo.com

Abstract. This research presents an experimental study to investigate the development of plug length and load capacity of the pipe piles foundation embedded within unsaturated cohesionless soil. The influence of matric suction (i.e., capillary stresses) in unsaturated zone is typically considered on load capacity investigation and plug length development within the pipe piles.

The experimental work consist of twenty models of open pipe piles, these models are divided into four different configurations; single pipe pile, group of double pipe piles, group of triple pipe piles and group of six pipe piles. All these models are loaded and tested under three different states of saturation; dry, fully saturated (i.e., matric suction 0 kPa) and unsaturated conditions with three different matric suction values of (6.0, 8.0 and 10.0) kPa, which are achieved by predetermined lowering of water table. The relationship between matric suction and depth of ground water table was measured in suction profile set by using three Tensiometers (IRROMETER). The soil, water characteristic curve (SWCC) estimated by applying fitting methods through the program (Soil Vision).

The results of experimental work demonstrate that the matric suction has great influence on the ultimate load capacity of all the pipe pile models and the variations of load capacity with respect to matric suction are similar to that of shear strength of unsaturated soils. The increasing value of the ultimate load capacity for different configuration of pipe pile models under unsaturated conditions is approximately (1.3 to 2.7) times than that of saturated condition. Also it shows that the plug length decrease with increasing the value of matric suction for the same configuration of pipe piles, and with increasing the number of pipe piles.

Keywords: Pipe pile · Partially saturated soil · SWCC · Soil suction · Soil plug length

M. Abu-Farsakh et al. (eds.), *Advances in Analysis and Design of Deep Foundations*, Sustainable Civil Infrastructures, DOI 10.1007/978-3-319-61642-1_3

1 Introduction

Steel open ended pipe piles uses have been gradually more popular during the recent years, because of a significant time savings were achieved than using closed-ended and the other types of piles. Also, open-ended piles would consume a lower driving resistance; and reduced vibrations and displacements developed during driving. The majority of the time savings comes from that the open-ended piles being more stable during driving and thereby requiring fewer adjustments, as well as, to the possibility of forming a soil plug during pile driving that would leads to extra-resistant for pipe pile (Tan and Lin 2012).

Conventional pile design methods according to soil mechanics theories treat soil as either fully saturated or dry. However, a large number of geotechnical problems involve the presence of partially saturated soil zones where the voids between the soil particles are filled with a mixture of air and water (Georgiadis 2003). Unsaturated soil is the most common material encountered in the field of geotechnical engineering. It has a complex multi-phase system consisting of air, water and solid material whose response is a function of the stress state, moisture condition and other internal variables present within the soil. During the last few years that theoretical frameworks and constitutive models have been proposed to describe the mechanical behavior of such soils (Berney et al. 2003).

Luking and Kempfert (2013), investigated the influence of different factors on the plugging effect and the change in the load-bearing behavior mainly in non-cohesive soils using experimental, numerical and statistical methods during jacking an open-ended non displacement pile. The soil is entering through the pile toe into the profile; this plug can close up the pile toe completely. Because of such reason, the pile can be treated approximately as a fully closed-ended displacement pile and is able to mobilize an additional base resistance. The load transfer in the plug takes place by compression arches, which are mainly influenced by the pile diameter and the soil density.

Karlowskis (2014), described the mechanism of plugging phenomenon at the toe of vertically loaded open-ended piles. The ground below the pile toe is deformed by pile penetration, the deformed and dilated soil intrudes inside the pile and friction is produced between the pile and the soil. If the inner friction resistance and self-weight balance with the bearing resistance of the ground below the pile toe, a plug is produced.

Al-Soudani and Fattah (2015), present study focuses on the determination of effect of soil plug on the ultimate capacity of single open-ended steel pipe pile, and makes a comparison with closed-ended pipe pile. Axial compression load tests were performed on model piles. It was concluded that the pile load carrying capacity in dense sand is several times greater than those in loose and medium sands. On the other hand, the removal of soil plug decreases the pile load capacity.

Abdulaziz et al. (2015), investigated the effect of matric suction on the load carrying capacity of piled raft foundation embedded within partially saturated sandy soil. The experimental work consists of 6 models of pipe piles; all these models are loaded and tested under both fully saturated and unsaturated conditions. The results demonstrated that matric suction has a significant influence on the load carrying capacity of all tested models and the variations of load carrying capacity with respect to matric suction are similar to that of shear strength of unsaturated soils.

2 Experimental Work

2.1 Soil Material

The soil used in this study is fine gray river sand obtained from Baghdad city (Abu Nawas) site. Soil standard laboratory tests were performed to determine physical and engineering soil properties as shown in Table 1. The grain size distribution of the sand used is shown in Fig. 1.

Table 1. Physical and engineering properties of the sand used in present study

Property	Value	Specification
Effective size, D_{10}	0.14	
Coefficient of uniformity, C_u	1.98	
Coefficient of curvature, C_c	1.60	
Classification (USCS)	SP	
Specific gravity, Gs	2.685	
Maximum unit weight, $\gamma_{d(max)}$, kN/m^3	16.16	ASTM D854-02
Minimum unit weight, $\gamma_{d\ (min)}$, kN/m^3	13.34	ASTM D4253-00
Maximum void ratio, e_{max}	0.97	
Minimum void ratio, e_{min}	0.63	

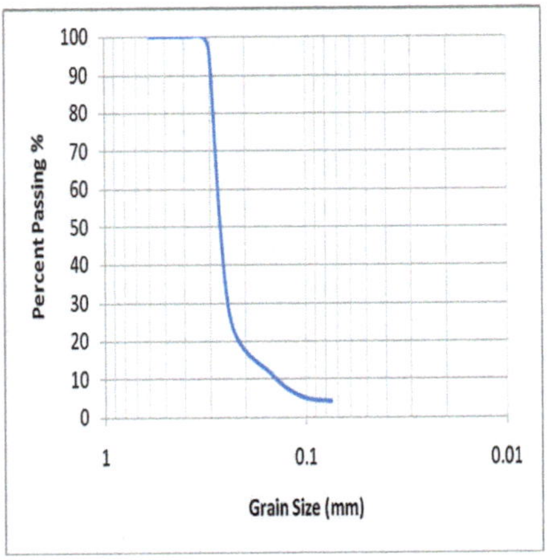

Fig. 1. Grain size distribution of the sand used

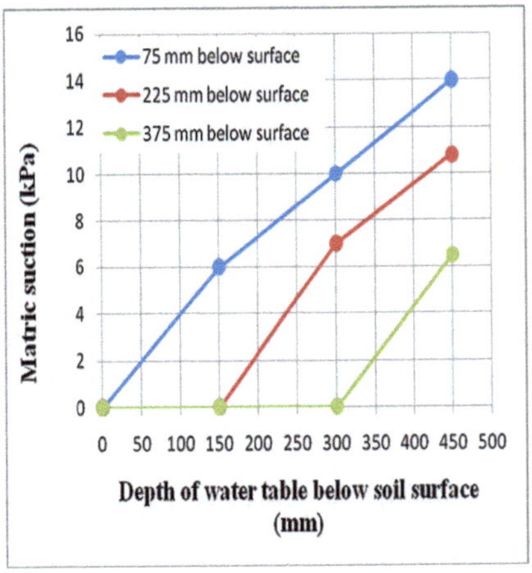

Fig. 2. Matric suction values for the three stages of water table below soil surface

Plate 1. Loading frame and soil container

2.2 Model Set-up Formulation

The model consists of steel container, steel loading frame, axial loading, system of (hydraulic jack), load cell, and digital weighing indicator.

The container was made of steel plate 4 mm thickness with dimensions of (60 × 60 × 70 cm height) used for testing all the pipe pile models. The dimensions of the tank were chosen to provide fully mobilized pressure within the soil media during the loading testing. The container has some devices used for saturation and de saturation of the soil bed. These devices were achieved by using 4 valves fixed along one side of the container, three at the same vertical distance of 150 mm between each other for de saturation and the fourth at the lower base of container used for saturation and lowering water table. The water table level in the tank was measured by using four Piezometers which fixed at each side of the container. Steel loading frame was manufactured to support the axial loading system (hydraulic jack). The axial load is applied through a hydraulic jack system of 10 ton capacity. A manual system is fixed to control hydraulic pressure at the right column of frame with pressure gage used to measure the axial pressure. Two dial gages with 0.01 mm precise have been used for measuring the displacements of the pipe pile model. A compression load cell of 2 ton capacity is used to measure the applied load. A digital weighting indicator was used for displaying the load measurement. Plate 1 shows the container with the loading frame.

Plate 2. Pipe pile models with different configurations with fixing rings

2.3 Models of Pipe Pile Groups and Pile Caps

The pipe pile models used in this study are of steel hollow pipe of 1.5 mm thickness and 500 mm length with the inner diameter of 23 mm (the recommended insertion depth within the soil bed 400 mm).

Four different configurations of pipe pile group models were used; these models consist of single pipe pile; double pipe piles (1 × 2), triple pipe piles (1 × 3) and six pipe piles (2 × 3). The pile caps used for pile groups were made of steel plate with smooth surface having a thickness of 14 mm. Four steel plats with smooth surface; of (80 × 75) mm, (160 × 75) mm, (220.5 × 75) mm and (230.5 × 150) mm were used as a cap for group pipe piles of (1 × 1), (1 × 2), (1 × 3), and (2 × 3) respectively. Circular groves of 27 mm diameter and circular rings of 30 mm height were made within the pile caps for fixing pipe pile models; these groves and rings were distributed according to the dimension and spacing of piles. Pipe pile model configurations and the fixing rings are shown in Plate 2.

2.4 Soil Bed Preparation

The steel container depth was divided into 7 lifts of (100) mm height for each layer of the soil used for controlling the compaction. The lower layer is a filter material that used to prevent the erosion of soil particle during dewatering and prevent excess pore water pressure. A double layer of Geo-mesh was placed above the filter material to protect this layer and prevent the mixing of the soil with the filter material.

The sand deposit was prepared by using a steel tamping hammer of 2 kg weighing to get the predetermined relative density of 65%.

2.5 Soil Suction Measurement

The Tensiometer (IRROMETER) as shown in Plate 3 was used to measure the negative pore water pressure. This Tensiometer is able to measure the soil suction in a range between (0–100) kPa. It is able to detect the changes in the soil moisture content with several hours after installation, in the case of poor drainage soil; The hand vacuum pump supplied with the Tensiometer will apply a vacuum pressure between (80–85 kPa Tensiometer) to release the air entrapped in the ceramic disc.

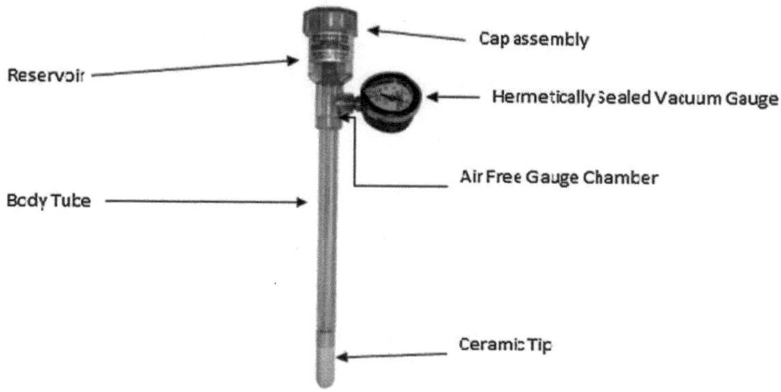

Plate 3. Tensiometer (IRROMETER Manual Book, IRROMETER company, Inc.)

2.6 Tensiometer Installation

After Saturation of Tensiometer as stated, by field solution prepared according to the instruction label, the Tensiometer was inserted with the aid of coring unit to the required depth.

A strong vacuum applied to the Tensiometer with the hand vacuum pump with removing the filler cap and submerging the tip in wet sand, this pump done vigorously until a reading of 80–85 kPa shown on the gauge.

2.7 Suction Profile Set

This technique was designed to measure the relationship between matric suction and depth of ground water table to achieve expected matric suction in the soil bed faster by simply adjusting the water table.

In this study three Tensiometers were installed at different depths below the soil surface to measure the suction profile in the sand after three lowering of water table below soil surface as shown in Plate 4. The period to achieve equilibrium conditions was 24 h (Li 2008). The first Tensiometer was installed at 7.5 cm below the soil surface, the second was installed at 22.5 cm below the soil surface and the third was installed at 37.5 cm below the soil surface.

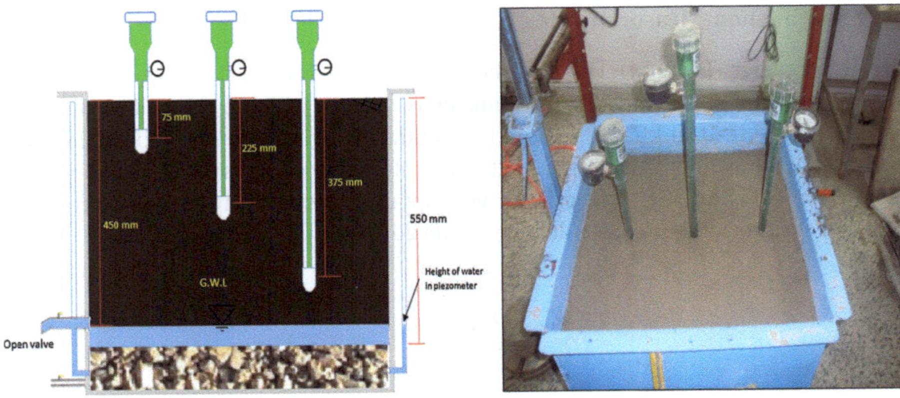

Plate 4. Inserting three Tensiometers for profile set suction

3 Results of the Suction Profile Set

The soil was initially saturated by raising the water level in the container from the bottom to ensure fully saturation and escape air from soil. The water table was then lowered to 15 cm below the soil surface (i.e. first stage) for a period of 24 h to let capillary suctions reached the equilibrium conditions. This procedure was repeated by varying depth of water table below the soil surface to another different depths (30 and 45) cm (i.e. second stage and third stage) respectively. Then the matric suction is measured after 24 h as described before. Figure 2 shows the measured matrix suction values above the water table for the three stages of lowering water table.

The suction profile of the soil used shows that the matric suction increases with the lowering of water table. The soil suction near soil surface (at depth of 7.5 cm below the soil surface) increases steeply as the depth of water table increases. According to Li (2008) and Sun (2010) the rapid increase of matric suction values at the soil surface of sand may be attributed to evaporation of water from the soil surface. Table 2 summarized the results of the corresponding average matric suction after lowering water table.

Table 2. Results of the corresponding average matric suction after lowering water table

Soil conditions	Lowering of Water table from soil surface in (cm)	Corresponding average matric suction in (kPa)
Fully saturated	0	0.0
Unsaturated	15	6.0
	30	8.0
	45	10.0

3.1 SWCC Estimated by Tensiometer

Figures 3 and 4 show the SWCC for the soil as estimated by the Fredlund and Xing (1994) and Van Genuchten (1980) equations respectively with the aid of Soil Vision program.

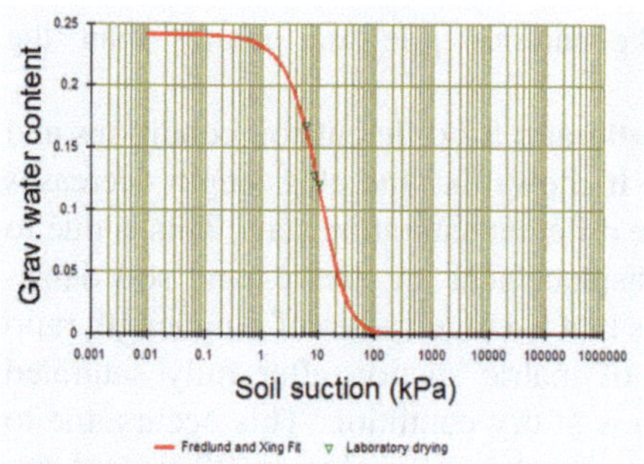

Fig. 3. Relationships between the gravitational water content and the matric suction obtained by the program Soil Vision by using Fredlund and Xing equation

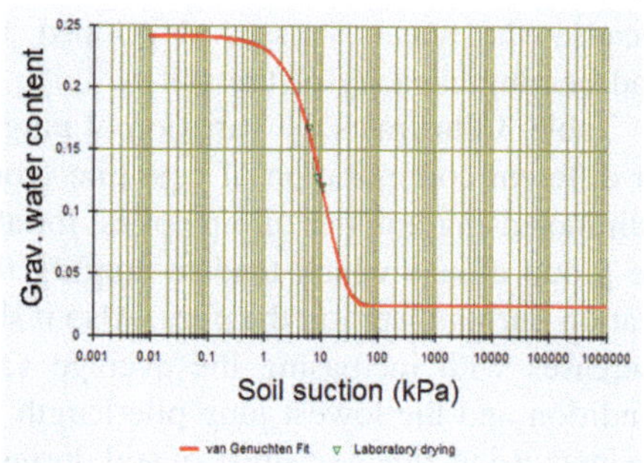

Fig. 4. Relationships between the gravitational water content and the matric suction obtained by the program Soil Vision by using Van Genuchten equation

The changes in slope defined two points that are pivotal to describe the SWCC (Fredlund et al. 2001). The first point is termed "air-entry value" which is the matric suction where air starts to enter the largest pores in the soil. The second point is defined as "Residual water content" which is the water content where a large suction change is required to remove additional water from the soil (Fredlund and Xing 1994). From the soil water characteristics curves, and fitting curves proposed by Fredlund and Xing (1994) and Van Genuchten (1980), the air-entry value $(u_a - u_w)_b$ (kPa) was found to be 2.37 kPa and 2.4 kPa respectively.

3.2 Insertion of Pipe Pile Models with Measuring Increment Filling Ratio

After preparing the soil bed under different stats of saturation (dry, fully saturated, and partially saturated conditions). Pipe pile model inserted within soil bed by using a manual hydraulic jack with a constant loading rate measured from a digital weighing indicator connected with a load cell of (100) kN capacity to the required penetrate depth of 400 mm.

Soil plug length was measured during pipe pile installation by insertion steel measuring tape from the top of the hollow pile every 2 cm of insertion to determine the plug length within the pipe piles. Where the soil plug attributes strongly influence the bearing capacity of open-end piles. The degree of soil plugging can be represented by the plug length ratio (PLR) which is defined by Paik et al. (2003), as the ratio of soil plug length to the pile penetration depth.

3.3 Formation of Soil Plug

At the beginning of pipe pile installation a soil column is formed within the pipe pile, its height typically equals to the penetration depth (full coring state). As the penetration depth increases, the developments of internal shear stresses become larger and reduce the amount of soil to penetrate within the pipe pile. This will continue for a short time, and then the soil column forms a plug, and does not move relative to the pile. This is because the shear stresses developed in the interior pile are greater than the load-bearing capacity of the soil.

Table 3 illustrates the variation of plug length ratio for different soil conditions and for different configuration of pipe pile groups. It shows that the plug length decreases with increasing number of pipe piles, for all the different saturation stats. This is due to the group effects which tend to amplify the displacement on stresses and soil densification during insertion the piles. Also it shows that average values of plug length ratio decreases with increasing the average value of matric suction after fully saturated condition and the lowest plug pile length ratio is at dry condition. This occurs due to the increase in effective stress of soil during lowering the water table. In other word, the friction resistance along the pile shaft will increase due to the increase in effective stress which agreed with Escario et al. (1989), who shows that shear strength for such soil increases when the degree of saturation approaches to zero value (which is the maximum increment of the soil suction).

Table 3. Plug length ratio at the end of insertion pipe piles for different soil stats and different configurations of pile group

Configurations of pile group	Plug length ratio (PLR) %				
	Dry state	Full saturation (0) kPa suction	Unsaturated 6 kPa suction	Unsaturated 8 kPa suction	Unsaturated 10 kPa suction
1 × 1	17.5	22.5	21.2	20.5	19.0
2 × 1	16.5	21.7	20.0	19.3	18.2
3 × 1	15.5	20.2	19.5	18.5	17.7
2 × 3	13.5	18.2	17.5	17.0	16.0

Figures 5, 6, 7 and 8 shows the relationship between average soil plug lengths for different configuration of pipe pile groups pressed in different soil stats of saturation. The figures show that a marginal increase in plug length with increasing soil suction accompanied with a little discrepancy during insertion, and this may attribute to the scale effect where small scale model used in testing.

3.4 Loading Test

After insertion pipe pile to the required depth static load test were carried out according to the quick test method of ASTM D1143M-07 by applying axial load with increment of 5% of the anticipated failure load, and keep the load constant for a time interval of

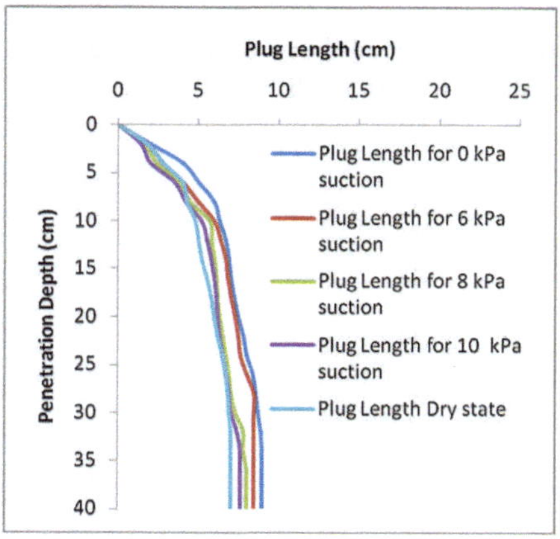

Fig. 5. Average soil plug length for single pipe pile pressed in different states of saturation soil

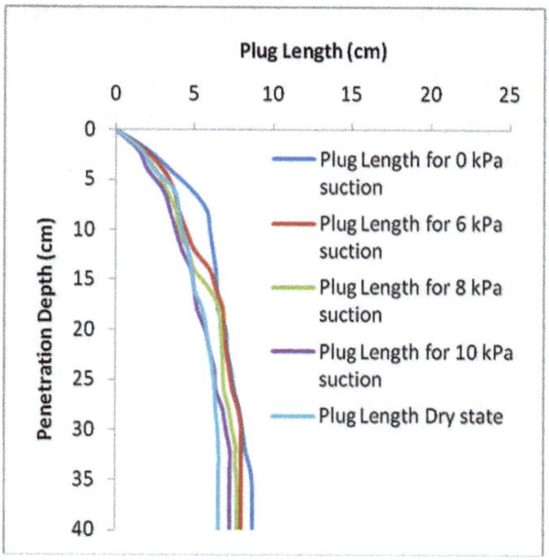

Fig. 6. Average soil plug length for group of double pipe piles pressed in different states of saturation soil

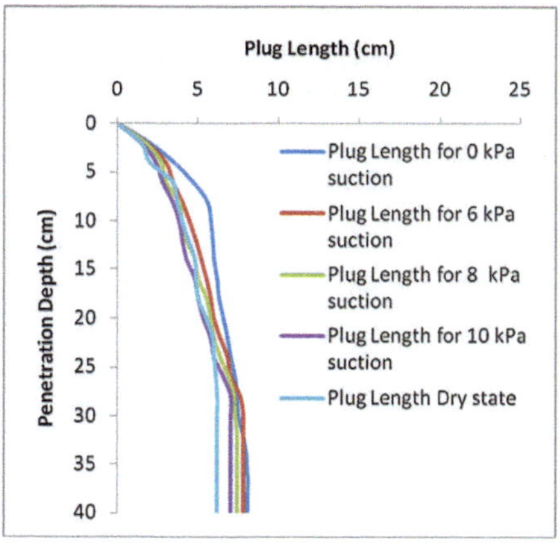

Fig. 7. Average soil plug length for group of triple pipe piles pressed in different states of saturation soil

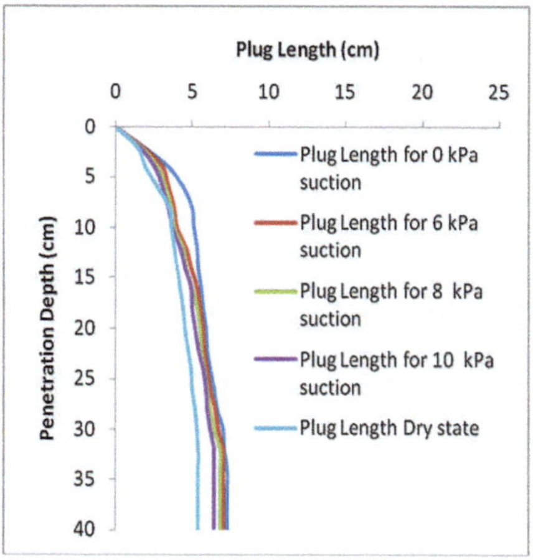

Fig. 8. Average soil plug length for group of six pipe piles pressed in different states of saturation soil

10 min during each increment in load. The test was continued until the recorded settlement exceeded 15% of the pile diameter. The constant loading rate adopted in the whole test program is measured from a digital weighing indicator connected with a load cell. The displacement of the pile was measured by taking the average of two dial gauges reading with an accuracy of (0.01 mm/division). The failure criteria adopted to determine ultimate load capacity is the Tangents proposal; at which, the definition of failure is based on the intersection of the two tangents of load-settlement curves.

3.5 Ultimate Load Capacity of Different Configuration of Pipe Piles

Figures 9, 10, 11 and 12 shows load settlement curves for the four configuration of pipe pile models embedded within different saturation stats of different matric suction values.

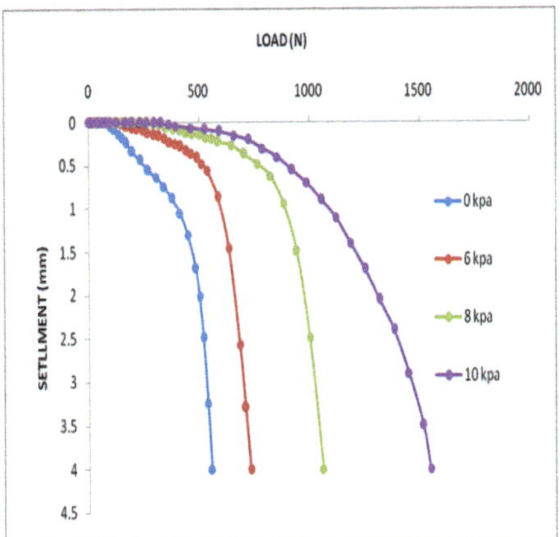

Fig. 9. Load settlement curve for single pipe pile models under different average matric suction values

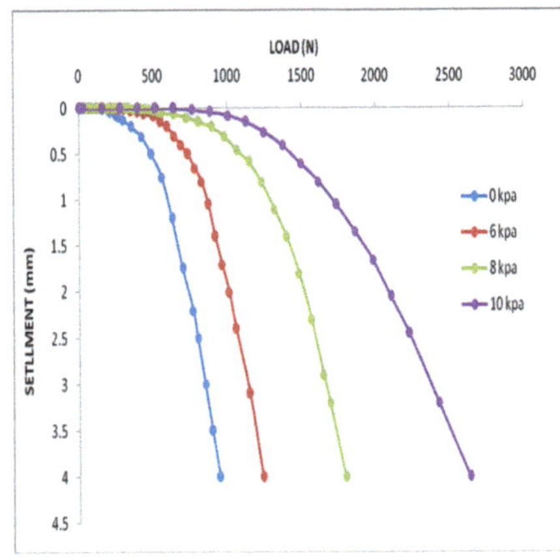

Fig. 10. Load settlement curve for double pipe piles models (2 × 1) under different average matric suction values

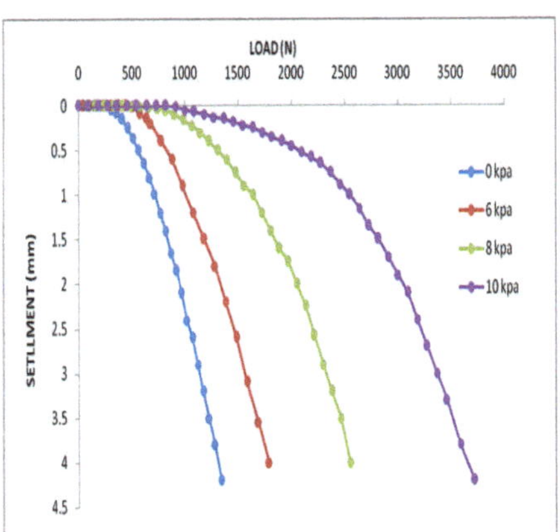

Fig. 11. Load settlement curve for triple pipe piles models (3 × 1) under different average matric suction values

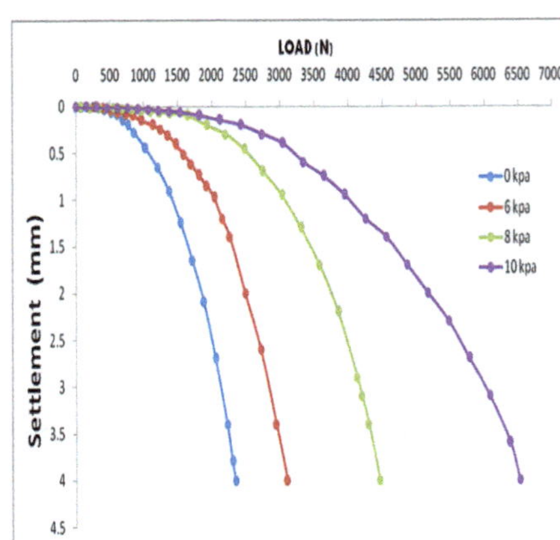

Fig. 12. Load settlement curve for six pipe piles models (2 × 3) under different average matric suction values

The figures show that the ultimate bearing capacity increases with increasing of matric suction for all different pipe pile configurations. The increment of load capacity was caused by the increase of effective shear strength of soil due to lowering of water table and increasing of matric suction.

Table 4 shows the ultimate load capacity values for different configuration of pipe pile groups embedded within partially saturated soil. It can be seen that the ultimate load capacity of pipe pile in partially saturation soil is more than that in a fully saturated by about (1.3–2.7) times, this increment in load capacity caused by the effect of soil suction along the shaft resistance. In the shallow lowering of water table the effect of partially saturated soil is less important from that of higher depth of lowering. The amount of increasing in the bearing capacity values due to lowering of water table is agreed with the amount of increasing of matric suction obtained by Mohammed and Vanapalli (2006).

Table 4. Ultimate load capacity values for different configuration of pipe pile groups under different saturation conditions

Configurations of pile group	Ultimate load capacity (N)			
	Full saturation 0 kPa suction	Unsaturated 6 kPa suction	Unsaturated 8 kPa suction	Unsaturated 10 kPa suction
1 × 1	450	590	990	1215
2 × 1	700	910	1540	1890
3 × 1	980	1275	2150	2650
2 × 3	1600	2100	3520	4320

Figure 13 show the improvement ratios in bearing capacity for different configurations under different values of matric suction.

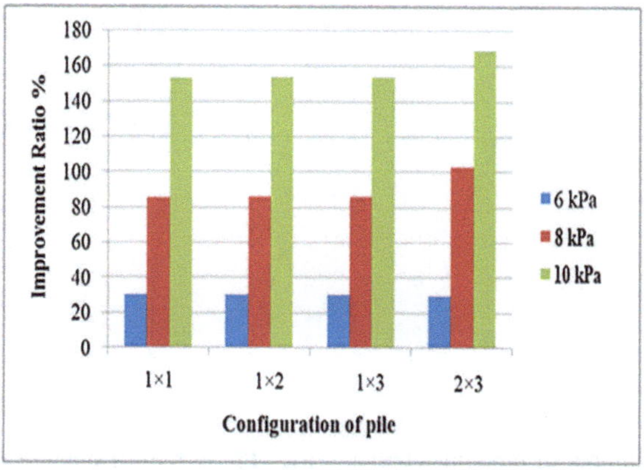

Fig. 13. Improvement ratio in bearing capacity for different matric suction

3.6 Group Action Effect of Pipe Piles

This part investigates the effect of group action of pipe pile on the ultimate load capacity under different stats of saturation (dry, fully saturated and unsaturated). Figures 14, 15, 16, 17 and 18 illustrate load-settlement curves for single, double, triple and six pipe piles.

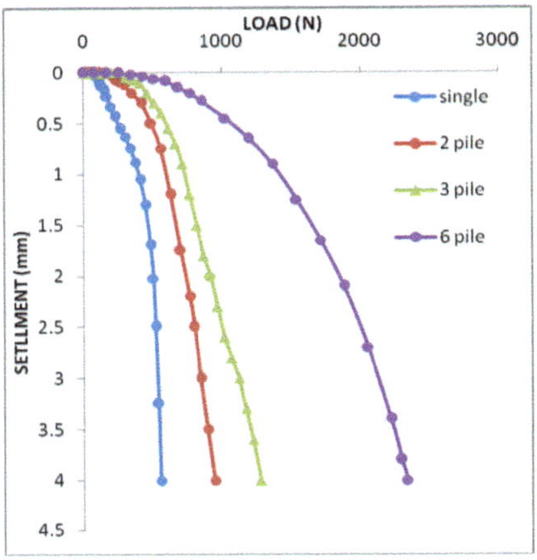

Fig. 14. Effect of number of pipe piles under fully saturated condition

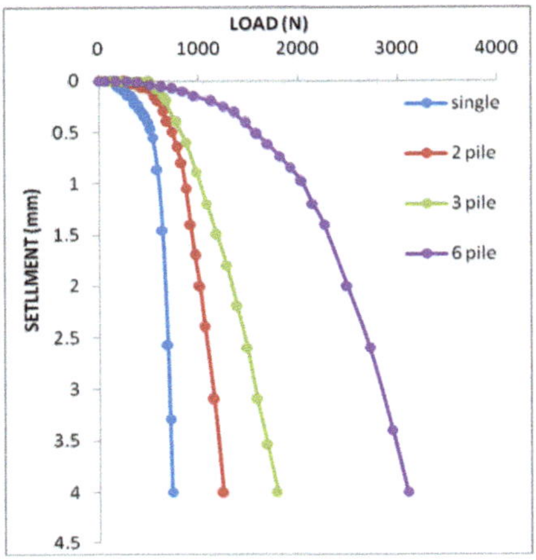

Fig. 15. Effect of number of pipe piles under the average of matric suction 6.0 kPa

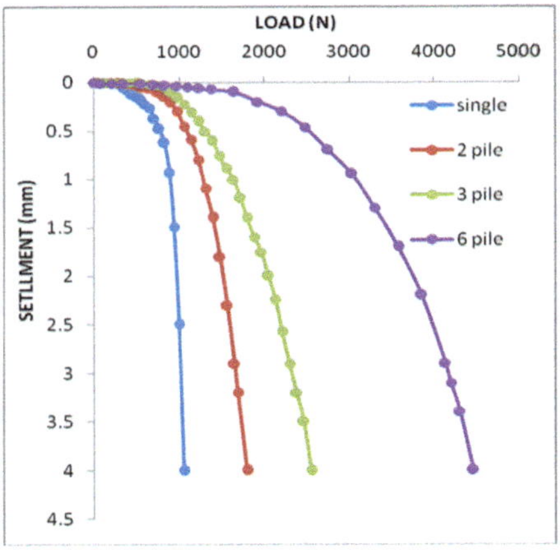

Fig. 16. Effect of number of pipe piles under the average of matric suction 8.0 kPa

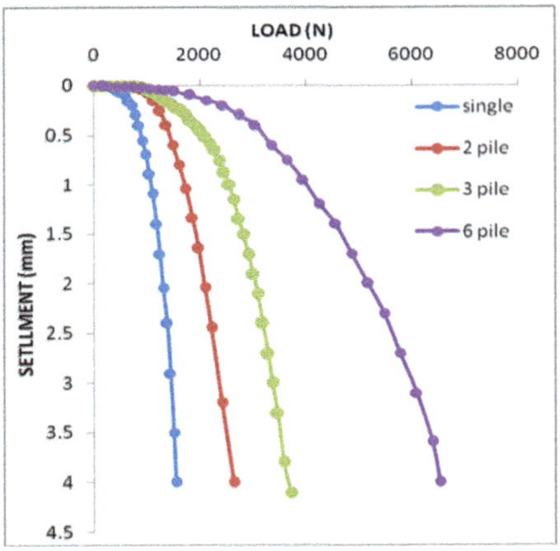

Fig. 17. Effect of number of pipe piles under the average of matric suction 10.0 kPa

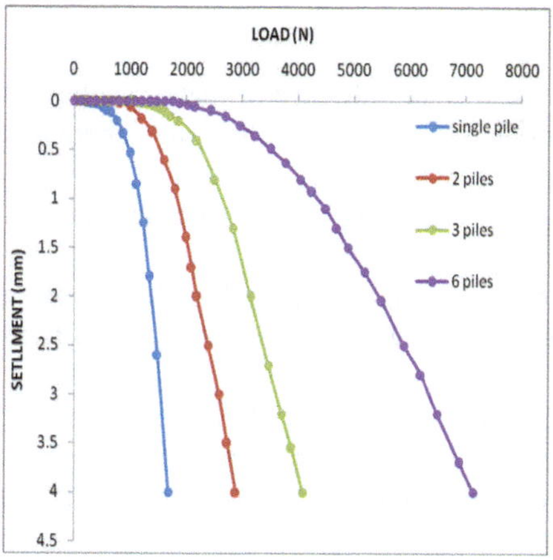

Fig. 18. Effect of number of pipe piles under dry state condition

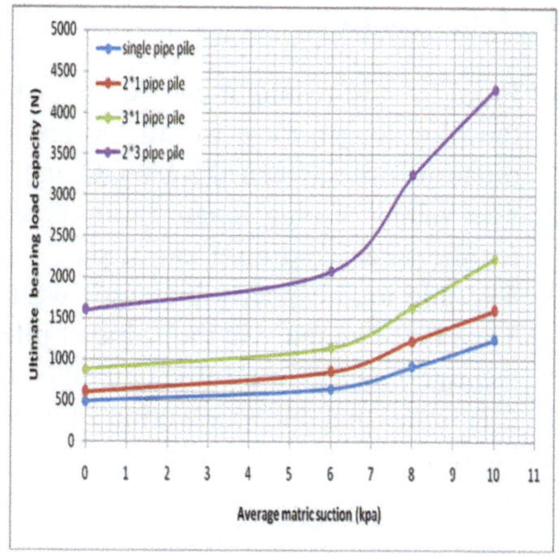

Fig. 19. Variation of the ultimate load capacity with respect to average matric suction for all the models

From the figures, it can be seen that the load capacity increases with the increasing number of pipe piles at the same wet condition of testing (dry, full saturation Unsaturated conditions).

Figure 19 shows the behavior of ultimate bearing capacity for different configuration of pile groups with different values of average matric suction. The trends of the testing curves demonstrate that the ultimate load capacity increases with increasing matric suction. It shows that also the bearing capacity approximately increases linearly with increasing matric suction up to the air–entry value and there is a non-linear increase in the bearing capacity with respect to matric suction beyond the air-entry value.

Table 5. Efficiency of group action for the configurations of pipe pile model due to partially saturation

Configuration of pipe pile	Efficient of group pile
Group of double pile (2*1)	77%
Group of triple pile (3*1)	72%
Group of six pile (2*3)	60%

The efficiency of group action for the three configurations is summarized in Table 5 by using the following equation.

$$E = \frac{load\ for\ group\ piles}{load\ of\ single\ pile \times number\ of\ pipe\ pile}$$

4 Conclusions

1. From the test results it can be noticed that there is an influence of matric suction on the plug length ratio for all the tested models. These results show that there is a relationship between the SWCC and the plug length ratio for pipe piles.
2. The values of plug length ratio decreased with increasing values of matric suction for the same configuration of pipe piles groups.
3. The pile load capacity increases with decrease in length of plug length ratio PLR. The rate of increase in the value of pile load capacity with PLR is greater in partially saturated soil than a fully saturated soil.
4. It is found that due to lowering of water table and contribution of matric suction, the bearing capacity for different pipe piles configuration in unsaturated soil is approximately (1.3–2.7) times the capacity of the same soil and pile configurations under saturated conditions.
5. It can be seen that the increase in number of pipe piles cause an increase in carrying load capacity with different values under the same test condition due to group action (dry, fully saturation with different matric suctions)
6. The steeply increase in soil suction near soil surface due to lowering of water table increases due to evaporation of water from soil surface which tends to this rapid increase of matric suction values. The procedure of set profile suction which proposed in this work is found to be a good procedure based on the encouraging results.

References

Abdulaziz A.K.: Experimental Study of Single Piled Raft Model Embedded within Partially Saturated Cohesionless Soils. M.Sc. Thesis, Department of Civil Engineering, University of Technology, Baghdad (2015)

Al-Soudani, H.S., Fattah, M.Y.: Bearing capacity of open-ended pipe piles with restricted soil plug. Ships Offshore Struct. (2015). doi:10.1080/17445302.2015.1030247

Berney, E.S., Peter, J.F., Smith, D.M.: The thermodynamics of three-phase unsaturated soil. In: 16th ASCE Engineering Mechanics Conference (2003)

Escario, V., Juca, J., Coppe, M.: Shear strength and deformation of partly saturated soils. Soil Mech. Found. Eng. Rio de Janeiro **1**, 43–46 (1989)

Fredlund, D.G., Xing, A.: Equations for the Soil Water Characteristic Curve. University of Saskatchewan, Saskatoon (1994)

Fredlund, D., Rahardjo, H., Leong, L., Ng, C.W.: Suggestions and recommendations for the interpretation of soil-water characteristic curves. In: Proceedings of the 14th Southeast Asian Geotechnical Conference, Hong Kong, pp. 503–508 (2001)

Georgiadis, K.: Development, Implementation and Application of Partially Saturated Soils Models in Finite Element Analysis. M.Sc. Thesis, University of London (2003)

Karlowskis, V.: Soil Plugging of Open-Ended Piles During Impact Driving in Cohesion-less Soil. Royal Institute of Technology (KTH), Stockholm (2014)

Luking, J., Kempfert, H.: Plugging effect of open-ended displacement piles. In: Proceedings of the 18th International Conference on Soil Mechanics and Geotechnical Engineering, Paris, pp. 2363–2366 (2013)

Li, X.: Laboratory Studies on the Bearing Capacity of Unsaturated Sands. M.Sc. Thesis, University of Ottawa (2008)

Mohamed, F., Vanapalli, S.: Laboratory investigations for the measurement of the bearing capacity of an unsaturated coarse-grained soil. In: Proceedings of the 59th Canadian Geotechnical Conference, Vancouver, BC, pp. 1–8 (2006)

Paik, K., Salgado, R., Lee, J., Kim, B.: Behavior of open-and closed-ended piles driven into sands. J. Geotech. Geoenviron. Eng. **129**(4), 296–306 (2003)

Sun, R.: Bearing capacity and settlement behavior of unsaturated soils from model footing tests. M.Sc. Thesis, Civil Engineering Department, University of Ottawa, Canada (2010)

Tan, Y., Lin, G.: Full-scale testing of open-ended steel pipe piles in thick varved clayey silt deposits along the Delaware River in New Jersey. J. Geotech. Geoenviron. Eng. **139**(3), 518–524 (2012)

Van Genuchten, M.T., Leij, F.J., Yates, S.R.: The RETC code for quantifying the hydraulic functions of unsaturated soils, USEPA Report 600/2-91/065. U.S. Environmental Protection Agency, Ada, Oklahoma (1991)

Bearing Capacity and Settlement of Pile Based on Cone Loading Test

Philippe Reiffsteck[1(✉)], Henk van de Graaf[2], and Catherine Jacquard[3]

[1] IFSTTAR, Paris, France
philippe.reiffsteck@ifsttar.fr
[2] Lankelma, Oirschot, The Netherlands
henkvandegraaf@lankelma-zuid.nl
[3] Fondasol, Avignon, France
Catherine.jacquard@fondasol.fr

Abstract. The Cone Loading Test (CLT) consists in stopping the static penetration at a desired level and carrying out a loading of the cone by successive steps. Measurement of the settlement of a cone submitted to incremental loading allows the determination of a modulus correlated to those obtained using laboratory tests or derived from other in situ tests. This test can be carried out with standard static penetrometer equipment. Considering that the cone of the penetrometer is a reduced model pile, a direct outcome of the CLT test is its ability to be a tool for foundation design. For this purpose, a direct method using the cone resistance and limit sleeve friction of the CLT test was proposed to calculate the bearing capacity and predicting the settlement of a pile. This method is a new approach to transform the CLT test curves of cone loading and friction mobilization in load-displacement curve of a pile (t-z curves).

Keywords: In situ testing · CPT · Penetrometer · Piezocone · Incremental loading test · Pile design t-z curve

1 Introduction

After several attempts to develop a test method using a penetrometer test based on either the plate bearing test or static pile loading test procedures to reach a stress strain relationship (Haefeli and Fehlmann 1957; Ladanyi 1976; Sanglerat 1974) Reiffsteck et al. (2009) proposed a procedure leading to possible improvement of direct design method. This test, called Cone Loading Test (CLT), can be carried out during a Cone Penetration Test (CPT) as a complementary, fast and economic test. After a dissipation test performed with the rods unclamped, the cone is loaded by a minimum of ten successive steps lasting 60 s or at constant very slow speed until the cone resistance of the soil is reached (Arbaoui et al. 2006; Reiffsteck et al. 2009). The stress-settlement curve links the pressure applied on the cone to the settlement of the top of the push rods at the end of each loading stage (Fig. 1). The curves consist of a linear part at small strain level, then a curved part at the onset of plasticity and a linear part at higher strain level.

M. Abu-Farsakh et al. (eds.), *Advances in Analysis and Design of Deep Foundations*,
Sustainable Civil Infrastructures, DOI 10.1007/978-3-319-61642-1_4

Furthermore, use of an electric cone allows derivation from sleeve friction measurement of the mobilization curve (t-z curve) of local side friction at each test level.

This test can be performed with a standardized electrical cone penetrometer (CEN/ISO 2005). In case of use of a piezocone, increase of pore pressure and shear resistance on the friction sleeve are measured.

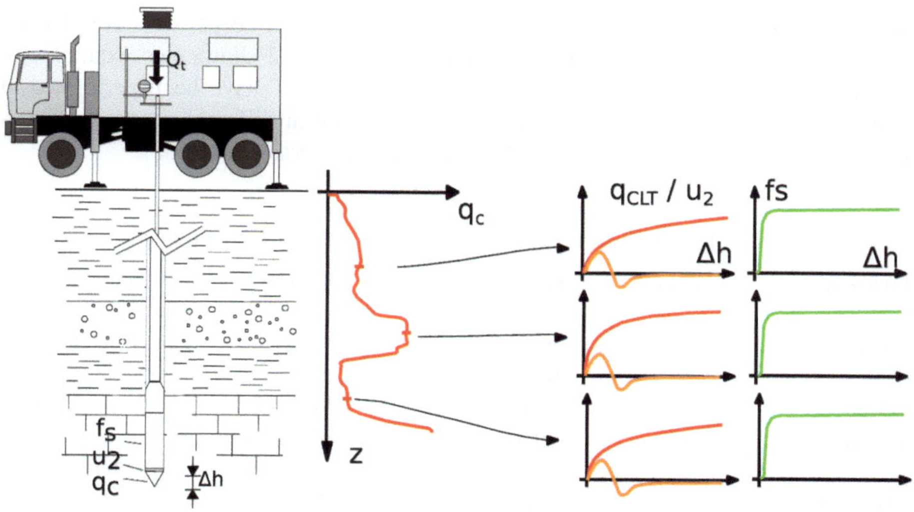

Fig. 1. Cone loading test principles (Reiffsteck et al. 2009)

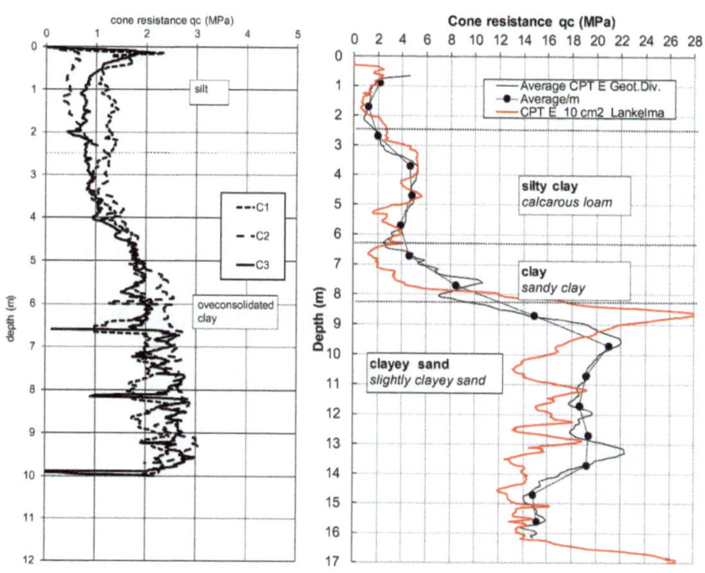

Fig. 2. CPT tests profiles for (a) Merville, France and (b) Limelette (Belgium)

The cone loading test requires several corrections resulting from the push rods compression effect or the initial conditions of the test. A comparison of correction methods of the measured displacement shows that the use of a friction reducer imposes a precise measurement of the thrust force applied at the top of the rods (Q_t). Generally a 15 cm^2 cone is used minimizing friction on the 10 cm^2 string of rods. The calculation of the correction, proved to be reliable, is to estimate of a mean friction obtained from the difference between the force measured at the top and at the cone level (Ali et al. 2008). At shallow depth deviation of rods is neglected but if deeper tests are planned use of inclinometer is suggested (CEN/ISO 2005). Figure 2 shows two examples of CPT profiles obtained on sites listed in Table 1 were CLT have been performed at several depths. Although loading tests were carried out at different depths (approximately every 1.5 m) no disturbance of the CPT profiles was observed.

Table 1. Test sites where CPT and CLT test have been performed

Location	Country	Soil profile	Depth	Cone type	Driving unit
Jossigny	France	Plateau silt	3	Mechanical	Lightweight DPT
Orleans	France	Sand	4	Mechanical	Lightweight DPT
Merville	France	Ypresian clay	4/9.8	Mechanical -piezocone	Lightweight DPT/mini-crawler
Choisy-au-bac	France	Silty sand/chalk	12	Piezocone	Mini-crawler
Oirshot	Netherland	Clay	8	Piezocone	Mini-crawler
Limelette	Belgium	Silt/sand	18	Piezocone	Mini-crawler
StMichielsgestel	Netherland	Sand	13	Piezocone	Mini-crawler
Les Moulineaux	France	Sandy soil/chalk	23	Piezocone	Truck

2 Pile Design

Determining the working load of a pile so as to be close to its actual bearing capacity is still very difficult.

In direct design method the ultimate bearing capacity R_u of a pile is computed from:

$$R_u = R_{pu} + R_{su} \tag{1}$$

where: R_{pu} is the ultimate pile tip capacity and R_{su} the ultimate shaft resistance capacity.

This separation of the pile capacity into two terms is a common feature of all the design methods used in practice: i.e. φ'-c', CPT, standard penetration test (SPT), Ménard pressuremeter test (MPT). The tip capacity is related to a mean value of the shear strength derived from these tests (in situ and lab) multiplied by a factor related to the failure mechanism and adjusted for the soil type and for the remoulding effect of the technique. The shaft term accounts for the change of soil properties in the vicinity of the pile after it has been installed, for the soil variability (vertically but not only) and for the (complex) pile-soil interaction. Figure 3 shows an example of the loading sequence of an instrumented pile. Hence, for every segment of the shaft, shaft resistance has to be computed from the shear strength times a factor depending of the influences above.

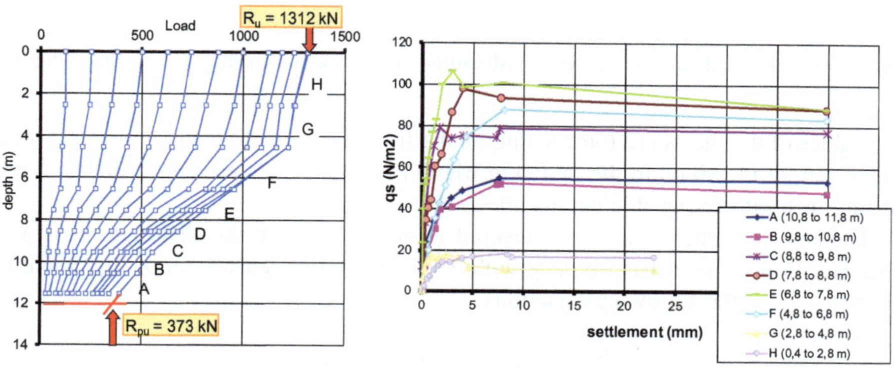

Fig. 3. (a) Evolution of load in pile segments (b) Measured shaft friction (Merville, France)

2.1 Cone Loading Test Direct Design Method

By considering that the cone penetrometer is a reduced pile model, and that the deformation of the soil around a cone indicates effects similar to the soil deformation observed around a full-scale pile during loading, then, the cone loading test curves reflect the interaction between a slender pile and the surrounding soil. A very useful and interesting repercussion of this statement is the potential of the cone loading test to be a foundations designing tool. As in a pile loading test, during a cone loading test, the applied cone pressure is recorded according to displacement in each loading stage. In the same manner, the curve of sleeve friction mobilization law can be drawn as shown on Figs. 4 and 5.

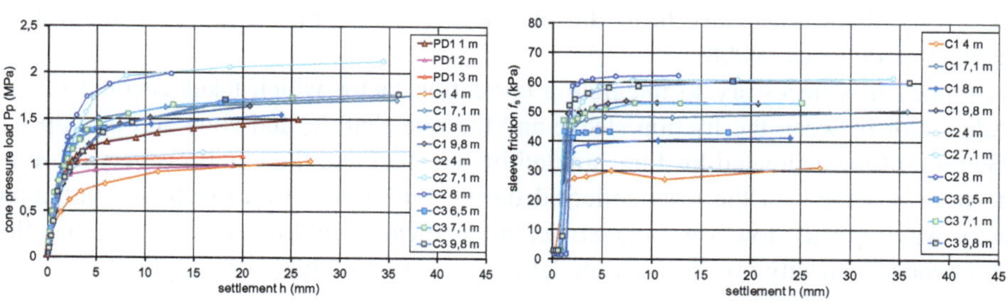

Fig. 4. (a) cone loading tests curves (b) Measured sleeve friction in clay (Merville, France)

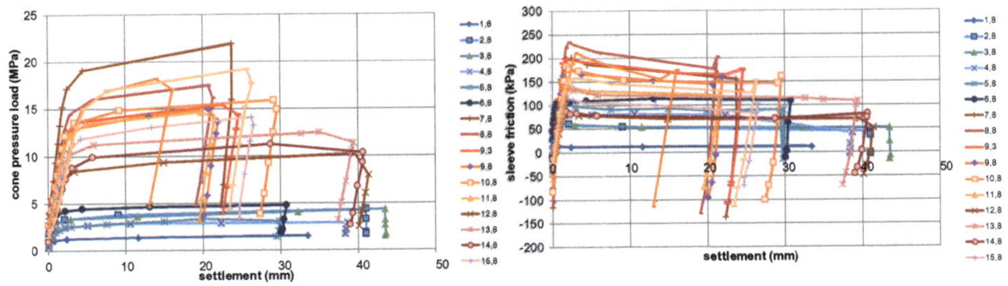

Fig. 5. (a) cone loading tests curves (b) Measured sleeve friction in sand (Limelette, Belgium)

Values of the measured force acting on the friction sleeve divided by the area of the sleeve can be identified to the unit skin friction resistance q_s determined during a pile loading test with retrievable extensometer (Fig. 3).

Then in the general case of a layered ground for which the distribution of penetrometer limit cone resistance q_{CLT} with depth are known, each term of Eq. 1 will be computed from the following equations:

$$R_{pu} = \left[q_o + k_p.(q_{CLT})\right]\pi.B^2/4 \tag{2}$$

$$R_{su} = \sum_{i}^{n} k_s.fs.\pi.B.l_i \tag{3}$$

where q_o is the total vertical pressure, k_p is the bearing factor, k_s is the pile soil interaction factor, B is the diameter of the pile, f_s is the limit unit shaft friction of the ith layer, l_i the thickness of the ith layer, q_{CLT} the limit cone resistance obtained near the tip of the pile.

The comparison and the correlation between the mobilization curve obtained from the cone loading test and that of the static pile loading test, for the same site and same ground type, give access to influence factors. However, there are some specificities which depend on the size and geometry of the tip used in the cone loading test (in order to limit the friction and thus limit the corrections on the measurement a 15 cm^2 cone is used), as well as the friction sleeve influence (steel used for friction sleeve is particularly smooth compared to some piles shafts). It must be noted that values derived from CLT test differs drastically form the one used for example in the very interesting keynote of de Cock (2008).

Indeed it is necessary to integrate a correction factor which takes account of the scale factor between the friction sleeve and the pile shaft k_s and cone and pile tip k_p. The ratio of the estimated load transmitted by the shaft and the pile tip is very close to the ratio observed on the tests collected during this work. As a result, no particular factor was applied. However, on average, the predicted total load remains less than the total bearing capacity observed. A model factor equal to 2 has to be applied to correct the calculated values.

Figure 7a shows a regression curve obtained for 20 piles in various soils (clay, silt, sand and chalk) and of different kind (screw, bored, driven piles). The pile-soil interaction factor is given by NF P94-262 standard (AFNOR 2012). A good prediction is observed with a mean value close to unity at 50% of the frequency. Further fitting of model factors is needed. The usual French CPT direct design method used q_c and not f_s to derived the shaft friction.

3 Deriving t-z Curves from CLT Tests Results

The goal of this research was to develop a method that would give a prediction of a complete load settlement curve for pile based on the analysis of the friction mobilization curves (t-z curves) at different levels of the cone loading test. The approach consists of transforming the CLT sleeve friction and cone resistance curves representative of the soil along the shaft and below the tip into a settlement (or t-z) curve (Fig. 6). To plot these curves, the forces (qc and fs) are normalized by their maximum observed value and the displacement h is normalized by the diameter of the cone. Then, exponential curves similar to those used in the NF P94-262 standard are calibrated on cloud of points (AFNOR 2012; Burlon et al. 2014). The way this new method has been developed is very close to one described by Briaud and also used in NEN6743 (Briaud 2007; Larsson and Bengtsson 2008; NEN 1991).

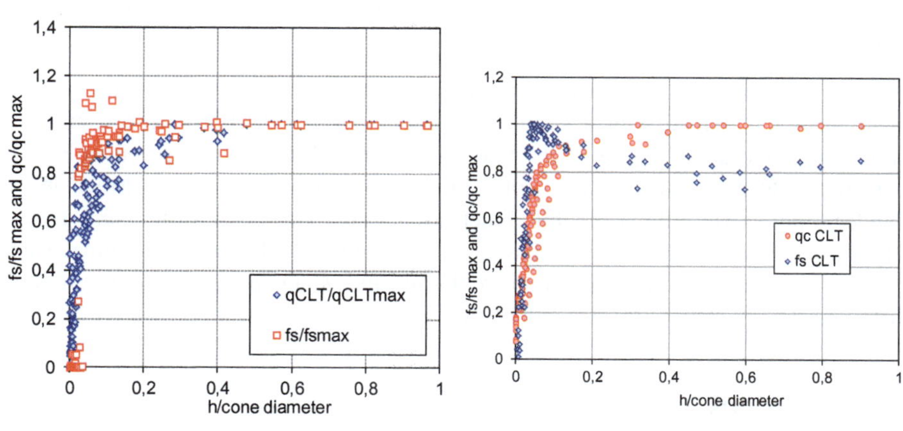

Fig. 6. Normalized t-z curves for tip and shaft for (a) clay and (b) sand

3.1 Prediction of Settlement

The settlement at the top of the foundation w_{found} can be determined as a function of the pile load as the sum of the value of the settlement of the upper end of the pile and the settlement due to compression of the soil layers situated below the pile point level. The settlement of the upper end of the pile includes the settlement relative to the pile point due to elasticity of the pile (elastic shortening).

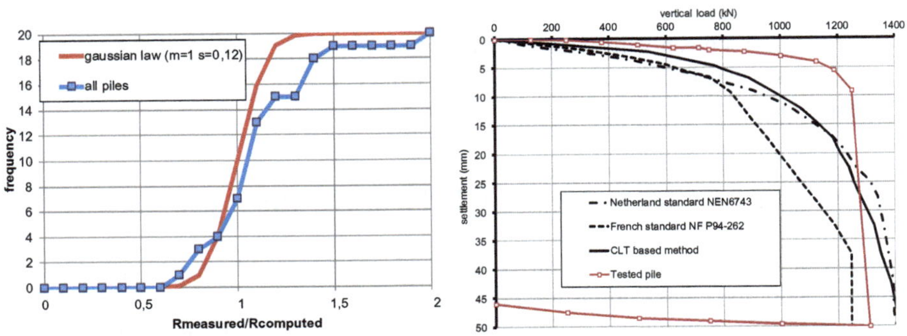

Fig. 7. Bearing capacity and settlement prediction for 20 piles; example of predicted load settlement curve in clay (Merville)

$$w_{found} = w_{skin} + w_{tip} + w_{el} \tag{4}$$

where

- w_{tip} and w_{skin} are obtained from relative load settlement curve constructed with the average cone resistance and sleeve friction versus the relative settlement s/B,
- w_{el} is function of the assumed linear variation of the vertical effort in the pile (Fig. 7b).

Figure 7b shows a comparison of the load settlement curve predicted with the CLT direct method obtained with a simple spreadsheet or the PIVER software, in comparison with the actual curve of the pile loading test. The example given in Fig. 7b corresponds to a pile with a depth of 12 m and a diameter of 0.5 m, drilled in the Flanders clay at Merville. The CLT dimensioning approach is well fitted and gives a very similar result to the methods based on the results of Ménard pressuremeter and cone penetration tests on which French (Frank and Zhao method used in AFNOR 2012) and Dutch (NEN 1991) standards are based.

4 Conclusions

The present paper has tried to describe a new direct design method based on the cone loading test taking advantage of the intrinsic quality of the CPT test: reproducibility, precision and furthermore giving access to strength but also stiffness. This new design method offers high potential application to spread footing design. Further research need to collect case studies on different sites and soils: loam and sand to define the adapted rheological factors.

Acknowledgements. The authors thank Fondasol, Lankelma Geotechniek Zuid, IFSTTAR, University of Clermont-Ferrand for the support they were able to provide during this study.

References

AFNOR: Justification des ouvrages géotechniques, Normes d'application nationale de l'Eurocode 7, Fondations profondes, NF P94-262, 204 pages (2012)

Ali, H., Reiffsteck, P., Bacconnet, C., Gourves, R., Baguelin, F., Van De Graaf, H.: Facteurs d'influence de l'essai de chargement de pointe, Journées Nationales de Géotechnique et de Géologie de l'Ingénieur JNGG' 08, Nantes, pp 467–474 (2008)

Arbaoui, H., Gourvès, R., Bressolette, Ph, Bodé, L.: Mesure de la déformabilité des sols in situ à l'aide d'un essai de chargement statique d'une pointe pénétrométrique. Can. Geotech. J. **43** (4), 355–369 (2006)

Briaud, J.-L.: Spread footings in sand: load settlement curve approach. J. Geotech. Geoenviron. Eng ASCE **133**(8), 805–920 (2007)

Burlon, S., Frank, R., Baguelin, F., Habert, J., Legrand, S.: Model factor for the bearing capacity of piles from pressuremeter test results—Eurocode 7 approach. Géotechnique **64**(7), 513–525 (2014)

CEN/ISO: Geotechnical investigation and testing—field testing—part 1: electrical cone and piezocone penetration tests, EN/ISO 22476-1, 41 p (2005)

De Cock, F.A.: Sense and sensitivity of pile load-deformation behaviour. In: BAP V: 5th International Symposium on Deep Foundations on Bored and Auger Piles, pp. 23–46 (2008)

Haefeli, R., Fehlmann, H.B.: Measurement of soil compressibility in situ by means of the model pile test. In: Proceedings of the 4th International Conference on SMFE, London, vol. 1, pp. 225–230 (1957)

Ladanyi, B.: Use of the static penetration tests in frozen soils. Can. Geotech. J. **13**(2), 95–110 (1976)

Larsson, R., Bengtsson, P.-E.: Field determination of stress-strain relations of clay till. In: 4th International Symposium on Deformation Characteristics of Geomaterials (IS-Atlanta 2008), pp. 875–882 (2008)

NEN: Calculation method for bearing capacity of pile foundation, compression pile, Dutch Standard NEN 6743, 31 p (1991)

Reiffsteck, P., Bacconnet, C., Gourvès, R., van de Graaf, H.C., Thorel, L.: Measurements of soil compressibility by means of cone penetrometer. Soils Found. **49**(3), 397–408 (2009)

Sanglerat, G.: The state of the Art in France, pp. 47–58. ESOPT, Stockholm (1974)

Experimental Study on Ultimate Capacity of Large Screw Piles in Beijing

Daping Xiao(✉), Chunqiu Wu, and Houshan Wu

Helical Foundations (Beijing) Technology Co. Ltd, Beijing, China
xiao@screwpile.com.cn

Abstract. This paper presents static load test results of ultimate uplift and compression capacities of screw piles with torque measurements performed in China, where the application of large size screw pile is currently not widespread. The 5 test piles have 114 mm shaft diameter and single/double helix of 406/508 mm diameter, and were installed in sandy silt and silty clay at a test field in Beijing, China. Torques during screw pile installation were measured by differential hydraulic pressure method which was calibrated with a wireless torque-meter. The piles were tested to their ultimate tension capacities, unloaded and then reloaded to their ultimate compression capacities. Load test results show that the screw piles have similar uplift and compression capacities. It is demonstrated that the empirical torque-capacity ratio can well predict the actual ultimate pile capacities of test piles. The "individual bearing" calculation method with consideration of shaft friction can also well predict the ultimate pile capacities of multi-helix screw piles.

1 Introduction

A screw pile consists of a steel shaft with one or multiple circular helices affixed to the central shaft. There is a wide range of shaft sizes available for design ranging from 76 to 219 mm for axially loaded piles. Screw piles of shaft diameter larger than 100 mm are normally classified as large-size piles. The pitch and spacing of the helices can be varied so that the upper helices will follow the lower one when advancing into the soil. The helix can be welded to the steel shaft, and the helical blades could be knife edged to facilitate their installation and minimize disturbance to the soil during installation. Screw piles are typically installed by rotating into the ground with a hydraulic torque drivehead mounted on an excavator (Perko 2009).

Screw piles, also known as helical piles/anchors, have been used in many applications in North America. However, currently their applications in China are limited to bracket foundations for solar power arrays with small-size screw piles of normally less than 2.5 m (Wang 2012). Field load test studies on large-size screw piles are still relatively rare, which to a certain extent has restricted their uses in buildings and infrastructures (Xiao and Wang 2012). Therefore, a field test program, including compression and tension pile load tests, was carried out on large-size screw piles installed in typical soils in Beijing. The test program was focused on examining the ultimate capacities of screw piles obtained from load tests with the predicated capacities by empirical torque-capacity ratio and formula calculation. These full scale field

M. Abu-Farsakh et al. (eds.), *Advances in Analysis and Design of Deep Foundations*,
Sustainable Civil Infrastructures, DOI 10.1007/978-3-319-61642-1_5

test results will be used to develop a more reliable design method for large-capacity screw piles.

2 Test Program and Load Tests

2.1 Field Test Program

Two types of screw piles were installed early in 2014 at Machikou town, Changping District, Beijing. One type is of single 508 mm diameter helice and another is of double diameter 406 mm helices, both types are of 114 mm shaft diameter. A total of 5 test piles were installed in a row, spaced at least 2 m apart. The test piles were installed to different depths of 4 m, 6 m, 8 m and 10.8 m respectively. Torques during screw pile installation were measured in each 0.25 m advance through differential hydraulic pressure method, which was later calibrated with a wireless torque-meter. Figure 1 shows a typical set up for installing test pile.

Fig. 1. Installation of screw piles

Stratified sediments (mostly sandy silt, silty clay and clayey silt) of low to medium plasticity occur at the test site to a depth of about 16 m. Two thin layers of sands, with thickness varied from 0.5 to 0.7 m, are found in the 16 m thick sediments. The Quaternary sediments are underlain by a thick dense sand layer. Ground water was not found in the maximum drill depth of 25 m during a detailed soil investigation conducted before test pile installation. Figure 2 shows the soil profile and test pile

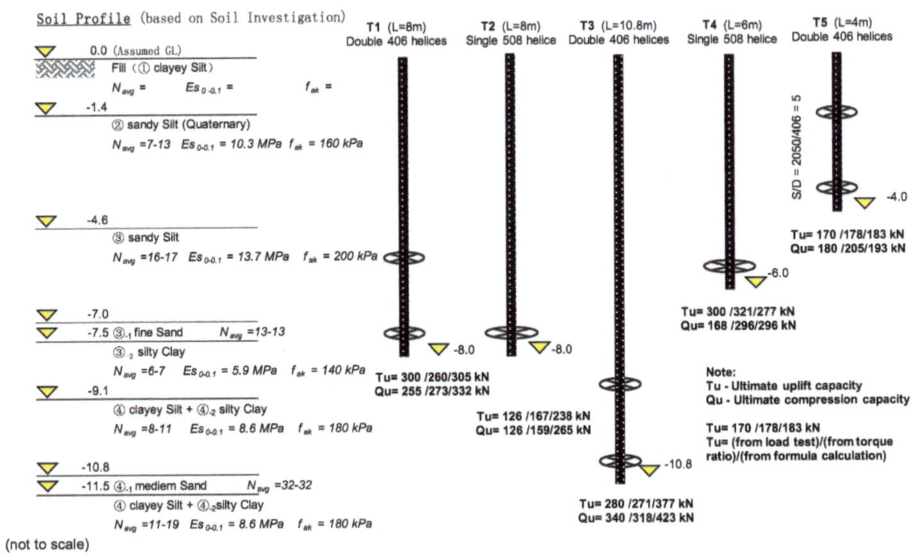

Fig. 2. The soil profile, test piles and ultimate capacities

arrangement, as well as ultimate uplift/compression capacities obtained from static load test, empirical torque-capacity ratio method and formula calculation method.

2.2 Pile Load Tests

Axial tension tests were firstly carried out on test piles, at least 3 days after their installation. A calibrated hydraulic jack with 150 tonnes capacity was placed on top of the test beam. High strength steel bars bolted to the reaction frame were used to tie the loading system with the test piles. The vertical pile movements were monitored by two dial gages, accurate to 0.01 mm, attached to the H-section steel reference beam.

The uplift load tests were conducted following the quick load test procedure, as described in Chinese Code GB5007-2011 Appendix T. Each pile was loaded to failure in increments of about 20% of the ultimate capacity of test pile. The loading increment was somehow too large due to an overestimate of ultimate pile capacity. Each loading increment was held for 2 h until "failure" has occurred, defined as continuous jacking is required to maintain the test load or cumulative deflection up to 100 mm occurs. Figure 3 shows a general setup for the axial tension test.

Axial compression tests were carried out on each test piles, at least 3 days after the completion of its uplift test. Kentledge reaction system was adopted for the axial compression pile load tests. A 5 tons preloading was first applied and maintained for 2 h in order to counterbalance the uplift influence from previous tension tests. Compression load tests were also conducted following the quick load test procedure, as described in Chinese Code GB5007-2011 Appendix Q. Each test pile was loaded to failure in increments of about 15% of the ultimate capacity of test pile.

Fig. 3. Axial tension test setup

3 Test Results and Analyses

3.1 Test Results

For all piles tested, measurements were taken to obtain the curves of load-deformation and deformation-logarithm of time for each load increment, which is the basis for determination of ultimate pile capacity according to Chinese Code GB5007-2011. Figure 4 shows the load-settlement (*Q-s*) curve from compression test and load-deflection (T-Δ) curve from uplift test for test pile T3.

Table 1 summarizes the geometry of all 5 test piles and the ultimate capacities obtained from the load tests. For easy reference, these results are also shown in Fig. 2

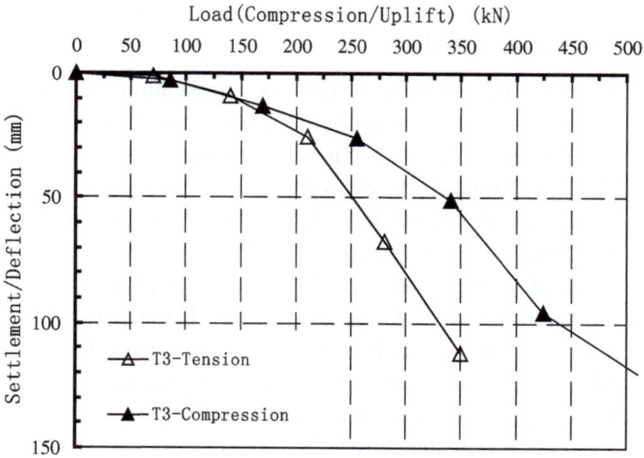

Fig. 4. Tension and compression load-deformation curves of T3

together with the soil profile. Table 1 also lists the comparisons of ultimate capacities obtained from empirical capacity-torque ratio and from calculation of individual bearing method. It should be pointed out that for the method of using capacity-torque ratio, the average of torques recorded in the last pile penetration of 4 times of helix diameter depth (about 2 m) was used for predication of the ultimate tension capacity T_u, and the final 0.5 m torque recorded was used for predication of ultimate compression capacity Q_u.

Load test results, as shown in Table 1, reveals that the test piles are of similar uplift and compression capacities. However, it shall be noted that the ultimate compression capacity Q_u might be somehow influenced by the test program; i.e., for each test pile, ultimate tensioning was first performed, followed by unloading, and then reloading to ultimate compression capacity of the pile.

Table 1. Summary of test screw piles and test results

Test piles	Pile depth (m)	Shaft Dia. (mm)	No. of helix	Helix Dia. (D, mm)	Helix spacing (S, mm)	Ultimate capacity (Q_u, T_u)		
						Based on load test (kN)	Based on torque ratio (kN)	Based on formula calculation (kN)
T1-Uplift.	8	114	2	406	2050	300	260	305
T1-Comp.	8	114	2	406	2050	255	273	332
T2-Uplift.	8	114	1	508	–	126	167	238
T2-Comp.	8	114	1	508	–	126	159	265
T3-Uplift.	10.8	114	2	406	2050	280	271	377
T3-Comp.	10.8	114	2	406	2050	340	318	423
T4-Uplift.	6	114	1	508	–	300	321	277
T4-Comp.	6	114	1	508	–	168	296	296
T5-Uplift.	4	114	2	406	2050	170	178	183
T5-Comp.	4	114	2	406	2050	180	205	193

3.2 Predication of Pile Capacities

Based on current researches and industry practice, predication methods for screw pile capacity can be categorized into "calculation method", and "torque method".

The "calculation method" can be further categorized into, according to helix spacing ration S/D, "cylindrical shear" method and "individual bearing" method. Previous experimental results demonstrate that for $S/D < 1.5$, screw pile capacity is a combination of the bearing of the bottom helical plate and the side shear along the cylinder of soil encased between the helical plates. For S/D larger than 3, each helix will act independently (Xiao and Wang 2012). For all the 3 test piles with double helices, the helix spacing ration S/D is 5, therefore "individual bearing" method was applied in the calculation of pile capacities. For example, the ultimate compression capacity of screw pile Q_u is the sum of the individual capacities of n helical bearing plates plus adhesion along the shaft, given by

$$Q_u = \sum Q_{ik} + \alpha H(\pi d) \tag{1}$$

It should be noted that in current practice, for example AC358, shaft friction for screw pile of shaft diameter <76 mm is often excluded in capacity estimation. As the pile shaft diameters increase, which in turn implies a deeper pile penetration, the contribution of shaft friction to pile capacity shall not be ignored anymore. This point was also revealed from the pile capacity calculation in Table 1.

In addition, an empirical method called the installation torque method is commonly used in screw pile industry. This method was developed based on empirical correlation, but lacks explicit definition related to traditional geotechnical concepts. However, it has been used successfully in the construction of thousands of screw piles over the past 30 years (Hoyt and Clemence 1989). For example, the ultimate tension capacity of screw pile T_u is related to the final installation torque T and the capacity-to-torque ration K_t, given by

$$\mathrm{T}_u = K_t T \tag{2}$$

It can be seen from Table 1 that the empirical torque-capacity ratio method gives a good predication of ultimate pile capacities, following $K_t = 16$ for tension capacity and $K_t = 18$ for compression capacity for the 114 mm shaft test pile (Perko 2009). Torque monitoring during installation provides a suitable method of production control, it is suggested that the required installation torque should be specified in the design.

4 Conclusions

A full scale field test program was successfully performed with the purpose of developing a more reliable design method for large-size screw piles. Static load test results show that the large-size screw piles are of similar uplift and compression capacities. It is demonstrated that the empirical torque-capacity ratio method can well predicate the actual ultimate pile capacities. For design of screw piles, the "individual bearing" calculation method with consideration of shaft friction can give a good predication of ultimate pile capacities. Torque monitoring during installation provides a suitable way for production control, and the required installation torque should be specified in the design.

References

Perko, H.A.: Helical Piles: A Practical Guide to Design and Installation. Wiley, New Jersey (2009)

Wang, D.: Experimental study on vertical bearing mechanism of screw pile (in Chinese). Master of Science Thesis, Tianjin University, Tianjin, China (2012)

Xiao, D., Wang, D.: Experimental study on optimum helix spacing of screw pile (in Chinese). Key Technologies in Design of Solar Power Station. China Electric Power Press, Beijing, China (2012)

Hoyt, R.M., Clemence, S.P.: Uplift capacity of helical anchors in soil. In: Proceedings on 12th International Conference on Soil Mechanics and Foundation Engineering, Rio de Janerio, Brazil, vol. 2, pp. 1019–1022 (1989)

Stability Analysis of Steel Pipe Pile Reinforced Inhomogeneous Slope by Using Strength Reduction Finite Element Method

Shi Baotong(✉) and Kong Xiangxing

The First Highway Survey and Design Institute of China Communications Construction Company Ltd., Xi'an, China
117021654@qq.com, 103121153@qq.com

Abstract. Taking an inheterogeneous slope reinforced by steel pipe piles as an object, the finite element strength reduction method, in which the connection of equivalent plastic strain zone is taken as failure standard, is used to analyze the stability coefficient of slope before and after reinforcement. From the results, it is shown that the safety coefficient increases from 1.3 to 1.75 as a result of the effect of steel pipe piles, which indicates that the reinforcement effect of steel pipe piles is feasible.

1 Introduction

In the slope, cracks have been appeared in the place with a distance of about 6-8 m from the trailing edge line, and the width of the widest crack is about 1.5 cm. The safety coefficient of the slope calculated by limit equilibrium method is about 1.15, which is far less than the local specification value of 1.43. In order to ensure the safety of the slope, reinforcement treatments is needed. As the steel pipe pile has the advantages of high degree of mechanization, quick and convenient construction, it is adopted to reinforce the slope. The proposed reinforcement plan is to constructing three rows of steel pipe pile in the top of the slope, of which the diameter is 200 mm and the row spacing is 1.5 m.

In order to study the effect of steel pipe piles, the stability of the slope is analyzed by using three-dimensional finite element method. As a commonly used method for slope stability analyses, the reliability of strength reduction method has been validated by many researchers [1–6]. So, in this paper the finite element strength reduction method is used to analyze the slope stability, and the safety factors of slope before and after reinforcement are computed and compared with each other in order to evaluate the reinforcement effect of steel pipe piles.

2 Theory of Finite Element Reduction Method

The mechanism of strength reduction is to decrease the shear strength parameters of rock and soil until it arrives to failure. In the reduction process, the slip surface can be obtained automatically and the safety factor of slope F is also obtained [7, 8], that is,

M. Abu-Farsakh et al. (eds.), *Advances in Analysis and Design of Deep Foundations*, Sustainable Civil Infrastructures, DOI 10.1007/978-3-319-61642-1_6

$$c' = \frac{c}{F}, \quad \tan\phi' = \frac{\tan\phi}{F} \tag{1}$$

where c, c′ and φ, φ' are cohesions and internal friction angles before and after reduction, respectively.

In the finite element strength reduction method, the constitutive model for geotechnical material is ideal elastic-plastic model, which relates closely to the safety factor. In traditional limit equilibrium method, Mohr-Coulomb criterion (referred to as M-C criterion) is used. However, the yield surface of M-C criterion on π plane in the principal stress space is irregular hexagonal cross section that spires exist, which makes it difficult to use in numerical simulations, while Drucker-Prager criterion (referred to as D-P criterion) is a circular on π plane, so it is much easier to be achieved in programming, and the calculation efficiency is high because there is no numerical spires existing. The expression of D-P criterion is as follows,

$$f = \alpha I_1 + \sqrt{J_2} = k \tag{2}$$

where I1 and J2 are the first invariant of stress tensor and the second invariants of deviatoric stress tensor, α and k are constants related to cohesion c and internal friction angle φ.

As the D-P criterion is the approximation of M-C criterion, different values of constants α and k in Eq. (2) represent different types of circles on π plane. Zheng Y.R. [7] has summed up the relations between each criterions as shown in Table 1. From Zheng's study, it is suggested that M-C criterion can be used to replace by DP3 criterion.

Table 1. Parameters for each D-P criterion

No.	α	k
DP1	$2\sin\phi/\sqrt{3}(3-\sin\phi)$	$6c\cos\phi/\sqrt{3}(3-\sin\phi)$
DP2	$2\sin\phi/\sqrt{3}(3+\sin\phi)$	$6c\cos\phi/\sqrt{3}(3+\sin\phi)$
DP3	$\frac{2\sqrt{3}\sin\phi}{\sqrt{2\sqrt{3}\pi(9+\sin^2\phi)}}$	$\frac{6\sqrt{3}c\cos\phi}{\sqrt{2\sqrt{3}\pi(9-\sin^2\phi)}}$

In the strength reduction FEM analysis, the solutions for slope failure are classified into two kinds: one is the non-convergence of force and displacement, the other is the run through of equivalent plastic strain from the bottom to the top. In this paper, the latter is adopted to evaluate the failure of slope.

3 FEM Model of Inhomogeneous Slope

In this paper, the slope is inhomogeneous and contain five layers of soil (Fig. 1), of which the mechanical parameters of each layer are shown in Table 2. When establishing the three-dimensional finite element model, the strike of slope is taken as axis

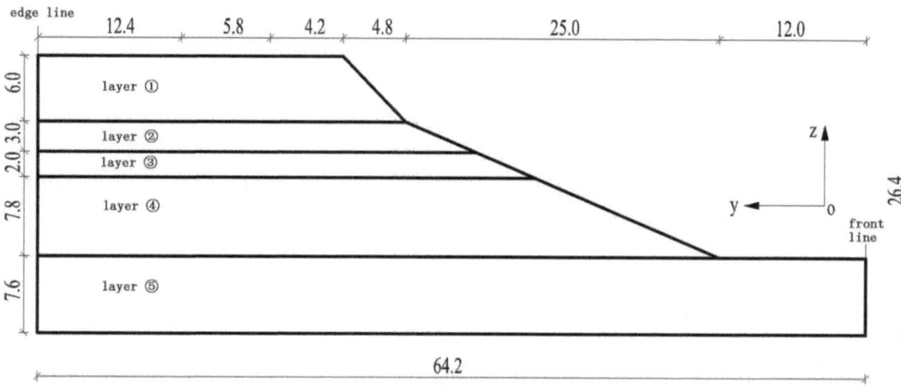

Fig. 1. Section of the slope

Table 2. Physical and mechanical parameters of soils

Soil layer	E (MPa)	μ	γ (kN/m^3)	γ' (kN/m^3)	C (kPa)	φ (°)
①	5.4	0.3	18.7	8.9	5	32
②	1.58	0.3	17.1	7.3	13	14
③	3.4	0.3	18.0	8.2	4	27
④	6.0	0.3	19.8	10.0	15.0	23
⑤	8.7	0.3	19.3	9.5	3	35

x and the positive axis y is perpendicular to strike and points towards the direction of the slope trailing edge, the vertical direction is z axis and upward is positive. The scopes of slope model are 104 m in x direction, 64.2 m in y direction and about 26.4 m in z direction.

In the finite element model, the displacement boundary conditions include displacement constraints in x, y, z directions, that is, the left and right sides of the slope are constrained with zero displacement, the front and back sides of the slope are also constrained with no displacement, while the bottom of the slope is restricted but the top of the slope is free and displacement is permitted. The stress boundary conditions of FEM model contain self-weight stress and hydro-static pressure. Self-weight stress is exerted in the slope in the form of volume force, while the hydrostatic pressure is acted on the slope surface in the form of equivalent surface load.

Furthermore, in the FEM model, 8 node hexahedral element is introduced to soil, 3D beam element with 2 nodes is used to simulate steel pipe pile. The total element number of the slope model is 11474, in which the entity elements are 9454, and beam elements are 2020.

4 Case Design

According to the purpose of this research, two cases are taken into consideration and the corresponding load conditions are shown in Table 3.

Table 3. Load conditions of each case

No.	Description	Load condition
Case 1	Before reinforced	Uniform load of 20 kN/m^2
Case 2	After reinforced	Uniform load of 20 kN/m^2 and equivalent uniform loads resulted from construction machinery

5 Results and Discussions

As mentioned above, the run through of plastic zone is used as the failure criterion when performing FEM numerical analyses of slope stability. Herein, the stability of slope in two cases have been analyzed by using strength reduction method. The plastic zone distribution in the slope under the two cases are shown in Fig. 2.

From the distribution of plastic zone in slope during failure, it is found that the development of plastic zone firstly started at the trailing edge and the third layer of soil, then it gradually developed into the deep area of slope and connected with the plastic zone in the third layer of soil, then the plastic zone continued to develop toward the front edge of the slope, finally, the run through plastic zone was formed from the toe of the slope to the trailing edge, which represented the failure of the slope. The corresponding reduction factor is the safety factor of the slope. According to the results, it was found that in case 1, the safety factor of slope before reinforcement is 1.3, which is less than the specified standard value of 1.43, which indicated that the slope is unstable and should be reinforced. However, the safety coefficient calculated by three-dimensional finite element method is greater than the safety factor of 1.15 calculated by using limit equilibrium method. The reason for this discrepancy is that the strength reduction finite element method used in this paper is three-dimensional. When the slope is reinforced by steel pipe piles, the safety factor increased to 1.75, which indicated that the reinforcement effect of steel pipe piles will increase the stability of the slope to a large extent.

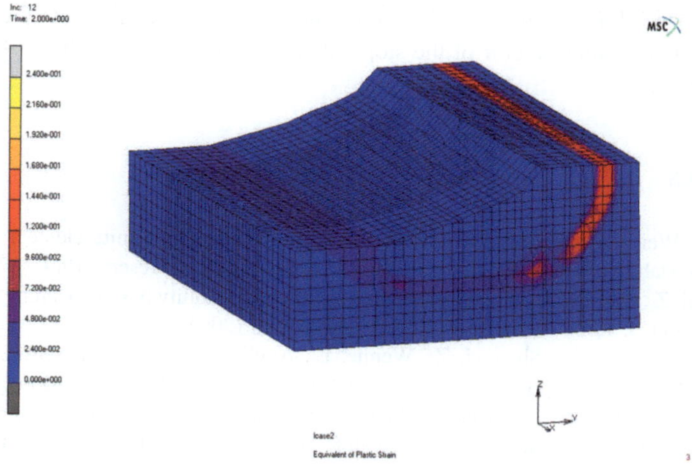

(a) Case 1: F=1.3

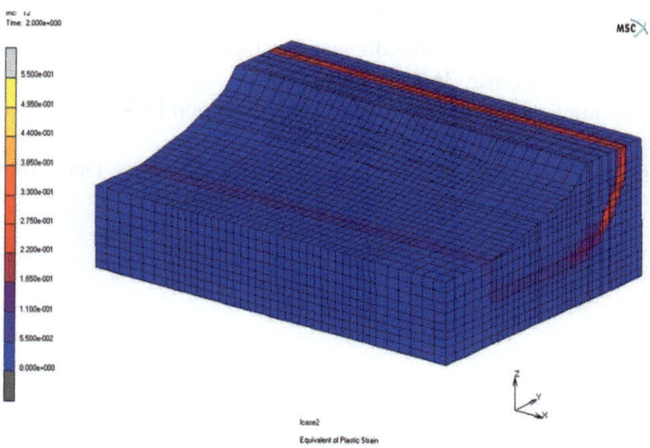

(b) Case 2: F=1.75

Fig. 2. Distribution of plastic zone when the slope failed

6 Conclusion

In this paper, in order to analyze the stability of an inhomogeneous slope and evaluate the effect of steel pipe pile, the finite element strength reduction method is introduced to investigate the stability of slope before and after reinforcement. Firstly, the theory of finite element strength reduction method was presented detailedly, and the Drucker-Prager criterion is adopted to perform numerical simulations. Then a three-dimensional finite element model of the slope was established to simulate the reinforcing process and two cases were taken into consideration to represent the conditions before and after reinforcement. From the calculation results, it was found that the safety factor of slope before reinforcement was about 1.3 while the safety factor of slope after reinforcement was 1.75, which indicated that the effect of steel pipe pile could promote the safety factor of the slope, thus the reinforcement effect of steel pipe pile is positive.

References

1. Song, Y., Wei, Y., Lin, M.: An analysis with three-dimensional finite element method for high slope stability to Kaiyuan road. J. Geol. Hazards Environ. Preserv. **19**(4), 87–91 (2008)
2. Ma, J., Lai, Z., Cai, Q., Xu, Z.: 3D FEM analysis of slope stability based on strength reduction method. Chin. J. Rock Mech. Eng. **23**(16), 2690–2693 (2004)
3. Yakun, S., Yingren, Z., Shangyi, Z., Wenjie, L.: Application of three-dimensional strength reduction FEM in slope[J]. Chin. J. Undergr. Space Eng. **2**(5), 822–827 (2006)
4. Chen, F.: Application of strength reduction technique to the three-dimensional stability analysis of landslides. Master thesis, Institute of Rock and soil Mechanics, the Chinese Academy of Sciences, Wuhan (2006)
5. Fei, C., Jianhui, D.: Three-dimensional stability analysis of rock slope with strength reduction method. Chin. J. Rock Mech. Eng. **25**(12), 2546–2551 (2006)
6. Jinghua, H.Z.L.: Analysis of three-dimension slope stability by strength reduction finite element method. Ind. Constr. **36**(6), 59–64 (2006)
7. Yingren, Z., Shangyi, Z.: Application of strength reduction FEM in soil and rock slope. Chin. J. Rock Mech. Eng. **23**(19), 3381–3388 (2004)
8. Zhao, S., Zheng, Y.R., Shi, W., Wang, J.: Analysis on safety factor of slope by strength reduction FEM. Chin. J. Geotech. Eng. **24**(3), 343–346 (2002)

The Evolution of Analysis Methods for Laterally Loaded Piles Through Time

Ahmed Moussa(✉) and Petros Christou

Civil Engineering Department, Faculty of Engineering,
Frederick University, Nicosia, Cyprus

Abstract. Today's structures impose a higher demand on their foundations as their height increases and their designs are more daring and complex. Naturally, the use of pile foundations, which provide the means for transferring the loads to greater depths, is increasing. Despite the advances in the construction of such foundations as well as the advances in software and technology, the analysis of pile foundations and especially those loaded laterally remains a challenge to the engineers. This paper reviews the literature regarding the development of the analysis methods through time going all the way back to 1867 and the subgrade reaction method by winkler and reaches today and the current state of the art. The analysis methods are grouped into the subgrade reaction methods, the ultimate limit state methods, the continuum methods and finally the finite element method. The main objective is to present the fundamentals of each method in an effort to make a contribution to the design of pile foundations by providing a reference and a critical discussion to the geotechnical engineers.

1 Introduction

Pile foundations are extensively used in bridges, offshore structures, high-rise buildings, wind turbines, etc. It is the oldest technique used for the construction on weak or soft soil stratums as pointed out by Pulikanti and Ramancharla (2013). Piles transfer the axial loads from the superstructure (i.e. dead and live loads) to the supporting soil by two mechanisms: (a) transferring the loads through the pile tip to the soil bed below and (b) the side friction between the pile interface and the surrounding soil. In addition, a significant amount of lateral loads is also applied due to the wind, earthquakes and the lateral movement of the soil, Kavitha et al. (2016). Hence, the lateral load transfer mechanism is different from those mentioned for the axial loads. A great number of research efforts are found in the literature presenting various methods for analyzing the laterally loaded piles (LLP). In general, these methods can be grouped into four main categories: (a) the ultimate limit state method (ULS), (b) the subgrade reaction approach, (c) the continuum method, and (d) the finite element method (FEM). The ultimate limit state (ULS) methods are used to obtain the maximum lateral load that a pile can carry. The methods in the subgrade reaction approach are used to calculate the allowable lateral pile deflection, i.e. serviceability limit states, and are based on the assumption that the soil is represented by a series of linear or nonlinear springs. Similarly, the continuum approach is used to obtain the allowable lateral displacement, however, the idealization of the soil is presented as a linear elastic infinite stratum.

M. Abu-Farsakh et al. (eds.), *Advances in Analysis and Design of Deep Foundations*,
Sustainable Civil Infrastructures, DOI 10.1007/978-3-319-61642-1_7

The boundary element method (BEM) and simplified continuum models are the main paths used in continuum approach and finally, the finite element method (FEM) is a numerical technique that is based on the concept of the continuum approach. In general, it can handle the complexity of the loading and the soil boundary conditions.

This paper provides a comprehensive review of the research for the analysis methods of laterally loaded single piles under static loading. It focuses on the evaluation of each method and highlights the pros and cons of each method.

2 Subgrade Reaction Approach

2.1 Winkler's Hypothesis

Winkler (1867) is the oldest and simplest single parameter model used to determine the bending moments and deflections along a pile. Emil Winkler who used Hooke's law in modelling the bedding of railway tracks first introduced the idea in 1867. He modeled the railway track as a continuous beam on an infinite number of unconnected linear-elastic springs that represented the soil, Kurrer and Melchers (2008). In this method, it is assumed that the soil bearing pressure is proportional to the ground settlement; in other words, the reaction at any point on the beam is influenced only by the deflection at the same point. This approach was later used for laterally loaded piles, since the general behavior is similar to that of the flexible beams against transverse loads, only rotated at 90°. The mathematical expression for the Winkler method is given by the following equation:

$$p = k_h.w \tag{1}$$

where p = the soil reaction per unit area (F/L^2)
k_h = coefficient of sub-grade reaction (F/L^3)
w = beam settlement (L)

The method has major drawbacks, such as the disregard of the spring coupling effect and soil continuity, Horvath (2002). The use of reinforced concrete for the foundations and the growing development of the foundation engineering theory in the early 1920s, led to the extension and improvement of the Winkler's hypothesis by using Euler-Bernoulli beam on top of the elastic soil springs. Biot (1937) and Hetenyi (1946) developed a fourth order linear differential equation (Eq. 2) that covers the deflection and bending moments for such a beam-foundation system. They also provided analytical solutions for various types and locations of loads and load distributions with constant subgrade reaction.

$$E_pI_p\frac{d^4u}{dz^4} + Q\frac{d^2u}{dz^2} + k_huD = 0 \tag{2}$$

where u = Lateral deflection of pile at point z along the length of the pile
E_p = Pile modulus of elasticity
k_h = horizontal coefficient of subgrade reaction
P = Soil pressure over the pile
D = Pile diameter

I_p = Pile cross section moment of Inertia

Q = Axial load on pile

Other researchers presented work accounting for the variation of k_h with depth. For instance, Barber (1953) proposed solutions to calculate the deflections and rotations at the ground surface when k_h is increasing linearly with depth. Terzaghi (1955) proposed rules for the selection of the recommended values of k_h that can be used in Eq. 2 for over consolidated clays, stiff clays and sandy soils. However, no experimental data or analytical procedure were given to validate his recommendations. Reese and Matlock (1956) obtained non-dimensional charts to be used in calculating the ground-line deflection, slope, shear and maximum bending moment on the pile and modified Eq. 1 for laterally loaded piles. This method was based on p-y curves (Eq. 3) that are manually created in addition to manual numerical solutions.

$$p = ky \tag{3}$$

where p = the soil reaction per unit length of the pile (F/L)

k = the modulus of subgrade reaction (F/L^2)

y = lateral deflection of the pile (L)

For this method, the value of k was assumed to be zero at the ground level and then increased linearly with depth. In addition, Matlock and Reese (1960) extended the earlier non-dimensional curves to include cases where k follows polynomial and power function distributions and they provided a solution for a soil profile, where k has a certain value at the ground surface, then linearly increasing with depth. Similarly, Poulos and Davis (1980), Prakash and Sharma (1990) provided curves and tables to find the non-dimensional coefficients for a constant value of k, pile deflections, moments and slopes as functions of depth.

Further Vesic (1961) had extended the work of Biot (1937) by adding the coupled loading case and introducing a new expression for k_h as a function of the soil and the pile properties, as shown in Eq. 4. However, the main shortcoming of Eq. 4 is its independence of pile diameter since the moment of inertia (I_p) for square and circular piles is also proportional to the pile width raised to the fourth power (i.e. D^4).

$$k_h = \frac{0.65E_s}{\left(1 - \mu_s^2\right)} \left[\frac{E_s D^4}{E_p I_p}\right]^{1/12} \tag{4}$$

where E_s = soil modulus of elasticity,

μ_s = Poisson's ratio of the soil

D = pile diameter

E_pI_p = flexural rigidity of the pile

Therefore, Bowles (1996) argued that the proposed value of k_h from Eq. 4 should be doubled. Further, Carter (1984), Ashford and Juirnarongrit (2003) improved Vesic's equations to achieve a more rigorous prediction of the laterally loaded piles' behavior by taking into account the effect of pile diameter.

The subgrade reaction approach has been extended by Davisson and Gill (1963) when they analyzed laterally loaded piles (free and fixed head conditions) embedded in a two layered soil system with a different constant value of k for each layer.

They concluded that the surface layer has a significant influence on pile response and provided the results in non-dimensional forms. Gill (1968) investigated further the appropriate values of k for the designing of laterally loaded piles in mud soils by performing lateral load tests on full-scale piles. Gill concluded that k is a nonlinear function of the pile displacement and can be back calculated from the load-deflection measurements. In addition, an empirical correlation was presented where k is related to the shear strength of the soil as well as other soil properties.

Alizadeh and Davisson (1970) performed field tests to define design criteria for piles in sandy soils. They concluded that a triangular distribution of k with depth is a valid approximation. Moreover, it was pointed out that k was insensitive to the low-pile head deflections and vice versa. Conversely, k was significantly increased, up to 100%, by the densification of the sand.

The following notes are meant to clarify the definition of the parameters E_s, k_h and k.

1. Soil modulus of elasticity (E_s) is a soil property. The rate of change in E_s with depth called the horizontal coefficient of subgrade reaction (k_h)
2. Modulus of subgrade reaction (k, from Eq. 3), and called spring stiffness, is dependent on pile diameter, pile's deflection, and soil properties
3. The slope of a p-y curve is also the spring stiffness (k), but for a particular soil-foundation system

2.2 p-y Curve Method

The p-y curve method relates the nonlinear behavior between the soil resistance (p) and the lateral pile deflection (y). The plot of the p and y variables at particular depth is called the p-y curve, as shown in Fig. 1, Haiderali and Madabhushi (2016). The p-y curves are derived and validated by performing field tests on fully instrumented piles. In 1950s, the full scale testing for laterally loaded piles was possible because of two main developments at that time: (a) the availability of digital computers to solve Eq. 2 and, (b) the ability for remotely reading the strain gauges to obtain the soil response, Reese and Van Impe (2010). The method implicitly accounted for the soil continuity and the nonlinearity of the pile-soil system, however, its main shortcoming is that the p-y curve is unique for a particular soil and pile properties, Horvath and Colasanti (2011). Owing to the popularity of the p-y curve method and its reasonable results, it was suggested to be used in engineering practice by API (1987, 2007). Furthermore, FHWA (2011) recommended the use of LPILE or FBPIER software that use the p-y curve method, in the analysis of laterally loaded piles, Favaretti et al. (2015).

The method was suggested by McClelland and Focht (1958). The authors used the finite difference method to solve the beam bending moment equations with applied nonlinear loads versus deflection curves to model soil response. They conducted a full-scale test on 60 cm (24 in.) steel pipe pile as well as laboratory tests on undisturbed clay samples. By using the output results, they recommended a procedure to correlate the consolidated-undrained triaxial laboratory test to p-y curves at various depths below the ground line, by using the following equations.

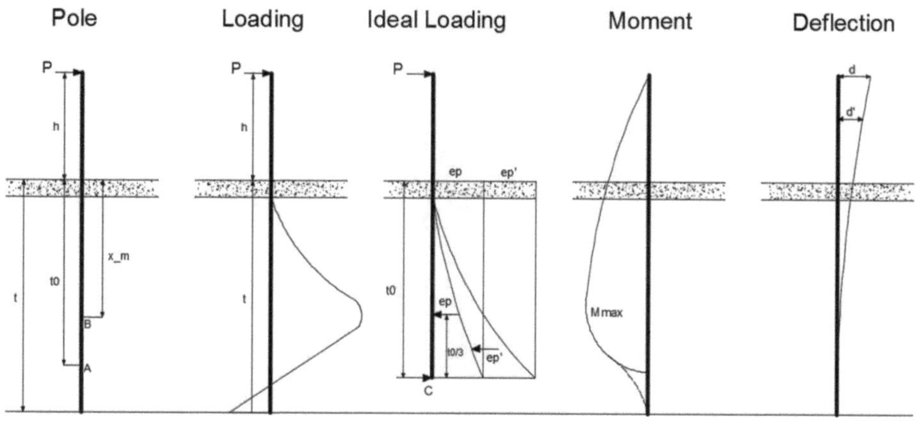

Fig. 1. Schematic representation of Blum method, Ruigrok (2010)

$$p = 5.5b\Delta\sigma \tag{5}$$

$$y = 0.5b\varepsilon \tag{6}$$

where b = diameter of pile

$\Delta\sigma = (\sigma_1 - \sigma_3)$ or deviatoric stress from the stress-strain curve

ε = strain from the stress-strain curve

Over the years, several field tests have been conducted as well as various equations has been derived, to develop p-y curves for various soil types. Matlock (1970) proposed a procedure for soft clay soil under static and cyclic loading based on results of full scale testing, Fig. 2. The proposed equation to construct the p-y curve for the static loading is the following.

$$\frac{p}{p_u} = 0.5\left(\frac{y}{y_{50}}\right)^{\frac{1}{3}} \tag{7}$$

where $p_u = N_p c d$

$$N_p = 3 + \frac{\sigma_z}{c} + J\frac{z}{d}$$

$$y_{50} = 2.5\varepsilon_{50} d$$

p_u is the ultimate soil resistance per unit length of the pile, y_{50} is the deflection at one-half of the ultimate soil resistance, c is the soil strength, d is the pile diameter, σ_z is the overburden pressure, z is the depth from ground surface to p-y curve, ε_{50} is the strain at one half of the maximum deviatoric stress in undrained test, J is an empirical constant and N_p is the ultimate lateral soil resistance coefficient.

Four years later, Reese et al. (1974) performed full-scale tests on two 60 cm (24 in.) diameter piles under static and cyclic loading to develop p-y curves for sandy soil.

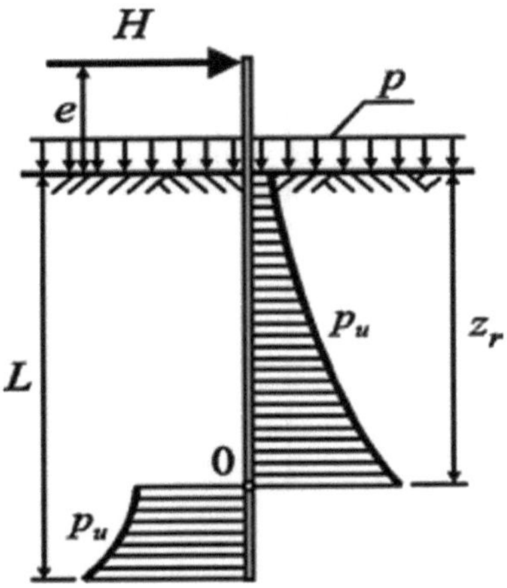

Fig. 2. Schematic representation of Brinch Hansen's method, Pula (2007)

The proposed p-y curve, as shown in Fig. 3, was divided into two portions, an initial straight line and a parabola, which are given, by Eqs. 8 and 9, respectively.

$$p = Kzy \tag{8}$$

$$p = Cy^{\frac{1}{n}} \tag{9}$$

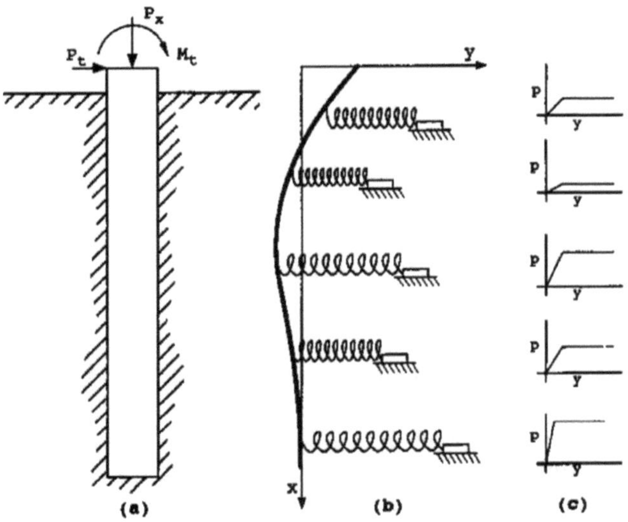

Fig. 3. Model of laterally loaded pile: (a) elevation view; (b) as elastic line; (c) p-y curves, Reese (1997)

Furthermore, Reese et al. (1975) proposed another set of p-y curves for laterally loaded piles embedded in stiff clay above and below the water table under static and cyclic loading. The characteristic shape for p-y curves above and below the water table are illustrated in Figs. 4 and 5, respectively. Another full-scale field test was conducted by Lee and Gilbert (1979) on a laterally loaded pile embedded in a very soft and highly plastic clay under large lateral loads, in an effort to simulate hurricane wind loads. The needed soil parameters to construct the new p-y curve were based on an unconfined compression test and the procedure for developing the new p-y curve were similar to the one proposed by Matlock (1970) for soft clay. The prediction of the pile behavior from the proposed p-y curve and the measured p-y curves from the test were in a good agreement. Also, Stevens and Audibert (1979) compared the work of Matlock (1970) and Reese et al. (1975) with seven field results on lateral loaded piles with diameters up to 1.5 meters. The authors concluded that Matlock and Reese equations for predicting p-y curves significantly overestimate the pile deflection at ground surface and underestimate the maximum bending moment. In addition, they recommended values for the parameters N_p and y_{50} with respect to pile diameter.

Georgiadis (1983) presented the development of p-y curves for a layered soil system. In his method, an equivalent depth of all soil layers existing below the upper layer was calculated based on actual depth, overburden pressure and strength properties of the overlying layers, as illustrated in Fig. 6.

The p-y curve method was integrated in software programs such as COM624P Reese (1984) and LPILE (Reese and Wang 1989) to predict the laterally loaded piles' response and to aid the engineers in the design process. Reese (1997) extended the p-y

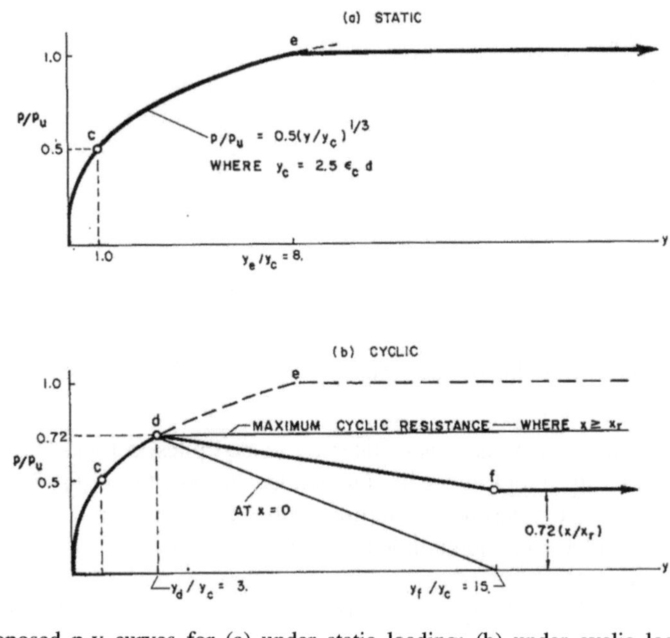

Fig. 4. Proposed p-y curves for (a) under static loading; (b) under cyclic loading, Matlock (1970)

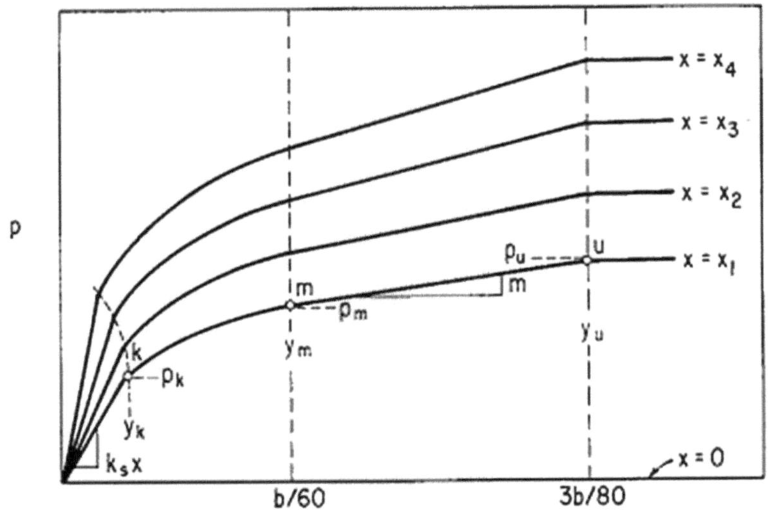

Fig. 5. Characteristic shape of a family of p-y curves for static loading in sand, Reese et al. (1974)

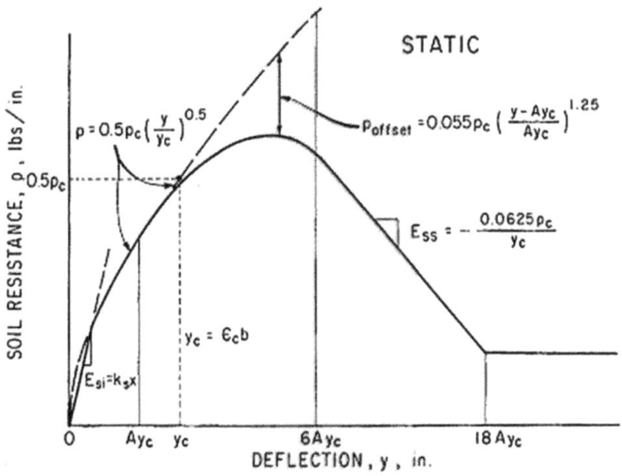

Fig. 6. p-y curve for stiff clay below water table (static loading) (Reese et al. 1975)

curve method to include the analysis of laterally loaded piles in rock. However, only two full-scale field tests in rocks were used to validate the equations of p-y curves, thus, this method was termed "interim". Therefore, in the case of the presence of rock and especially if the rock contains joints filled with weak soil, the method should be used with great caution.

A study presented by Chong et al. (2011) aimed to investigate the effect of the secondary structural effect on laterally loaded piles socketed in jointed mudstone rock.

The interaction between the pile-soil system was modelled by using the three-dimensional distinct element code (3DEC), where the Coulomb slip joint model was used to model the interface element. The pile was modelled with a linear elastic and isotropic material while the soil stress-strain relationship was modelled as a Mohr–Coulomb constitutive model. The p-y curves were developed for one, two and four joints' sets to capture their effect and it was concluded that a single p-y curve could reasonably capture the interaction of pile-soil system. The number of joint sets affect considerably the pile head deflection.

Zhang et al. (1999) developed nonlinear p-y curves for laterally loaded single battered piles, at any angle, embedded in dry sand, based on the results from centrifuge tests. The predicted results from the proposed p-y curves were in good agreement with centrifuge tests results, however the proposed method was limited to the dry sand soil type and a meager free length span range. Mokwa et al. (2000) developed another set of p-y curves for partially saturated silts and clays. The p-y curves formulation were based on Brinch-Hansen's theory as well as 20 lateral load tests at 5 different sites. A spreadsheet was created to calculate the proposed p-y curves for both piles and drilled shafts of any size in cohesion and friction soils. Thereafter, they used the results as an input in the software LPILE Plus 3.0, Reese (1997) to compare the soil resistance and deflections between the developed p-y curves and the default p-y curve for silt, in LPILE Plus 3.0. It was found that the default p-y formulation underestimated the soil resistance and overestimated the deflection for loads exceeding 50% of the ultimate.

Rollins et al. (2005) proposed p-y curves for laterally loaded piles in liquefied sand based on a full scale field tests for single pile and pile group for pore water pressure ratio, $r_u = 95\%$ under blast loading. The shape of the proposed p-y curves was very different than the standard p-y curves. It looked like an inverted S shape as shown in Fig. 7. This was attributed to the increased soil stiffness with larger displacement, which related to the dilative behavior of the soil and the decrease of r_u around the pile. The comparison of the proposed p-y curves with other methods yielded to a better prediction of the measured bending moments and deflection for a wide range of loads and pile diameters.

Chang and Hutchinson (2012) developed nonlinear p-y curves for partially saturated sand that is based on a lab experiment of a steel pile embedded in a partially saturated Nevada sand and the test was conducted by using a large 1 g laminar soil box. This paper provided not only insights for the back-calculated and normalized p-y curves characteristics at different levels of liquefaction but also a detailed description of the design and construction of the test. The laminar soil box was subjected first to dynamic loading to achieve three pore water pressure ratios, i.e. $r_u = 25$, 50 and 90%, then immediately followed by lateral loading on pile head. The empirical equations for the p-y curves were based on the proposed equations for Rollins et al. (2005), but they were modified to fit the test conditions. Also, it was concluded that even for r_u values range from 10 to 15%, the p-y curve will follow the inverted S shape not the standard p-y curve shape, Fig. 7.

Yang et al. (2012) proposed a p-y curve for laterally loaded piles in frozen silt based on a full scale field test. The soil resistance (p) and pile deflection (y) were back calculated based on the field test. The p-y curves proposed by Matlock (1970) and

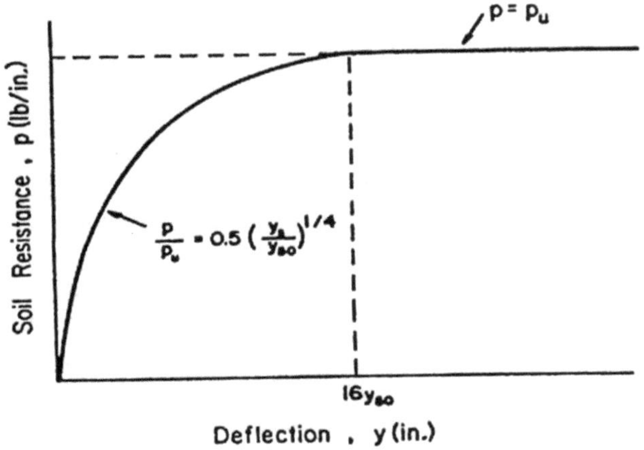

Fig. 7. p-y curve for stiff clay above water table (static loading) (Reese et al. 1975)

Reese (1997) were adjusted to construct the proposed p-y for the frozen silt soil, then used as an input in LPILE software to predict the pile response and to compare the result with field tests results. It was concluded that the proposed p-y curve is capable of predicting the laterally loaded piles behavior in frozen silts as well as permafrost with good accuracy.

In order to investigate and understand the soil-pile interaction under passive loading that was applied at several stages, p-y curves were also developed for a passive pile embedded in well-graded sand by Suleiman et al. (2014). A test was conducted in an advanced lab that is equipped with advanced sensors and stereo digital image correlation to capture the 3D soil movements, pile lateral movements and soil pressure along the pile shaft. The soil box was divided into two parts, i.e. top and bottom box, with a sliding surface between them to simulate the moving soil. A selected p-y curve for a load stage was compared with methods proposed by Broms (1964b) and Ito and Matsui (1975). It is found that the Broms' method overestimated the maximum lateral soil resistance, while Ito and Matsui method underestimated it. Also, the authors noted that the p-y curves for passive piles are dependent on the soil type and properties, therefore, the proposed p-y curves are valid only for the conducted test conditions. The work on developing more rigorous p-y curves has been continuously updated until now.

A comprehensive evaluation of the offshore guidelines for wind turbines (OGL) and other modified p-y methods for a homogenous sand soil had been presented by Thieken et al. (2015). A 3D finite element model has been created with an advanced soil constitutive model to carry out the evaluation. The results of the evaluation showed that the OGL underestimated the pile head displacement for large diameter monopiles under extreme loads, however, overestimated it under small operational loads. Moreover, the evaluated p-y methods was not sufficient for large diameter monopiles or piles with arbitrary dimensions and load levels. Therefore, the proposed p-y method for piles with arbitrary dimensions and load levels was presented based on a parametric study of 250 pile-soil systems, which gave a better prediction for arbitrary soil-pile system in homogenous sand compared to the evaluated methods.

Another study for laterally loaded short rigid monopiles with large diameters that are used heavily in offshore wind turbines is carried out by Zhu et al. (2015). Six lab tests were performed on a short rigid pile embedded in sandy silt, with two relative densities. The lateral load was applied on the pile head at three different heights. The results showed the formation of a rotational center at depth of (0.75 – 0.85)L and a continuous increase in the pile capacity after a displacement of (0.1 – 0.3)B. This was attributed to the strain-hardening of the soil, where L and B are the embedded length and the outside diameter of the pile, respectively. Based on these conclusions, a new set of work-hardening p-y curves were proposed based on horizontal coefficient of subgrade reaction (k_h) that captures the local pile displacements. The new p-y curves were in a very good agreement with the test measurements compared to the p-y curves that were based on ground level displacement.

The common practice in considering the effect of scour on laterally loaded piles is by ignoring the scour-hole dimensions. Lin et al. (2016) proposed a modified p-y curve that takes into account the 3D scour-hole dimensions (i.e. depth, width and slope angle) in soft clay soil. The p-y curve formulation presented by Matlock (1970) was modified by substituting the parameter z from Eq. 9 with another depth that considers the scour-hole dimensions. To validate the proposed simplified approach, a 3D finite difference model was created based on the results from a field test. The modified p-y curve was used as an input in LPILE, in order to be compared with the numerical results. It was found that the scour depth affected significantly the pile behavior more than the width and the slope. By increasing the scour depth from 0 to 8D (pile diameter), the pile head displacement and the maximum bending moment increased significantly. Also, due to the increase of the slope angle, the pile lateral resistance increased. Therefore, it was concluded that the current practice in dealing with scour conditions is definitely conservative and uneconomical in design.

2.3 Strain Wedge (SW) Models

The strain wedge method overcame the limitations of the p-y curve method by taking into account the variation of the soil profile and pile properties. Norris (1986) was the pioneer of implementation of the strain wedge (SW) method on the analysis of laterally loaded piles, Fig. 8. The main concept of the SW approach is to characterize the pile response parameters in terms of the 3D soil-pile behavior. The author analyzed a free pile-head embedded in either clay or sand uniform soil. The results were in very good agreement with the Mustang Island field test results (Cox et al. (1974)). Ashour et al. (1998, 2002, 2004) contributed significantly to Norris' work. They extended the SW method to obtain the response of laterally loaded isolated long piles, drilled shafts and pile groups in layered soils and rock deposits. Further, the developed SW models covered critical factors that affect significantly the analysis of laterally loaded piles such as the nonlinear behavior of pile and soil, liquefaction of the soil and the interaction between piles in a pile group. According to Ashour et al. (1998), the general procedure for the SW model which was validated through other methods and full scale results, are as follows:

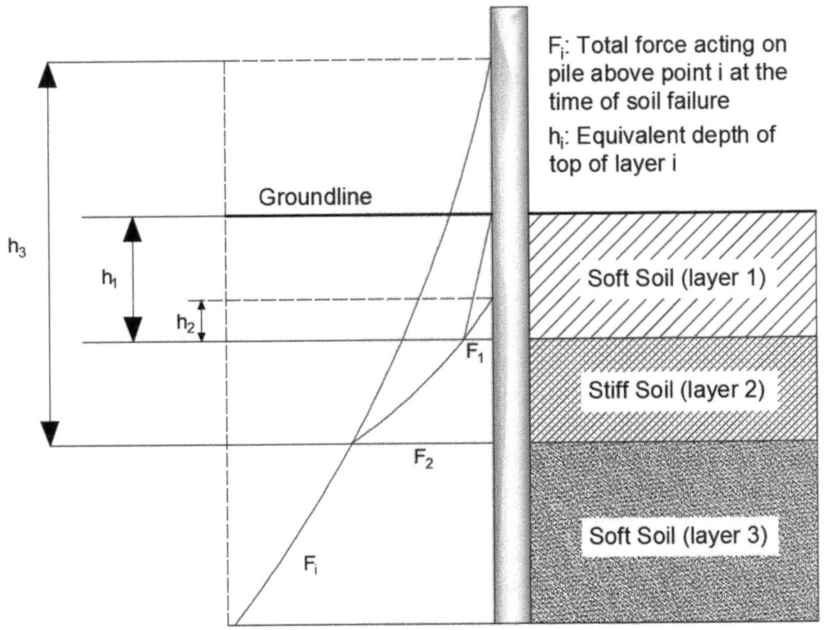

Fig. 8. Typical determination of equivalent depths in a layered soil profile (Georgiadis 1983)

1. Increase in horizontal strength ($\Delta\sigma_h$), stress level (*SL*) and young's modulus (*E*) values are found for a certain value of ε, according to the soil stress-strain relationship
2. The related geometry of the passive wedge of the soil is evaluated based on a postulated initial value *h* of the passive wedge depth. The steps 1 and 2 are applied to the sublayers of the soil layers of depth *h*
3. The current change in soil-pile line load (*p*) along *h* is found in terms of soil and pile parameters and the pile-cross section shape. Then, the values of pile head-head deflection
4. The pile deflection along the depth is obtained in terms of ε, Poisson ration, SL and the size of the passive wedge
5. According to the current pattern of E_s, the loaded pile is assessed as a BEF with a random pile-head lateral load. The values of pile-head deflection (Y_0) and the depth X_0 evaluated using BEF are compared to the values of the SW analysis
6. Iterative processes are used on the same value of soil strain $\dot{a}$ to determine the converged values of *h* and Y_0
7. Finally, the value of $\dot{a}$ is increased and used in the steps 1 to 6 repeatedly

The SW models are used for the development of p-y curves taking into account the pile bending stiffness, the shape of the cross section, the pile head condition and the pile soil interaction effects. Other methods are used to derive the p-y curves based on experimental data; such procedure imposes the constrain that the p-y curves are site depended and therefore can be used for sites with similar conditions. This distinct difference is the main advantage of the SW method when compared to the others.

3 Ultimate Limit State (ULS) Methods

3.1 Blum Method

Blum (1932) proposed a simple method that is applicable for short rigid piles embedded in sandy soil. Blum assumed a fixed theoretical penetration depth (t_0), Fig. 9, where the deflection, moment and rotation are restrained at t_0. A lateral load was applied at point C to maintain equilibrium. A lateral load at the top of the pile (P) and the passive resistance of the soil (E_p and $E_{p\prime}$) are applied above t_0. Blum calculated the ultimate displacements by using simple rules of mechanics at ultimate loads. In general, Blum derived a set of equations to either find the ultimate load (P) applied on the pile if the pile dimensions are known, or the pile minimum dimensions if P and its location are known. This method is used in engineering practice with the aid of spreadsheets to provide a quick estimate of the pile design.

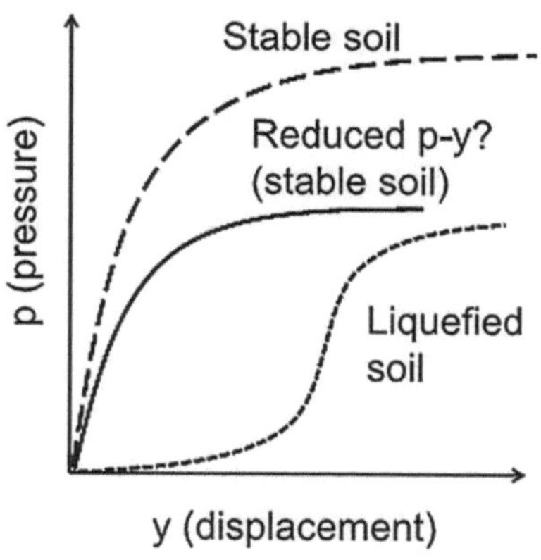

Fig. 9. General schematic comparing p-y curves in stable and liquefied soil, Chang and Hutchinson (2012)

3.2 Brinch Hansen's Method

Hansen and Christensen (1961) developed a model based on earth pressure theory to determine the ultimate resistant of laterally loaded piles. Unlike Blum (1932), the penetration depth is variable along the pile shaft and the soil resistance was separated into active and passive, Fig. 10. Hansen argued that the ultimate lateral load applied (H) is significantly affected by the location of the point of rotation (O) which is defined by the sum of the moments of the active and passive soil resistance above and below point O. The derived equations can either result in the minimum penetration depth of the pile or the ultimate applied lateral force (H). The major advantage of this method over the Blum method is its applicability for composite soils as well as layered soil systems. In addition, both active and passive earth pressures are considered. However, it is limited for short piles only. Christensen (1961) validated the Brinch Hansen's method by performing 26 pile load tests on 5 × 5 cm wooden piles. Cristensen concluded that the Brinch Hansen's method provided reasonable results compared to the load tests values.

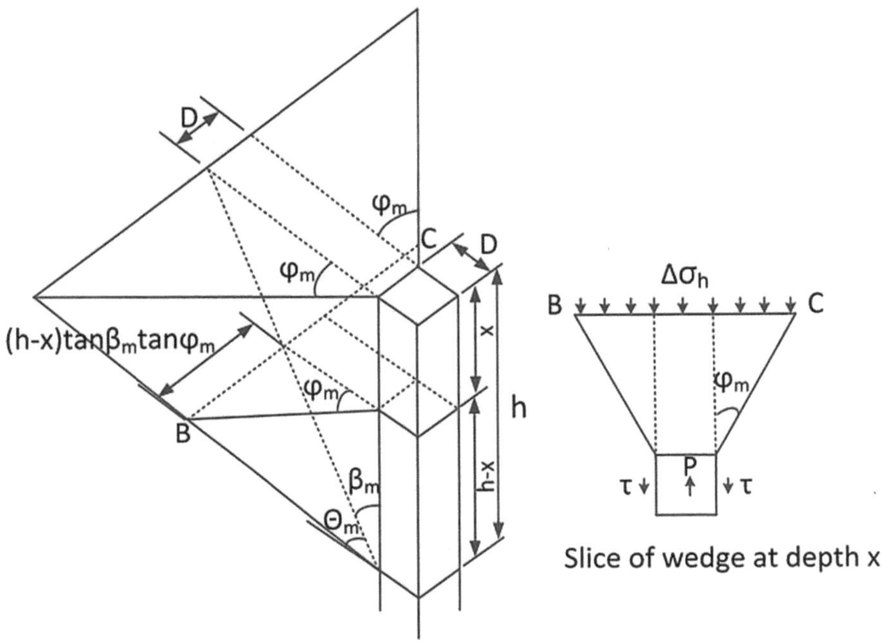

Fig. 10. Strain wedge model in uniform soil, Xu et al. (2013)

3.3 Broms' Method

Broms (1964a, b, 1965) derived equations to solve for the deflections at working loads, the ultimate lateral resistance and moment distribution for laterally loaded single piles and pile groups, for cohesive and cohesionless soils. Broms developed non-dimensional curves to find t_0, given that the soil strength, pile diameter and applied load are known. In addition, he used the concept of the subgrade reaction approach to predict the lateral displacement. Regarding the selection of the horizontal coefficient for the subgrade reaction values, Broms used his own suggested values in addition to the values proposed by Terzaghi (1955) and Vesic (1961). The author suggested that a constant value of k_h could be used for cohesive soils if the applied load was within the range of one-third to one-half of the ultimate lateral capacity of the pile. For cohesionless soils, the value of k_h is assumed to increase linearly with depth. Broms' method can take into account short and long piles for both cohesive and cohesionless soils. On the other hand, it cannot consider a layered soil system as well as composite soils. The model was validated by numerous field tests however, it is not reliable to solve for the deflections at serviceability limit state due to the observed large variation between the measured and calculated deflections.

4 Continuum Method

In the continuum approach, the assumption of modeling the soil by linear or nonlinear springs is replaced by a more appealing and realistic representation. The soil is represented by an infinite, linear and elastic medium and its behavior is described by the deflection along the vertical soil profile, Kavitha et al. (2016). However, this approach is hard to be used by the engineers in practice due to its mathematical complication. The works in this field and they are grouped in two main categories: (a) the boundary element method (BEM), and (b) the simplified continuum models.

4.1 Boundary Element Method (BEM)

The early work on BEM for laterally loaded piles goes back to Douglas and Davis (1964) who introduced a simple theory for a laterally applied load and a moment on the top of a thin vertical plate buried in elastic media. The mathematical theory for the proposed method included two steps: (a) the integration of the Mindlin (1936) equation for a horizontal displacement induced by a lateral point load within a semi-infinite mass, and (b) the calculation of the distributed normal stress by employing a numerical solution. In addition, it was assumed that the plate is perfectly smooth and the adhesion between the plate and the soil is neglected. Similarly, Spillers and Stoll (1964) proposed a simplified theoretical solution based on BEM by introducing a limiting lateral pressure value for the soil-pile interaction to account for the soil nonlinearity for a pile in both elastic half-space and plastic half space.

Poulos (1971a, b) utilized the previous work and contributed significantly in this field. The author introduced the stress for single and group pile-soil system using the BEM with various boundary conditions at the top and bottom of the pile. The pile was modelled as a thin rectangular strip, which interacted with an ideal linearly elastic media. As shown in Fig. 11, the pile was divided into equal segment lengths, except at the top and bottom of the pile. The response of the soil-pile interaction system was calculated by considering an applied normal uniformly distributed load on each pile segment, whereas, the shear stress at the interface of the pile sides was neglected. Further, the horizontal displacement at each segment can be calculated by equating the displacements of the pile and soil at the center of each segment. In this process, the soil horizontal displacement was calculated through the Mindlin equation, while the pile deflection was calculated by the differential equation, which is expressed as a finite difference expression for bending of a thin beam. Poulos (1973) had improved his model by considering the soil yielding and nonlinearity, i.e. the elastic-perfectly plastic soil behavior.

Moreover, Banerjee and Davies (1978) used the BEM to analyze the laterally loaded piles in a two-layered elastic half space system with soil modulus that is linearly increasing with depth. The results of the proposed theory were compared with the

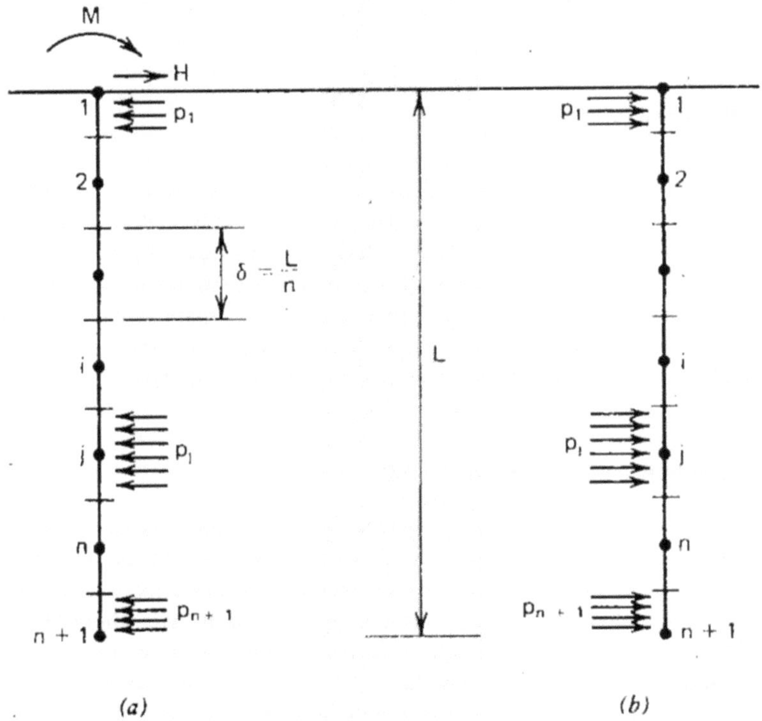

Fig. 11. Schematic diagram of Poulos and Davis (1980) method

behavior from full-scale tests; they showed that their method could provide a more rigorous prediction of the response, due to the linear variation of soil modulus. Thereafter, Davies and Budhu (1986), Budhu and Davies (1987, 1988) extended the linear algorithm of Banerjee and Davies (1978) to account for the nonlinear behavior of laterally loaded piles embedded in over consolidated clays, cohesionless soils and soft clays. The soils were modelled as linear elastic at small strains, while the soil plastic behavior was considered when the soil pressure at the pile surface was equal to certain values.

Randolph (1981) introduced the concept of characteristic soil stiffness (G_c) which is the average of soil stiffness over the critical length of the pile. In his parametric study, the deflection, moment and shear for laterally loaded piles were insignificant after a certain depth, which was smaller than the actual pile length, measured from the pile head. Therefore, in his analysis, he considered the use of critical depth (l_c) for a pile embedded in ideal elastic half-space. In addition, he provided simple equations, Fig. 12 obtained by the finite element method (FEM) to calculate the pile lateral displacement and rotation at ground level.

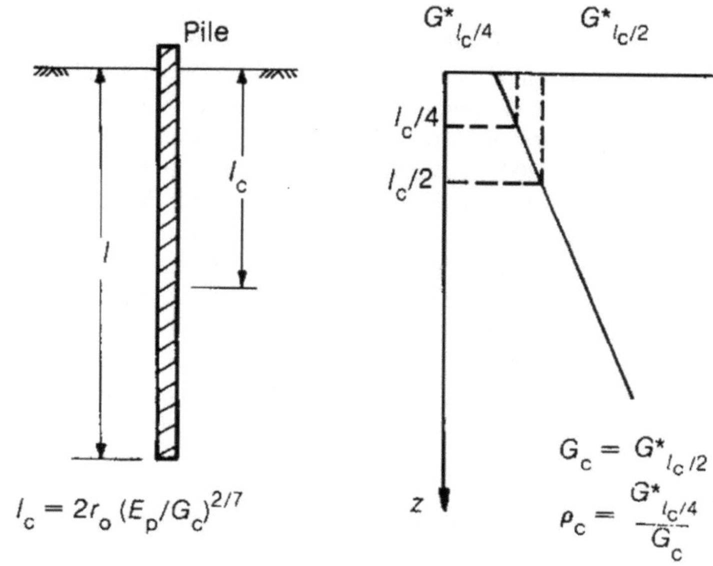

Fig. 12. Definition of p_c and Gc, Randolph (1981)

4.2 Simplified Continuum Models

4.2.1 Energy and Variational Calculus Method

Sun (1994) proposed a numerical approach for predicting the behavior of laterally loaded piles based on the modified Valsov model that was proposed by Vallabhan and Das (1988). In this model, the principles of variational calculus and the minimum of potential energy were used to obtain the governing differential equation for the pile-soil system. The numerical approach depended on a principal non-dimensional parameter γ, which represented the decaying of stress within the soil away from the pile and depended on a non-dimensional displacement $\bar{F}$. Moreover, Sun adapted the iterative technique given by Vallabhan and Das (1988) to evaluate γ, and went on to propose a closed form solution for calculating γ. The soil was modelled as an ideal linear homogenous elastic material and the gapping between the soil and pile was neglected. The soil properties included a constant elastic modulus and Poisson's ratio. A parametric study was then conducted to investigate the relationship between the main model parameters such as γ, slenderness ratio, Poisson's ratio and flexibility factor. Sun's model was extended by Zhang et al. (2000) to consider the nonlinear behavior of laterally loaded drilled shafts embedded in a layer of soil overlying a rock mass as well as the yielding of the soil and/or rock mass. Guo and Lee (2001) presented a new load transfer model similar to Sun (1994), however, the parameter γ was presented in a simple statistical expression and the need for the iterative technique was eliminated. In addition, it was argued that Sun's model does not provide reliable results at a high Poisson's ratio, i.e. $\nu_s \geq 0.3$. Therefore, the effect of variation of Poisson's ratio was considered and given by the parameter G_c that was proposed by Randolph (1981). Further improvements on Sun's model were suggested by Basu and Salgado (2007), when it was extended to consider the behavior of laterally loaded piles embedded in an arbitrary number of soil layers, in addition to the 3D interaction of pile-soil system.

Shen and Teh (2002) presented a new model for the analysis of laterally loaded pile groups based on the variational approach along with the assumption that the displacement and reaction pressures along the pile shaft were represented by a finite series. Furthermore, Shen and Teh (2004) extended the previous method to analyze the laterally loaded piles in soils with stiffness increasing with depth. More importantly, a spreadsheet was created based on the proposed method to calculate the pile displacement and bending moments along the pile shaft. Yang and Liang (2006) extended the work of Shen and Teh (2002, 2004) by proposing a similar variational approach for the laterally loaded piles in two-layered soils, mainly sands and sedimentary rocks. Four different variations of the soil profiles were considered with a constant soil stiffness or stiffness linearly changing with depth for each soil layer. It was observed that the assumption of a constant k_h underestimated the pile deflection.

4.2.2 Modified Reissner Models

Reissner (1958) solved the problem of an isotropic, homogeneous elastic continuum of infinite lateral extent but finite thickness, H, as shown in Fig. 13. This layer was supported by a rigid foundation and was subjected to pressure, P. In order to solve the problem, Reissner assumed that the plane stresses and the horizontal displacements of the foundation layer were zero. The relationship between the foundation surface displacement (w) and the applied pressure (P) was given by following equation.

$$C_1 w - C_2 \nabla^2 w = P - C_3 \nabla^2 P \tag{10}$$

where C_1, C_2 and C_3 are constants related to Young's modulus (E), shear modulus (G) and H

∇^2 is the Laplace operator

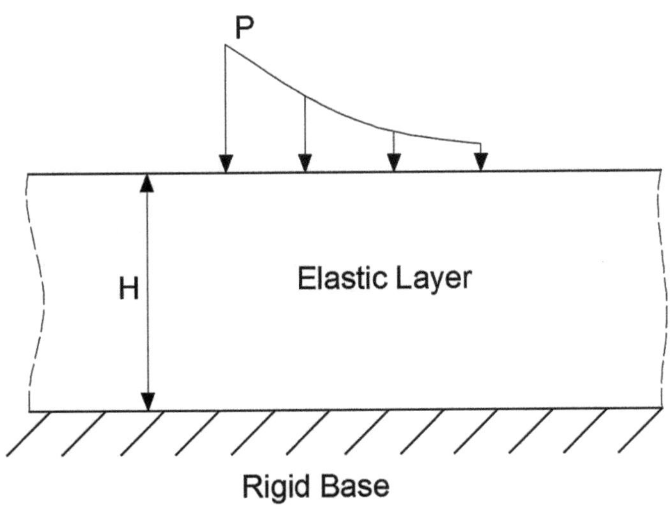

Fig. 13. Reissner's simplified continuum model

Horvath (1983) introduced Winkler-Type Simplified Continuum (WTSC) subgrade model that is based on Reissner subgrade model. The idea was to express k_h the Winkler subgrade model (Eq. 1) in the continuum approach. The author solved the same problem that is shown in Fig. 13, however, he only considered the vertical normal stresses and strains. Moreover, three different cases for the variation of the soil modulus with depth, e.g. constant, linear and nonlinear, were considered. Eventually, Horvath proposed the parameter k_{sc}, which is the equivalent modulus of subgrade reaction for a simplified continuum approach. The evaluation of k_{sc} is primarily dependent on the selection of the thickness (H) in the layer and the value and variation of soil modulus. Subsequently, Horvath (1984) extended the previous work to be applicable for the analysis of laterally loaded piles as shown in Fig. 14. The model was referred to as the Reissner simplified continuum (RSC) and the following assumptions had been made to solve the problem:

1. The vertical and out of plane stresses were equal to zero
2. The E was constant in the lateral direction at any depth
3. The variation of E with depth was handled by assigning different values for E at each depth

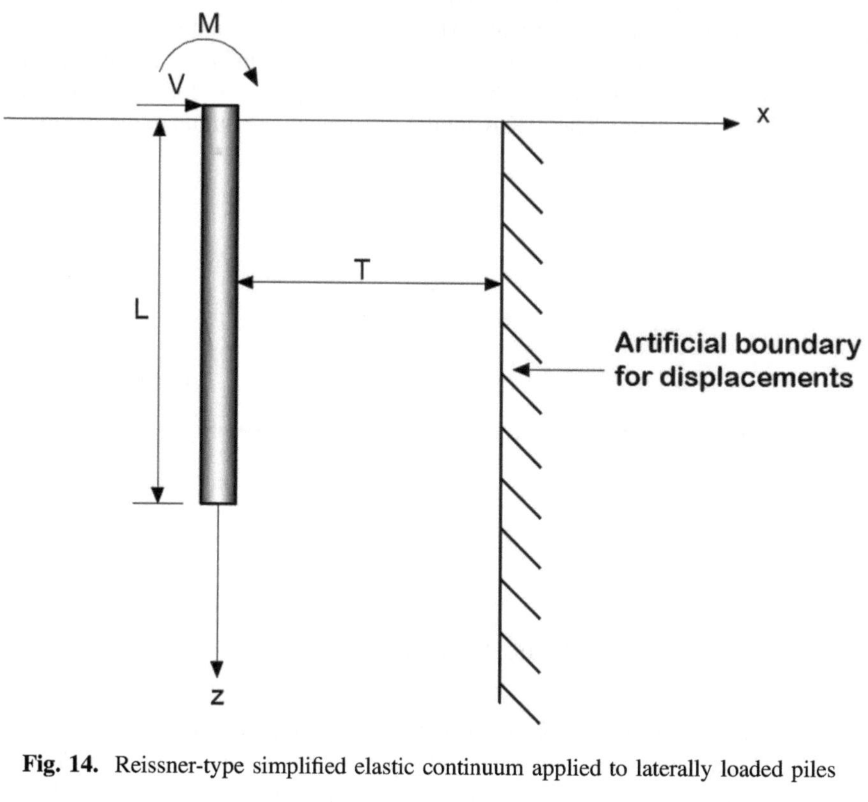

Fig. 14. Reissner-type simplified elastic continuum applied to laterally loaded piles

The relation between the applied pressure (P) and lateral pile displacement (U) was given by the following differential equation.

$$P = K_1 U_x - K_2 \nabla^2 U_x + K_3 \nabla^2 P \tag{11}$$

where K_1, K_2 and K_3 are constants and function of E, G and the distance to the artificial boundary (T).

U_x is the lateral displacement

The main advantage of the proposed RSC model is its ability to account for the soil continuity in addition to the consideration of varying E with depth.

Colasanti and Horvath (2010), Horvath and Colasanti (2011) proposed a new hybrid subgrade model (e.g. Modified Kerr/Reissner (MK/R) model) that utilizes the advantages of RSC model and modified the Kerr mechanical model. The authors argued that each model would compensate the drawbacks of the other. The modified Kerr mechanical model was first proposed by Kerr (1964, 1965), Fig. 15. It consists of an upper spring layer (K_u) tensioned membrane and a lower spring layer (K_l). Both upper and lower spring layers are series of linear, independent, axial springs acting only in the vertical direction. The tension membrane is an incompressible layer which has zero flexural stiffness. Its role is to account for the spring coupling mechanism in the model. Therefore, the advantage of using the modified Kerr mechanical model is its ease for the implementation in commercial software. The authors implemented the MK/R model in the ANSYS software (Ver.11.0) and they used the mat foundation as an application because of its simplicity. However, the model is also applicable for deep foundations (e.g. laterally loaded piles). The MK/R model parameters are summarized below:

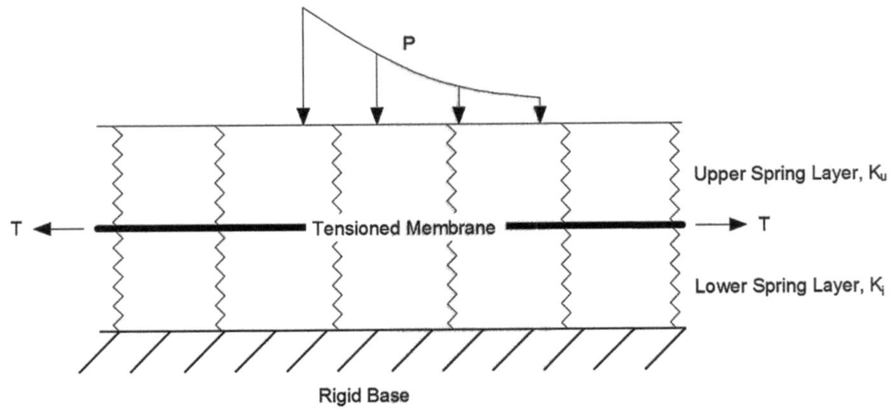

Fig. 15. Modified Kerr Model

1. Characterization of the soil profile to define the important soil properties (i.e. unit weight, elastic parameters) for each soil layer as shown qualitatively in Fig. 16a
2. Idealizing the soil profile into n arbitrary linear-elastic soil layers as shown if Fig. 16b

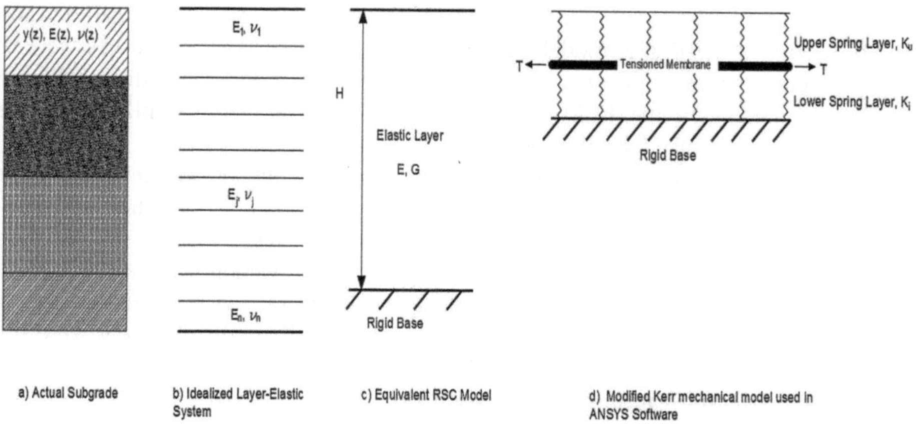

Fig. 16. MK-R Model parameters' evaluation

3. Conversion of the arbitrary linear-elastic soil layers to equivalent RSC model with constant elastic parameters as in Fig. 16c
4. After the identification of the parameters of the equivalent RSC model, a set of equations will be used to calculate the parameters for the modified Kerr mechanical model that is shown in Fig. 16d.

5 Finite Element Method (FEM)

The FEM method is a mathematical technique used to calculate an approximate solution to the partial differential equations based on the continuum theory. Desai and Appel (1976) started the early work on the 3D FEM by proposing a general finite element procedure to study the behavior of laterally loaded piles and their interaction with the soil. The soil behavior was assumed to be linear-elastic and the pile was modelled as an 1D beam element. The interaction between the pile and the soil was modelled using a thin-layer element, which was capable to consider different modes of deformations, e.g. stick mode, gapping or slippage. It was found that the use of the interface element considerably affected the pile-head displacements. Faruque and Desai (1982) included the material and geometric nonlinearity in their 3D finite element model for axially and laterally loaded piles. The soil nonlinear stress-strain relation was modelled with the Drucker-Prager plasticity theory. The authors concluded that the geometric nonlinearity had a significant effect on the analysis of pile-soil interaction.

Brown and Kumar (1989) developed p-y curves by creating 3D finite element models in ABAQUS considering elasto-plastic behavior for the soil. The p-y curves and the field results were not in a good agreement due to the simplicity of the soil model. However, Brown and Shie (1990, 1991), concluded that the 3D finite element models are an effective tool to capture the 3D the interaction between the piles and the soil, the soil nonlinearity, the variation of the soil strength with depth, and the interface friction. The model for the interface frictional element accounted for two soil constitutive laws, i.e. a simple elastic-plastic model with a Von Mises yield surface and an extended Drucker-Prager model with non-associated flow. The p-y curves were in good agreement with those obtained from the COM624 program. Additionally, the authors recommended the FEM for the evaluation of the factors influencing the behavior of the laterally loaded piles.

Trochanis et al. (1991) who studied the nonlinearity in pile-soil interaction for axially and laterally loaded piles presented another important work. Their soil constitutive models included a linear elastic and a Drucker–Prager model. The pile was modelled as a 1D-beam element and a special 2D interface element was used to capture the pile-soil interaction. The interface element was characterized by a friction coefficient, its geometry and an elastic stiffness. The main findings of this study were the factors that affect significantly the axially and laterally loaded piles which are the interface slippage and gap formation, respectively.

Additional work on FEA was presented by Yang and Jeremi (2002, 2005). They studied the effect of layered soils on the behavior of laterally loaded piles and provided accurate p-y curves. A simple von Mises material model and a Drucker–Prager model respectively modeled the clay and sand soils. In addition, the pile-soil interaction was simulated by the use of a thin layer element. The authors concluded that a simple elastic-plastic soil model could predict the pile head deflection with reasonable accuracy. Moreover, it was found that the upper soil layers influence the lower layers and vice versa, in addition to the antisymmetric effect of layered soils.

Yang and Liang (2006) investigated the effect of the push-pull resistance for laterally loaded drilled shafts in rocks using a 3D finite model. The modified Drucker-Prager model was used for the modelling of the soil, while the piles were modeled as an elastic material. The interaction behavior was simulated with a surface based contact element, where the master and slave surfaces were the shaft surface and rock surfaces, respectively. The results from the 3D model showed the insignificance of the pull-push resistance compared to the resistance to the applied moments at the ground surface, which is less than 5%.

Ahmadi and Ahmari (2009) created a 3D finite element model to study the effect of the shear strength anisotropy and the soil mass secondary structure on laterally loaded piles. The results of two full-scale case studies were considered in their analysis. Matlock reported the first case study referring to a steel pipe in soft clay and Reese and Welch reported the second case study for a cast-in-place pile in over consolidated clay. The maximum moment and moment distribution of the analyses were compared to the field results. The soil model considered the von Mises constitutive law which did not take into account the effect of soil anisotropy. Therefore, the soil shear strength and modulus of elasticity had to be back-calculated by fitting the pile head load-deflection curve to the field results. The back-calculated shear strength was used as an input to the proposed model to predict the pile-head load vs maximum moment and moment distribution curves. Then a comparison was made between the measured and the predicted curves. The p-y curves from the model were compared to the two p-y curves proposed by Matlock and Wu et al. A sensitivity analysis was carried out using the ANSYS software to study the model dimensions, mesh fineness, contact stiffness, and pile-soil adhesion. It was concluded that using back calculated shear strength and modulus of elasticity provided a good correspondence between the finite element analysis and the field measurements.

Peng et al. (2010) investigated the effect of fin dimensions on the lateral capacity of the pile in sand. The Mohr-Coulomb constitutive model was used in the numerical analysis using the software PLAXIS-3D. The sand was assumed to be a linear elastic perfectly plastic and the soil Young's modulus was assumed to increase linearly. The ultimate load was defined when the slope of the p-y curve was less than 0.05 MN/m, i.e. when the curve was straight and the pile failure load was taken as the displacement equal to 10% of the pile diameter. The numerical analysis results were compared to laboratory test results (1G model tests) for verification purposes. The conclusions were:

1. The lateral resistance increased when the length of the fins increased
2. Fins placed near the pile head provided more resistance than those near the pile tip
3. The lateral resistance varied with the direction of loading in relation to the direction of the fins but the difference was acceptably small for design purposes making the direction of the piles immaterial

Mardfekri et al. (2013) assessed the existing simplified methods (e.g., beam on elastic foundation, p-y method, and SALLOP) for predicting the deflection of laterally loaded piles in clay and sand by using linear and nonlinear FE analyses. The first objective was to compare the results of different analysis methods for laterally loaded piles and illustrate the variation of the results. The second objective was to show the effect of the pile diameter on the accuracy of the simplified methods. Three pile diameters (1 m, 2 m and 4 m) were presented in the paper. Four different models using linear analysis were created and the influence of the pile diameter was investigated. These models were:

1. A 3D ABAQUS finite element model
2. A model in which the soil was reproduced with 3D elements and the pile with 1D beam-column elements
3. A model using a consistent boundary matrix
4. A Winkler foundation model

In addition, three models were used for the case of nonlinear analysis as follows:

1. The same 3D ABAQUS finite element model of the previous runs. However the soil and pile-soil interaction were nonlinear
2. A model using the p-y curves (sand, hard clay p-y curves by Reese)
3. A model implementing the simple SALLOP approach

The results of linear analysis showed that the pile head deflection is not only a function of EI but also a function of the pile diameter. Furthermore, modeling the pile as 1D beam-column element leads to a smaller contribution of the surrounding soil to the lateral stiffness of the pile. Comparing the nonlinear results from the 3D ABAQUS finite element model to the results of the other two models, it is found that both the p-y model for sand and the SALLOP approach, provide reasonable results for the pile of 4 m in diameter. However, the accuracy deteriorates for smaller diameter piles. Further, the p-y curve model for clay as well as the SALLOP provide reasonable results for the pile of 1 m diameter but deteriorate for the piles of larger diameters.

Tuladhar et al. (2013) used the results of full scale field tests as the basis for the investigation of the capability of 3D finite element models in simulating pile-soil interaction. The soil was modeled as 20-node solid elements. The piles were modeled as either 3-node fiber-based beam elements or as 20-node solid elements. Two types of loads were considered (monotonic and reverse cyclic loading) as shown on Table 1. The model FB-Mon1 significantly underestimated the lateral load capacity of the pile, compared to the field results, because this model ignored the volume of the pile. However, using a rough mesh the accuracy was reasonable. The model SL-Mon1 extremely overestimated the lateral capacity of the pile because of the assumption of perfect bond between the soil and the pile. On the other hand, the load displacement curve from Model SL-Mon2 showed a good agreement with the field results because of the consideration of the interface elements. The stiffness reduction factor was not considered in the model SL-Rev1 which lead to the over estimation of the lateral load capacity of the pile. The load displacement curves were in good agreement with field results when a stiffness reduction factor of 0.2 was considered in model SL-Rev2.

Table 1. 3D finite element analysis models, Tuladhar et al. (2013)

Pile element type	Loading	Name	Descriptions		
			Mesh division	Interface element	Stiffness reduction factor
3-node fiber-based beam element	Monotonic	FB-Mon1	Fine	No	-
		FB-Mon2	Rough	No	-
20-node solid element	Monotonic	SL-Mon1	-	No	-
		SL-Mon2	-	Yes	-
	Reversed cyclic	SL-Rev1	-	Yes	No
		SL-Rev2	-	Yes	Yes

Khodair and Abdel-Mohti (2014) compared the analysis results from a finite difference (FD) model and two 3D finite element models used to analyze the pile-soil structure interaction under axial and lateral loads of piles in stiff clay soil. The LPILE software was used for the FD model, while the ABAQUS and SAP2000 softwares have been used for FE models. The soil in ABAQUS was modeled with solid continuum elements whereas nonlinear springs were used in SAP200. The applied lateral displacement and the applied axial load was 2 cm and 298 KN, respectively. This study showed that the obtained bending moment along the pile in ABAQUS was higher than that for LPILE. The reason is that FE model in ABAQUS accounted for the soil continuity while in LPILE it was not. Furthermore, a parametric study investigated the effect of the variation of the modulus of elasticity, the amount of soil surrounding the pile, the number of soil springs and the effect of axial load. A range of 5 to 50 MPa for the soil modulus of elasticity was used to study the effect of soil stiffness under a lateral deformation of 2 cm. The results showed that when the modulus of elasticity is within the range of 20-25 MPa (medium to stiff clay), the difference for the bending moments and lateral displacements obtained from ABAQUS and LPILE were steadily reduced. In the ABAQUS model, the soil was surrounding the pile with a cylinder radius which varied from 0.5 m to 4 m. When the radius of the soil cylinder increased, the positive bending moments and lateral displacements along the pile shaft decreased and their magnitude approached to those obtained by LPILE. However, the lateral displacements obtained at the radius of 4 m were extremely different from the LPILE results. There was a very good agreement between the bending moments and the lateral displacements obtained from SAP2000 and LPILE, when using the largest possible number of soil springs. There was no considerable effect from the application of axial loads.

Kampitsis et al. (2015) verified the accuracy of an efficient beam formulation for inelastic analysis of soil-pile interaction in dry sand by a several laboratory pushover tests. The results were compared to a 3D FE model created in ABAQUS. The formulation was based on the boundary element method (BEM) and it accounted for the shear deformation effect. The proposed formulation was efficient and verified as it accurately represented the response of a single pile embedded in dry sand. Furthermore, the initial stiffness was over estimated compared to experimental pushover results and 3D FE model, however the ultimate capacities were accurately predicted. Also, the results from BEM were in very good agreement with those from the 3D FE model. More importantly, the proposed formulation required less computational effort and produced accurate results.

6 Conclusions

A detailed literature review on the analysis methods of laterally loaded piles is presented in this paper and the following conclusions are highlighted:

1. The ULS methods are used to calculate the maximum lateral loads that the piles can support. As such, the ultimate loads are associated with lateral displacements and the developed mathematical formulations do not consider the soil deformation. Therefore, the ULS methods are not suitable to apply to displacement control problems and do not provide realistic results for the pile-soil interaction problem.
2. The idealization of the soil in Winkler Hypothesis as linear independent springs overlooks the soil continuity, shear coupling between the springs and the effect of strength characteristics of the piles on the subgrade reaction. However, the method is still used until today because of its simplicity and its acceptable results; it is not recommended to be used when the soil profile is highly nonlinear. However, the availability of finite element software can overcome these drawbacks without significant computational effort.
3. The p-y curves are back calculated from empirical test results, therefore, they depend on the empirical test environment and the variation in soil and pile properties. Hence, it is critical to select the most appropriate p-y curve in the analysis of laterally loaded piles to obtain accurate and realistic results. Furthermore, a review

for the applicability of the selected p-y curve is crucial since they are experimental in nature.

4. The ability to take into account the 3D nature of the problem, the soil continuity as well as the soil-pile interaction make the use of the SW method very appealing to the practicing engineers. Conversely, the two main limitations of the method are: (1) Few empirical data were used to develop the stress-strain relationship, and (2) The determination of strain wedge depth and the value of the subgrade reaction modulus below the strain wedge are not a simple task.
5. In general, the continuum methods involve complex mathematical formulations to model the continuity of the soil, its nonlinearity and the boundary conditions. Furthermore, determining a suitable soil modulus for use is not an easy task. Thus, the continuum methods do not provide a practical solution for practicing engineers.
6. The advancement in the computational power of computers facilitated the use of FEM in the analysis of LLP. The FEM can provide rigorous results in relatively reasonable time with the consideration of material nonlinearity and heterogeneity, in addition to the interface modeling. However, it is important to validate the finite element model prior to using the results for the design.

References

Ahmadi, M., Ahmari, S.: Finite-element modelling of laterally loaded piles in clay. Proc. Inst. Civil Eng.-Geotechn. Eng. **162**(3), 151–163 (2009)

Alizadeh, M., Davisson, M.: Lateral load tests on piles—Arkansas River project. J. Soil Mech. Found. Div. **96**(5), 1583–1604 (1970)

American Petroleum Institute RP 2A Planning, designing, and constructing fixed offshore platforms, 17th edition. Washington, DC: API (1987)

American Petroleum Institute. Recommended Practice for Planning, Designing and Construction Fixed Offshore Platforms – Working Stress Design. API RP 2A-WSD, 21st Ed. Washington, DC (2007)

Ashford, S., Juirnarongrit, T.: Evaluation of pile diameter effect on initial modulus of subgrade reaction. J. Geotech. Geoenviron. Eng. **129**(3), 234–242 (2003)

Ashour, M., Norris, G., Pilling, P.: Lateral loading of a pile in layered soil using the strain wedge model. J. Geotech. Geoenviron. Eng. **124**(4), 303–315 (1998)

Ashour, M., Norris, G., Pilling, P.: Strain wedge model capability of analyzing behavior of laterally loaded isolated piles, drilled shafts, and pile groups. J. Bridge Eng. **7**(4), 245–254 (2002)

Ashour, M., Norris, G., Pilling, P.: Laterally loaded isolated piles, drilled shafts and pile groups using the strain wedge model. GeoSupport 2004: Innovation and Cooperation in the Geo-Industry (2004)

Banerjee, P.K., Davis, T.G.: The behaviour of axially and laterally loaded single piles embedded in non-homogeneous soils. Geotechnique **28**, 309–326 (1978)

Barber, E.: Discussion to paper by SM Gleser. ASTM, STP, vol. 154, pp. 96–99 (1953)

Basu, D., Salgado, R.: Elastic analysis of laterally loaded pile in multi-layered soil. Geomech. Geoeng. Int. J. **2**(3), 183–196 (2007)

Biot, M.: Bending of an infinite beam on an elastic foundation. Zeitschrift fiir Angewandte Maihematik und Mechanik **2**(3), 165–184 (1937)

Blum, Wirtschaftliche dalbenformen und deren berechnung. Bautechnik, Heft 5 (1932)

Bowles, J.E.: Foundation Analysis and Design, 5th edn. McGraw-Hill, New York (1996)

Broms, B.B.: Lateral resistance of piles in cohesionless soils. J. Soil Mech. Found. Div. **90**(3), 123–158 (1964a)

Broms, B.B.: Lateral resistance of piles in cohesive soils. J. Soil Mech. Found. Div. **90**(2), 27–64 (1964b)

Broms, B.B.: Design of laterally loaded piles. J. Soil Mech. Found. Div. **91**(3), 79–99 (1965)

Brown, D.A., Shie, C.-F.: Three dimensional finite element model of laterally loaded piles. Comput. Geotech. **10**(1), 59–79 (1990)

Brown, D.A., Shie, C.-F.: Some numerical experiments with a three dimensional finite element model of a laterally loaded pile. Comput. Geotech. **12**(2), 149–162 (1991)

Brown, D., Kumar, M., et al.: PY curves for laterally loaded piles derived from three-dimensional finite element model. Numerical models in geomechanics. In: Numog III. Proceedings of the 3rd International Symposium Held in Niagara falls, Canada, 8–11 May 1989. Publication of Elsevier Applied Science Publishers Limited (1989)

Budhu, M., Davies, T.G.: Nonlinear analysis of laterality loaded piles in cohesionless soils. Can. Geotech. J. **24**(2), 289–296 (1987)

Budhu, M., Davies, T.G.: Analysis of laterally loaded piles in soft clays. J. Geotech. Eng **114**(1), 21–39 (1988)

Carter, D. P.: A non-linear soil model for predicting lateral pile response. Ph.D. dissertation, ResearchSpace@ Auckland (1984)

Chang, B., Hutchinson, T.: Experimental evaluation of p-y curves considering development of liquefaction. J. Geotech. Geoenviron. Eng. **139**(4), 577–586 (2012)

Chong, W., Haque, A., Ranjith, P., Shahinuzzaman, A.: Effect of joints on p–y behaviour of laterally loaded piles socketed into mudstone. Int. J. Rock Mech. Min. Sci. **48**(3), 372–379 (2011)

Christensen, N.: Model tests with transversally loaded rigid piles in sand. The Danish Geotechnical Institute (1961)

Colasanti, R.J., Horvath, J.S.: Practical subgrade model for improved soil-structure interaction analysis: software implementation. Pract. Period. Struct. Des. Constr. **15**(4), 278–286 (2010)

Cox, W.R., Reese, L.C., Grubbs, B.R., et al.: Field testing of laterally loaded piles in sand. In: Offshore Technology Conference (1974)

Davies, T., Budhu, M.: Non-linear analysis of laterally loaded piles in heavily overconsolidated clays. Geotechnique **36**(4), 527–538 (1986)

Davisson, M., Gill, H.: Laterally loaded piles in a layered soil system. J. Soil Mech. Found. Div. **89**(3), 63–94 (1963)

Desai, C., Appel, G.: 3-D analysis of laterally loaded structures. In: Proceedings of the 2nd International Conference on Numerical Methods in Geomechanics, Blacksburg, pp. 405–418 (1976)

Douglas, D., Davis, E.: The movement of buried footings due to moment and horizontal load and the movement of anchor plates. Geotechnique **14**(2), 115–132 (1964)

Faruque, M., Desai, C.: 3-D material and geometric nonlinear analysis of piles. In: Proceedings of the Second International Conference on Numerical Methods in Offshore Pilling, pp. 553–575 (1982)

Favaretti, C., Lemnitzer, A., Stuedlein, A.W., Turner, J.: Recent discussions of py formulations for lateral load transfer of deep foundations based on experimental studies. In: Proceedings of the International Foundations Congress and Equipment Expo, March 17–21, 2015, San Antonio, Texas, pp. 388–399 (2015)

Federal Highway Administration (FHWA). LRFD seismic analysis and design of transportation geotechnical features and structural foundation, Report No. FHWA-NHI-11-032, National Highway Institute (NHI), U.S. Department of Transportation (2011)
Georgiadis, M.: Development of p-y curves for layered soils. In: Proceedings of the Geotechnical Practice in Offshore Engineering, ASCE, pp. 536–545 (1983)
Gill, H.: Soil behavior around laterally loaded piles. Technical Report, DTIC Document (1968)
Guo, W.D., Lee, F.: Load transfer approach for laterally loaded piles. Int. J. Numer. Anal. Methods Geomech. **25**(11), 1101–1129 (2001)
Haiderali, A.E., Madabhushi, G.: Evaluation of curve fitting techniques in deriving p–y curves for laterally loaded piles. Geotech. Geol. Eng. **34**(5), 1453–1473 (2016)
Hansen, J.B., Christensen, N.H.: The ultimate resistance of rigid piles against transversal forces. Copenhagen: Geoteknisk Institut (1961)
Hetenyi, M.: Beams on elastic foundation. The University of Michigan Press, Ann Arbor (1946)
Horvath, J., Colasanti, R.: New Hybrid Subgrade Model for Soil-Structure Interaction Analysis: Foundation and Geosynthetics Applications. In: ASCE, Proceedings of the Geo-Frontiers 2011 Conference, March 2011, 10 pp (2011)
Horvath, J.: Soil-structure interaction research project: basic SSI concepts and applications overview. Technical report, Manhattan College School of Engineering, Bronx, New York, USA (2002)
Horvath, J.S.: Modulus of subgrade reaction: new perspective. J. Geotech. Eng. **109**(12), 1591–1596 (1983)
Horvath, J.S.: Simplified elastic continuum applied to the laterally loaded pile problem—part 1: theory. In: Laterally Loaded Deep Foundations: Analysis and Performance, STP 835, E. T. Mosley and C.D. Thompson, eds., Philadelphia, American Society for Testing and Materials, pp. 112–121 (1984)
Horvath, J.S., Colasanti, R.J.: Practical subgrade model for improved soil-structure interaction analysis: model development. Int. J. Geomech. **11**(1), 59–64 (2011)
Ito, T., Matsui, T.: Methods to estimate lateral force acting on stabilizing piles. Soils Found. **15** (4), 43–59 (1975)
Kampitsis, A.E., Giannakos, S., Gerolymos, N., Sapountzakis, E.J.: Soil–pile interaction considering structural yielding: numerical modeling and experimental validation. Eng. Struct. **99**, 319–333 (2015)
Kavitha, P., Beena, K., Narayanan, K.: A review on soil–structure interaction analysis of laterally loaded piles. Innov. Infrastruct. Solut. **1**(1), 1–15 (2016)
Kerr, A.D.: Elastic and viscoelastic foundation models. J. Appl. Mech. **31**(3), 491–498 (1964)
Kerr, A.D.: A study of a new foundation model. Acta Mech. **1**, 135–147 (1965)
Khodair, Y., Abdel-Mohti, A.: Numerical analysis of pile–soil interaction under axial and lateral loads. Int. J. Concr. Struct. Mater. **8**(3), 239–249 (2014)
Kurrer, K.-E., Melchers, R.: The History of the Theory of Structures: From Arch Analysis to Computational Mechanics, vol. 8. Taylor & Francis, New York (2008)
Lee, P., Gilbert, L., et al.: Behavior of laterally loaded pile in very soft clay. In: Offshore Technology Conference (1979)
Lin, C., Han, J., Bennett, C., Parsons, R.L.: Analysis of laterally loaded piles in soft clay considering scour-hole dimensions. Ocean Eng. **111**, 461–470 (2016)
Mardfekri, M., Gardoni, P., Rosset, J.M.: Modeling laterally loaded single piles accounting for nonlinear soil-pile interactions. J. Eng. **3**, 1–7 (2013)
Matlock, H.: Correlations for design of laterally loaded piles in soft clay. In: Offshore Technology in Civil Engineering's Hall of Fame Papers from the Early Years, pp. 77–94 (1970)

Matlock, H., Reese, L.C.: Generalized solutions for laterally loaded piles. J. Soil Mech. Found. Div. **86**(5), 63–94 (1960)
McClelland, B., Focht, J.A.: Soil modulus for laterally loaded piles. Trans. Am. Soc. Civ. Eng. **123**(1), 1049–1063 (1958)
Mindlin, R.D.: Force at a point in the interior of a semi-infinite solid. J. Appl. Phys. **7**(5), 195–202 (1936)
Mokwa, R.L., Duncan, J.M., Helmers, M.J.: Development of py curves for partly saturated silts and clays. In: Proceeding to GeoDenver, Denver, Colorado, August 5–8, Geotechnical special publication, pp. 224–239 (2000)
Norris, G.: Theoretically based BEF laterally loaded pile analysis. In: Proceedings on 3rd International Conference on Numerical Methods in Offshore Piling, pp. 361–386 (1986)
Peng, J.-R., Rouainia, M., Clarke, B.: Finite element analysis of laterally loaded fin piles. Comput. Struct. **88**(21), 1239–1247 (2010)
Poulos, H.G.: Behavior of laterally loaded piles. J. Soil Mech. Found. Div. ASCE **97**(5), 711–731 (1971a)
Poulos, H.G.: Behavior of laterally loaded piles. II: Pile groups. J. Soil Mech. Found. Div. ASCE **97**(5), 733–751 (1971b)
Poulos, H.G.: Analysis of piles in soil undergoing lateral movement. J. Soil Mech. Found. Div. **99**(5), 391–406 (1973)
Poulos, G., Davis, H.: Pile Foundation Analysis and Design. John Wiley & sons, Inc., New York (1980)
Prakash, S., Sharma, H.D.: Pile Foundations in Engineering Practice. Wiley, Hoboken (1990)
Pula, W.: Reliability of laterally loaded rigid piles. In: Probabilistic Methods in Geotechnical Engineering. vol. 491, pp. 169–183. Springer (2007)
Pulikanti, S., Ramancharla, P.K.: SSI analysis of framed structures supported on pile foundations: a review. Front. Geotech. Eng. (FGE) **2**, 28–38 (2013)
Randolph, M.F.: The response of flexible piles to lateral loading. Geotechnique **31**(2), 247–259 (1981)
Reese, L.C., Cox, W.R., Koop, F.D.: Field testing and analysis of laterally loaded piles in stiff clay. In: Proceedings of the VII Annual Offshore Technology Conference, pp. 672–690. Houston, Texas, 2(OTC 2312) (1975)
Reese, L.C.: Analysis of laterally loaded piles in weak rock. J. Geotech. Geoenviron. Eng. **123** (11), 1010–1017 (1997)
Reese, L.C.: Handbook on Design of Piles and Drilled Shafts under Lateral Load. FHWA Report FHWA-IP-84/11, US DOT, Washington, DC, 360 pp (1984)
Reese, L.C., Matlock, H.: Nondimensional solutions for laterally loaded piles with soil modulus assumed proportional to depth. In: Proceedings of the VIII Texas Conference on Soil Mechanics and Foundation Engineering, University of Texas, Austin (1956)
Reese, L.C., Van Impe, W.F.: Single Piles and Pile Groups Under Lateral Loading. CRC Press, Boca Raton (2010)
Reese, L.C., Cox, W.R., Koop, F.D. Analysis of laterally loaded piles in sand. In: Offshore Technology in Civil Engineering Hall of Fame Papers from the Early Years, pp. 95–105 (1974)
Reese, L.C., Wang, S.T.: LPILE PLUS Computer Program Documentation. Ensoft Inc., Austin (1989)
Reissner, E.: A note on deflections of plates on a viscoelastic foundation. J. Appl. Mech. **25**(1), 144–145 (1958)
Rollins, K.M., Gerber, T.M., Lane, J.D., Ashford, S.A.: Lateral resistance of a full-scale pile group in liquefied sand. J. Geotech. Geoenviron. Eng. **131**(1), 115–125 (2005)

Ruigrok, J.: Laterally loaded piles, models and measurements. Ph.D. dissertation, TU Delft, Delft University of Technology (2010)
Shen, W., Teh, C.: Analysis of laterally loaded pile groups using a variational approach. Geotechnique **52**(3), 201–208 (2002)
Shen, W., Teh, C.: Analysis of laterally loaded piles in soil with stiffness increasing with depth. J. Geotech. Geoenviron. Eng. **130**(8), 878–882 (2004)
Spillers, W.R., Stoll, R.D.: Lateral response of piles. J. Soil Mech. Found. Div. **90**(6), 1–10 (1964)
Stevens, J., Audibert, J.: Re-examination of py curve formulations. In: Offshore Technology Conference (1979)
Suleiman, M.T., Ni, L., Helm, J.D., Raich, A.: Soil–pile interaction for a small diameter pile embedded in granular soil subjected to passive loading. J. Geotech. Geoenviron. Eng. **140**(5) (2014). doi:10.1061/(ASCE)GT.1943-5606.0001081
Sun, K.: Laterally loaded piles in elastic media. J. Geotech. Eng. **120**(8), 1324–1344 (1994)
Terzaghi, K.: Evalution of conefficients of subgrade reaction. Geotechnique **5**(4), 297–326 (1955)
Thieken, K., Achmus, M., Lemke, K.: A new static p-y approach for piles with arbitrary dimensions in sand. Geotechnik **38**(4), 267–288 (2015)
Trochanis, A.M., Bielak, J., Christiano, P.: Three-dimensional nonlinear study of piles. J. Geotech. Eng. **117**(3), 429–447 (1991)
Tuladhar, R., Mutsuyoshi, H., Maki, T.: Numerical modelling and full-scale testing of concrete piles under lateral loading. Aust. J. Struct. Eng. **14**(3), 229–242 (2013)
Vallabhan, C.G., Das, Y.: Parametric study of beams on elastic foundations. J. Eng. Mech. **114** (12), 2072–2082 (1988)
Vesic, A.: Beams on elastic subgrade and the Winkler's hypothesis. In: Proceedings, 5th International Conference on Soil Mechanics and Foundation Engineering, vol. 1, pp. 845–850 (1961)
Winkler, E.: Die Lehre von der Elastizit at und Festigkeit. Domimicus, Prague (1867)
Xu, L.-Y., Cai, F., Wang, G.-X., Ugai, K.: Nonlinear analysis of laterally loaded single piles in sand using modified strain wedge model. Comput. Geotech. **51**, 60–71 (2013)
Yang, K., Liang, R.: A 3D FEM model for laterally loaded drilled shafts in rock. In: GeoCongress 2006@ sGeotechnical Engineering in the Information Technology Age, pp. 1–6 (2006)
Yang, K., Liang, R.: Numerical solution for laterally loaded piles in a two-layer soil profile. J. Geotech. Geoenviron. Eng. **132**(11), 1436–1443 (2006)
Yang, Z., Jeremi, B.: Numerical analysis of pile behaviour under lateral loads in layered elastic–plastic soils. Int. J. Numer. Anal. Methods Geomech. **26**(14), 1385–1406 (2002)
Yang, Z., Jeremi, B.: Study of soil layering effects on lateral loading behavior of piles. J. Geotech. Geoenviron. Eng. **131**(6), 762–770 (2005)
Yang, Z., Li, Q., Horazdovsky, J., Hulsey, J., Marx, E.: Analysis of laterally loaded piles in frozen soils. In: State of the Art and Practice in Geotechnical Engineering (GeoCongress2012), Geotechnical Special Publication no. 225, pp. 215–224 (2012)
Zhang, L., Ernst, H., Einstein, H.H.: Nonlinear analysis of laterally loadedrock-socketed shafts. J. Geotech. Geoenviron. Eng. **126**(11), 955–968 (2000)
Zhang, L., McVay, M.C., Lai, P.W.: Centrifuge modelling of laterally loaded single battered piles in sands. Can. Geotech. J. **36**(6), 1074–1084 (1999)
Zhu, B., Sun, Y., Chen, R., Guo, W., Yang, Y.: Experimental and analytical models of laterally loaded rigid monopiles with hardening p–y curves. J. Waterw. Port, Coast. Ocean Eng. **141** (6), 04015007 (2015)

Reliability of Load-Transfer Approach in the Design of Large Diameter Bored Piles

Hayel El-Naggar[1(✉)], Mostafa Ahmed[2], and Sherif AbdelSalam[2]

[1] Geo-Institute, Housing and Building National Research Center, Cairo 11511, Egypt
hayelelnaggar@gmail.com

[2] Civil Engineering Department, Faculty of Engineering, The British University in Egypt, Cairo-Suez Road, El Shorouk 11837, Egypt
mhassanein992@gmail.com, sherif.abdelsalam@bue.edu.eg

Abstract. Pursuing current global trend to practice a reliability-based pile design methodology, most codes have calibrated resistance factors accounting for the strength limit, while the serviceability limit is still being assessed following a deterministic approach. Recently, the Load and Resistance Factor Design (LRFD) parameters were regionally developed by AbdelSalam et al. (2015) for the strength limit utilizing the electronic database "EGYptian Pile Test". This database contained results from more than 320 pile load test, most of them for large diameter bored piles. In this study, the database was upgraded to include load-transfer outcomes for all the available data using finite difference program Allpile v.6.5, which provided separate skin- and end-bearing behaviors. In addition, the total load-displacement acquired from the load-transfer analysis was employed to develop the LRFD resistance factors for groups of piles sorted by pile diameter, length, and soil conditions. These resistance factors were calculated based on limited total settlement of 1% and 2.5% of the pile diameter, which indirectly accounts for the serviceability limits. A comparison between the strength- and the serviceability-based resistance factors was conducted, and it was found that designs based on the serviceability is more efficient in case of large diameter bored piles.

1 Introduction

Several international codes such as Eurocode 7 (2004), Australian Standard 5100 (2004), AASHTO (2007), etc., shifted the design of deep foundations from the Working Stress Design (WSD) to the Limit State Design (LSD) approach. The process towards full acceptance and implementation of LSD in the Middle East and North Africa (MENA) region is still ongoing, yet emergent (AbdelSalam and El-Naggar 2014; AbdelSalam et al. 2016). This is due the increasing demand for more efficient and sustainable geotechnical designs that are compatible with the structural designs. The LSD is simply a process to calculate partial or global factors for loads and resistances (Becker 1996) where using these factors account for design uncertainties to achieve a defined target reliability, which also means a more reliable pile design compared with the WSD. There are two limit states that must be considered in the

M. Abu-Farsakh et al. (eds.), *Advances in Analysis and Design of Deep Foundations*, Sustainable Civil Infrastructures, DOI 10.1007/978-3-319-61642-1_8

design of foundations, first is the Ultimate Limit State (ULS), and second is the Serviceability Limit State (SLS). The ULS relates to capacity of the foundations, while the SLS relates to conditions such as total or differential settlement that may directly affect the function and usability of a structure under service load. According to Phoon and Kulhawy (2008), usually the SLS governs the failure criteria for large-diameter pile foundations; nevertheless, the ULS has considerably received more attention while developing the LSD recommendations for various codes. This is due to several uncertainties associated with modeling or predicting a pile load-displacement response under axial loading, making accurate prediction of the pile limiting displacement a relatively complicated task. According to Paikowsky et al. (2004), the development of LSD resistance factors that consider both the ULS and the SLS requires large pile load tests database and an accurate mean of prediction for the pile load-displacement response.

Several researches collected large databases and used the load-transfer model to conducted a LSD reliability-based assessment that considers the ULS and SLS together (e.g. Phoon and Kulhawy 2008; Misra and Roberts 2009; Abu-Farsakh et al. 2016; Haque et al. 2016; Misra et al. 2007). In the MENA region, the main problem was the non-availability of an electronic database that includes sufficient pile load test information and soil data, until recent. In 2015, Abdelsalam et al. (2015) developed a comprehensive electronic database called EGYptian Pile Test database (or EGYPT database), which was initially used to calculate the parameters required in the Load and Resistance Factor Design (LRFD) approach for large diameter bored piles considering the ULS.

The primary goal of this study was to develop the LRFD resistance factors for large diameter bored piles to fulfill both, the ULT and SLS together, using information from EGYPT database and by utilizing the load-transfer model, following the basic calibration procedure provided by Paikowsky et al. (2004). This required a significant upgrade in the existing database to include load-transfer outcomes for all the available test piles, using outcomes from the finite difference program Allpile v.6.5 (Civil Tech 2014) to acquire the total load-displacement response of a pile, in addition to separated skin- and end-bearing responses. Data were grouped afterwards according to the pile diameter, embedment length, and soil conditions, then the reliability-based LRFD resistance and efficiency factors were calculated for the load-transfer model based on limited total pile settlement criteria of 1% and 2.5% of the pile diameter. The study also included a comparison between the strength- and serviceability-based resistance factors to stand on the governing factor in the design.

2 Database Update

Results from a large number of pile Static Load Tests (SLTs) conducted in Egypt were collected and revised by AbdelSalam et al. (2015) to produce the first electronic database, EGYPT database. The database contained information about the soil profile at each test pile and their corresponding cyclic and/or monotonic SLT outcomes. The pile collection consisted of Large Diameter Bored Piles (LDBP), driven Vibro piles, and Continues Flight Augur (CFA) piles. The total number of records available at that time was 320, where LDBP (diameter, D $\geq$ 60 cm) formed the majority of this data with percentage 74%, then the Vibro piles with percentage 13%. The 8% of the remaining data was for CFA piles and 5% for small diameter bored piles (D < 60 cm). The distribution of all the available data was summarized on a map provided by El-Naggar (2016).

The diameters of the test piles in the database ranged from 60 cm to 150 cm with average value 93 cm. Test piles with diameter 100 cm has the largest percentage of 33% followed by 30% for the diameter of 80 cm, and a small percentage of about 0.45% was for the 90 cm, as well as the 70 cm as presented in Fig. 1a. A distribution indicating the various lengths of the LDBP test piles is provided in Fig. 1b, for which the highest value for the length ranged from 15 m to 20 m and the lowest value for the length ranged from 65 m to 70 m. The soil profile at each pile was also classified as sand, clay, or mixed soil following the "70% Rule" suggested by AbdelSalam et al. (2011). The 70% rule is a generalization for the soil profile along the pile shaft and ensures that the effect of this generalization would not considerably change the pile total skin-friction. In this rule, a pile would be embedded in sand if at least 70% of the soil layers along the shaft are classified as a sand based on the Unified Soil Classification System (USCS). Same is considered for clay, otherwise the soil layers are classified as mixed—knowing that sand, clay, and mixed soils are the classifications used for pile grouping in the AASHTO 2007, which did not provide a clear approach for soil generalization. The distribution of the SLTs by soil type is provided in Fig. 1c, for which most of the test piles were embedded in mixed soil by 57%, while the smallest number was for piles embedded in clay.

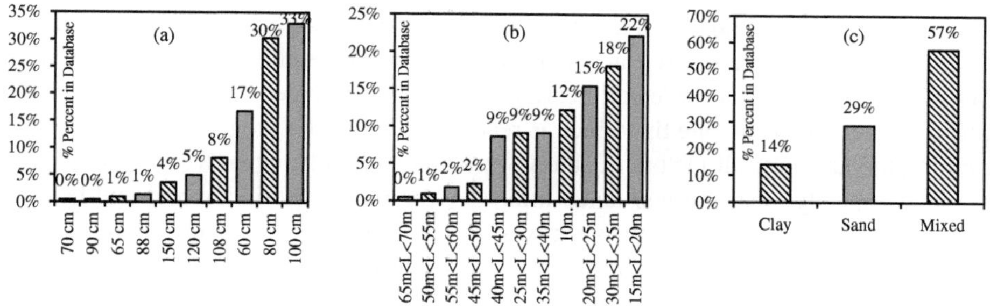

Fig. 1. LDBP distribution in the database by: (a) diameter; (b) length; and (c) soil type

The main difficulty reported by authors of EGYPT database was the deficient load–displacement curves for most LDBP, which prevented the curve from fulfilling the criterion of Davisson (1972)—knowing that Davisson's criterion was selected herein because it was previously indicated by AbdelSalam et al. (2015) that it provides a comparatively accurate and consistent results related to other approaches such as Chin (1970) and Hansen (1963), and also yields more efficient LRFD resistance factors. The problem of deficient load–displacement curves was solved by extrapolation following the method of approach (MoA) recommended by Paikowsky and Tolosko (1999). AbdelSalam et al. (2015) extrapolated the load–displacement curves for bored piles using the MoA and proved that this method can be confidently used for such type of piles after making necessary comparison with direct load–displacement curve extrapolation using a polynomial function from the third degree. Another deficiency in EGYPT database was excluding information about separate skin-friction and end-bearing resistances for the available data set. An effective way to overcome this deficiency was to run the load-transfer equations for the entire data, and include the outcomes in the database to be more comprehensive and serve for the LRFD calibration that accounts for both limits, the ULT and SLS together.

3 Load-Transfer Analysis

Design of LDBP depends on calculating a factored capacity that satisfies the ULS, while the settlement is typically checked afterwards. This procedure requires conducting several design iterations to satisfy both limit states, especially that settlement governs the design of LDBP (Misra and Roberts 2006). Hence, a pile design methodology that is based on a predicted total load-displacement response can simultaneously incorporate both limits within the design process in one single step (Alawneh 2006; Roberts et al. 2008).

Several methods can be used to predict the load-displacement response of axially loaded piles, one of them is the load–transfer method that is a simple approach, frequently used, and referred to as "t–z" analysis (Misra and Chen 2004). Allpile v.6.5 (Civil Tech 2014) is finite difference program that is capable of running the t-z analysis for piles depending on corrected number of Standard Penetration Test (SPT) blow counts along the pile embedded length. In the analysis, the pile material was determined as concrete, while the effect of pile reinforcement was neglected. Soft clay layers (if exist) were assumed to exert negative skin friction on the pile, while the effect of fill layers was neglected. The pile end-bearing stratum was assumed to extend 10 times the pile diameter under the pile tip, which is recommended in the program manual. Before running the analysis, all factors of safety were assumed to be equal to unity to acquire nominal capacity, also the resistance limits for the pile skin-friction and end-bearing were set unlimited and the settlement was allowed freely up to 5 cm. Figure 2a represents the predicted load-displacement curve for test pile ID #1 in EGYPT database after running the load-transfer analysis using Allpile program. On the same figure, the predicted curve was compared with the measured response from the SLT after extrapolation using the MoA and verification via a polynomial function from the third degree as per the recommendations by AbdelSalam et al. (2015) for bored piles. Pile ID #1 had a diameter of 100 cm, and was imbedded in a mixed soil profile. As can be seen from the figure, there is a good match between the measured and the predicted curves along both the elastic and yield portions. Figure 2b represents another comparison for pile ID #132 with a smaller diameter (60 cm), and similarly a good match between results was observed.

Upgrading the database was carried out to include outcomes from the load-transfer analysis such as the t-z curves along the pile length, which describe the stress–displacement relationships along the soil–pile interface. Also q-w curve was added to the database, which describes the penetration load versus displacement at the pile tip.

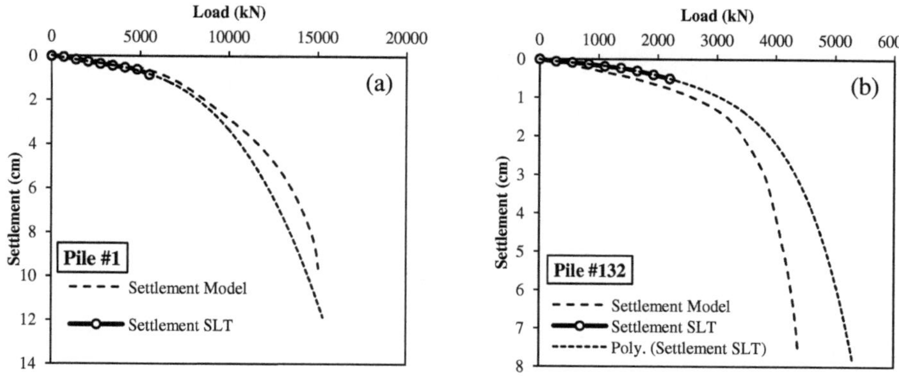

Fig. 2. Load-displacement curves from t-z and SLT: (a) Pile ID#1; and (b) Pile ID#132

In addition, the total load-displacement curve at the pile head was separated for the skin-friction and the end-bearing resistances as can be seen in Fig. 3 (a figure that represents a snap shot for the record detail view for the test pile ID #1 in EGYPT database after updates). More information about the record view components can be found in AbdelSalam et al. (2015). However, it is important to indicate that the LRFD calibration conducted in this paper was based on the predicted total response at the pile head (i.e., total load-displacement curve from the load-transfer analysis), not the separated responses for skin- and end-bearing.

4 LRFD Calibration

4.1 Grouping Criteria

The load–transfer analysis was performed for all the piles in EGYPT database, and accordingly data was divided into different groups sorted according to the pile diameter, embedded length, and soil profile. The number of main groups was four as follows: G_{D1}, a group for piles with diameter from 60 to 90 cm; G_{D2}, a group for piles with diameter from 900 cm to 150 cm; G_{L1}, a group for piles with length from 13 m to 30 m; and G_{L2}, a group for piles with length greater than 30 m. Each group was then sorted again based on the soil profile along the shaft and at the pile tip; therefore, group G_{S1} included piles in sand (along shaft and at tip), group G_{S2} included piles in mixed soil (along shaft) and sand (at tip), group G_{S3} included piles in mixed soil (along shaft) and clay (at tip), and group G_{S4} included piles in clay (along shaft) and sand (at tip). Figure 4 is a flowchart that summarizes all groups used in the analysis.

4.2 Serviceability Limits

The 2001 Egyptian Code of Deep Foundations (ECDF 2001) adopted a method of prediction from the German specification DIN 4014 – Part 2 (DIN 1990) for LDBP. This method suggests the relation between total load-displacement at pile head

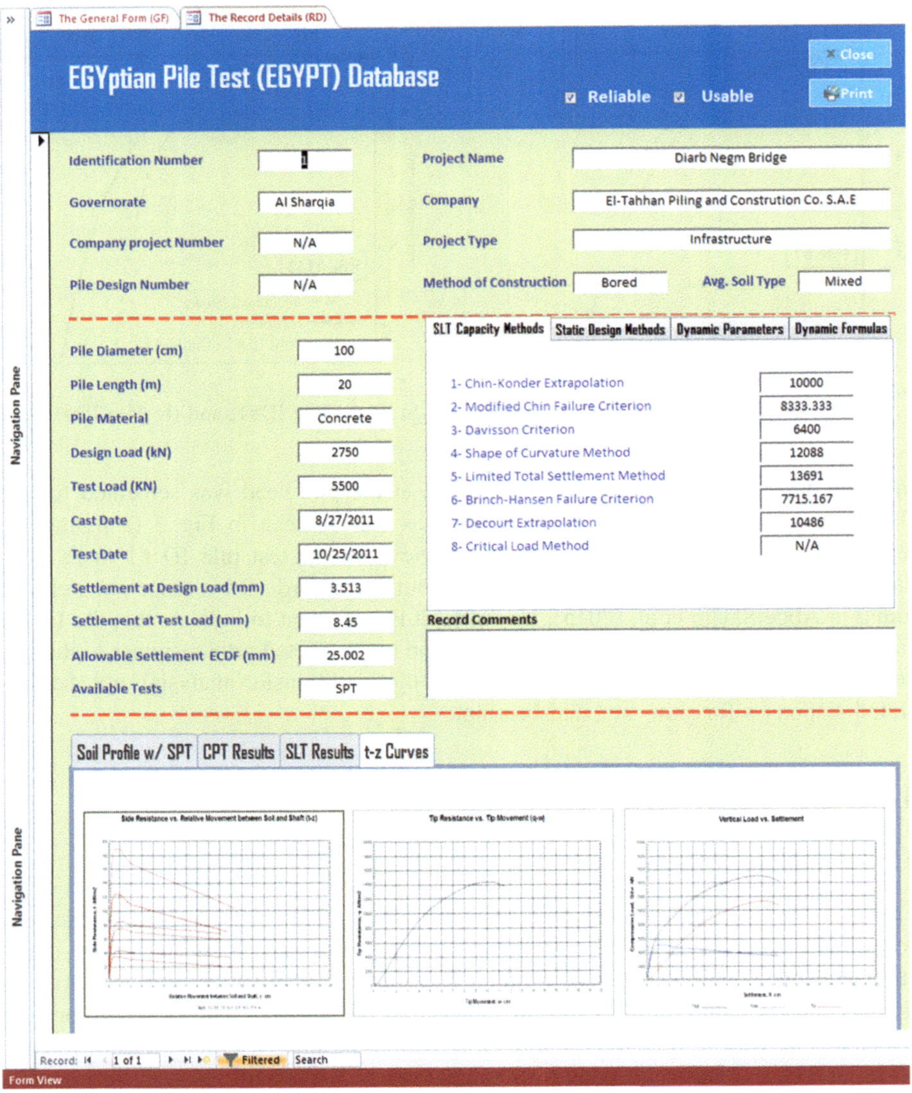

Fig. 3. Screen shot for an updated record in EGYPT database including t-z outcomes

by assuming empirical values for the pile skin-friction and end-bearing responses. The DIN 1990 method assumes that the performance of LDBP is more affected by settlement, as full mobilization of skin-friction occurs at settlement corresponding to 1% of the pile diameter, while full mobilization of end-bearing occurs at settlement of 5% to 10% of the pile diameter. Therefore, the pile settlement limit was determined in this study based on the aforementioned values, to be with a minimum of 1% of the pile diameter, and maximum of 2.5% of the pile diameter, and these two limits were indicated to in this paper as K1%, and K2.5%, respectively.

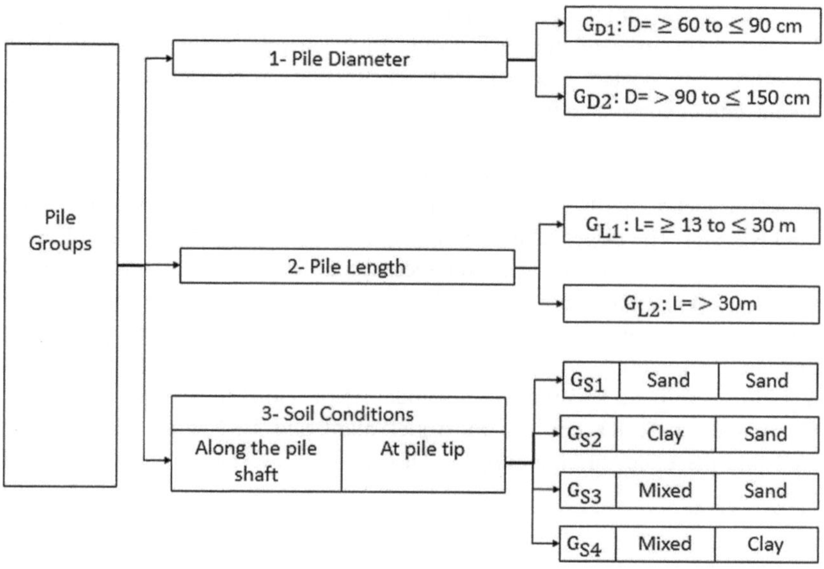

Fig. 4. Flowchart for the pile groups of EGYPT database

4.3 Statistical Approach

The calibration of the LRFD resistance factors was performed using the closed form solution of the First Order Second Moment (FOSM) approach presented in Eq. 1. In FOSM, the Probability Distribution Functions (PDFs) for the applied loads and resistance of pile are assumed to follow a log-normal distribution according to AASHTO (2007) and using mean bias for Dead and Live Loads (DL and LL), Coefficient of Variation (COV), load factors, and DL to LL ratios summarized in Table 1. The goodness-of-fit tests were performed to confirm the normality of PDFs for each group as presented in the sub-section below.

$$\varphi_R = \frac{\lambda_R\left(\frac{\gamma_{DL}Q_{DL}}{Q_{LL}} + \gamma_{LL}\right)\sqrt{\frac{(1 + COV^2_{Q_{DL}} + COV^2_{Q_{LL}})}{(1 + COV^2_{Q_R})}}}{1_R\left(\frac{\gamma_{DL}Q_{DL}}{Q_{LL}} + \gamma_{LL}\right)\exp\left\{\beta_T\sqrt{\left[(1 + COV^2_{Q_R})(1 + COV^2_{Q_{DL}} + COV^2_{Q_{LL}}\right]}\right\}} \quad (1)$$

Table 1. Summary of DL and LL bias, COV, factors, and ratio from AASHTO 2007

Bias		COV		Load factors		DL/LL ratio
λ_{QD^*}	λ_{QL^*}	Ω_{QD}	Ω_{QL}	γ_D	γ_L	$E(Q_D)/E(Q_L)$
1.05	1.15	0.10	0.20	1.25	1.75	2

The actual nominal pile capacity was acquired from the measured load-displacement curve of the SLT results for each pile in the database. Then, the corresponding predicated capacity of each pile was determined from the load-displacement curve calculated by Allpile program. For both, the measured and the predicted capacities, the ultimate values were determined on the curves at K1% and K2.5%. The product of dividing the measured and the predicted pile capacities was defined as the mean bias factor, K, which was calculated for each pile twice (i.e., for K1% and K2.5%). The target reliability index, β, was used in the analysis for redundant piles, five or more per pile cap where the recommended probability of failure is 1% and corresponds to $\beta = 2.33$ as per Paikowsky et al. (2004). Resistance factors were also calculated at β for non-redundant piles, four or fewer piles per pile cap where the recommended probability of failure is 0.1% and corresponds to $\beta = 3.00$. For the DL/LL ratio, there is no specific recommendation provided in regional code, so it was assumed equals to 2.0 according to AbdelSalam and El-Naggar (2014) who also indicated that the effect of changing the DL/LL on the resistance factors of LDBP is negligible.

4.4 Goodness-of-Fit

The normality test Anderson Darling (AD) and 95% Confidence Interval (95% CI) were used to check the best distribution for a PDF representing a specific group of piles in the database. Results for all available PDFs are presented in Fig. 5, which includes the bias factor K obtained at limited settlement 1% of pile diameter, or K1%. To determine which distribution was satisfactory (i.e., normal or log-normal distributions, or both), the probability of 95% confidence interval must be greater than 0.005, while the Anderson Darling value should be smaller for the best-fitting distribution. From the tests, it was found that all PDFs followed the lognormal distribution, which means that the FOSM can be used to calculate the LRFD resistance factors. Before substitution in the FOSM equation, and due to small numbers of available test piles in some groups, the calibration parameters needed for the FOSM equation were first undergo a Monte

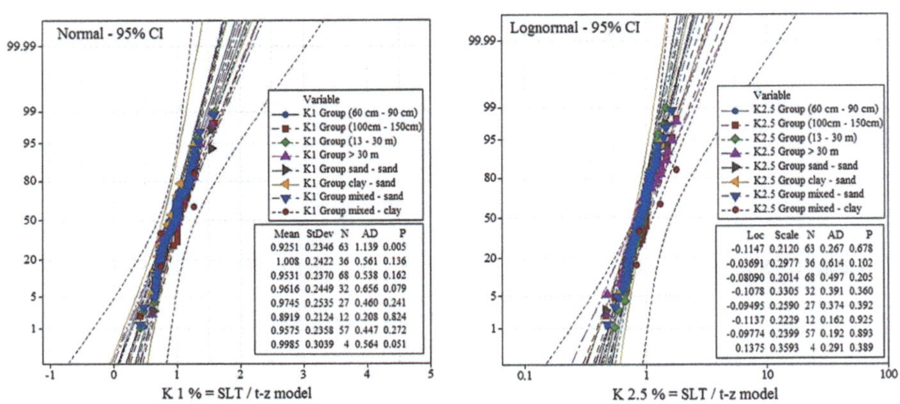

Fig. 5. AD and 95% CI results for all groups at K1%: (a) normal; and (b) log-normal

Carlo Simulation (MCS). The MCS was performed for all the available groups that were determined based on the soil profile, pile diameter, and length to obtain a more accurate bias and COV, using Minitab 17 statistical software (Minitab 2010). The simulations were performed until the number of iterations reached 10,000 for each group as per the recommendation provided by Abu-Farsakh et al. (2013). PDFs are presented in Fig. 6 for all pile groups just before running the MCS. The values of the calibration parameters were slightly adjusted after running the MCS, whereas the differences did not exceed 5%, and these parameters were then used in the FOSM to develop the resistance factors.

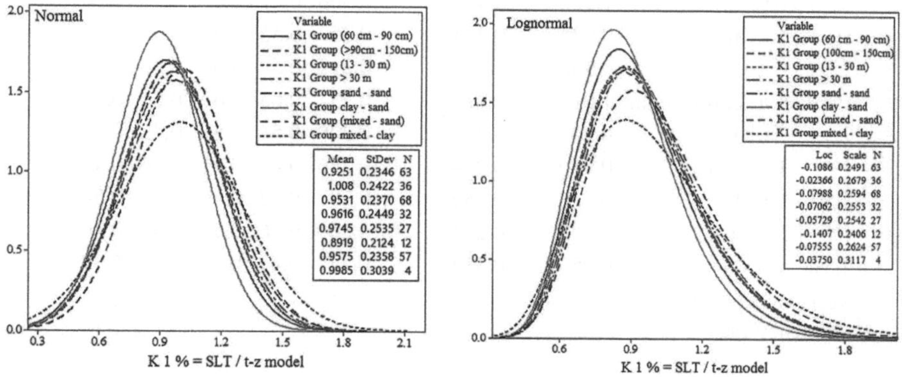

Fig. 6. PDFs at K1% before using MCS: (a) normal; and (b) log-normal

4.5 Resistance Factors

The LRFD resistance factors were calculated using the statistical parameters obtained from MCS and based on the FOSM. Table 2 includes all calibrated resistance factors for different pile groups. Efficiency factors were calibrated to be able to judge the performance of each resistance factor. For pile groups based on soil profile (i.e., G_{S1}, G_{S2}, G_{S3}, and G_{S4}), the resistance factors were in a close range with no significant variations, as the resistance factors ranged from 0.54 to 0.58 for redundant piles, and for non-redundant piles the resistance factors ranged from 0.42 to 0.47. The highest efficiency was for G_{S2} and equal to 0.63 at K2.5%. For pile groups based on pile diameter (i.e., G_{D1} and G_{D2}), the resistance factors for redundant piles ranged from 0.54 to 0.63. For non-redundant piles, the resistance factors ranged from 0.42 to 0.51. The highest efficiency for these groups was obtained was obtained for G_{D1} and equal to 0.64 at K2.5%. For pile groups based on pile length (i.e., G_{L1} and G_{L2}), the resistance factors for redundant piles ranged from 0.47 to 0.62. For non-redundant piles, the resistance factors ranged from 0.36 to 0.51, and the highest efficiency was for GL1 and equals to 0.66 at K2.5%. It is important to indicate that the resistance factors presented in Table 2 ensures the design reliability of a large diameter bored pile and accounts for both, the strength and the serviceability limit states together if the K1% or the K2.5% are limiting the pile settlement. Using these outcomes also requires adopting the

Table 2. Calibrated LRFD resistance factors

K	Group	Mean (λ)	St. Dev. (σ)	COV	$\beta = 2.33$		$\beta = 3.00$	
					φ^{a}	φ/λ^{b}	φ	φ/λ
1%	Sand – Sand	0.978	0.255	0.261	0.58	0.59	0.46	0.47
2.5%	(G_{S1})	0.940	0.239	0.254	0.56	0.60	0.45	0.48
1%	Clay – sand	0.888	0.214	0.241	0.54	0.61	0.44	0.49
2.5%	(G_{S2})	0.913	0.208	0.228	0.57	0.63	0.46	0.51
1%	Mixed – Sand	0.958	0.236	0.246	0.58	0.61	0.47	0.49
2.5%	(G_{S3})	0.935	0.226	0.241	0.57	0.61	0.46	0.49
1%	Mixed – Clay	0.998	0.301	0.301	0.54	0.54	0.42	0.42
2.5%	(G_{S4})	1.206	0.444	0.368	0.57	0.47	0.43	0.36
1%	60 cm–90 cm	0.925	0.233	0.252	0.55	0.60	0.44	0.48
2.5%	(G_{D1})	0.908	0.193	0.213	0.58	0.64	0.48	0.52
1%	>90 cm–150 cm	1.012	0.237	0.234	0.63	0.62	0.51	0.50
2.5%	(G_{D2})	0.994	0.298	0.300	0.54	0.55	0.42	0.43
1%	13–30 m	0.950	0.234	0.246	0.57	0.61	0.46	0.49
2.5%	(G_{L1})	0.938	0.185	0.197	0.62	0.66	0.51	0.54
1%	>30 m	0.963	0.257	0.266	0.56	0.58	0.45	0.46
2.5%	(G_{L2})	0.955	0.332	0.347	0.47	0.49	0.36	0.38

[a]Resistance factor
[b]Efficiency factor

load-transfer equations to predict the pile total load-displacement response, and for bored piles with diameters, lengths, and soil conditions similar to those of concern in this study.

4.6 Resistance Factors Comparison

AbdelSalam et al. (2015) provided LRFD resistance factors for traditional static analysis methods using the same database. Their calibration was not related to the pile settlement, either directly or indirectly; however, Davisson's criterion was used to determine the pile capacity from the SLT measurements. Hence, the resistance factors provided by AbdelSalam et al. (2015) only accounts for the ULS. By comparing the resistance factors developed for groups of piles embedded in similar soil conditions, it was found that ULS-factors are lower compared with the resistance factors developed in this study (i.e., K1% and K2.5%, or SLS-factors). This is because the ULS-factors were calibrated for static methods at pile full load mobilization, which corresponds to settlement of around or more than 10% of the pile diameter. But for the SLS-factors, the capacity was determined at a lower value of settlement to satisfy the structural requirements of bridges and other super structures. Some results of this comparison are summarized in Fig. 7, which shows the resistance and efficiency factors for pile groups with same soil profile at $\beta = 2.33$ and $\beta = 3.00$.

For sand at β = 2.33, the resistance factors ranged from 0.38 to 0.40 with mean of 0.39, these values were lower compared with the case of K1% and K2.5% by 33% and 30%, respectively. For clay at β = 2.33, the resistance factors ranged from 0.46 to 0.52 with mean of 0.49, these values were lower than those for K1% and K2.5% by 9% and 14%, respectively. For mixed soil, the resistance factors ranged from 0.39 to 0.42 with mean of 0.40, which is lower than those calculated for K1% and K2.5% as well. Additionally, the efficiency factors were generally lower in the case of ULS-factors compared with the SLS-factors. For instance, in case of sand the efficiency factors ranged from 0.40 to 0.45 with mean of 0.425, whereas the efficiency for the K1% was 0.59 and for K2.5% was 0.60. Hence, the average difference between the ULS and SLS-factors was around 28%.

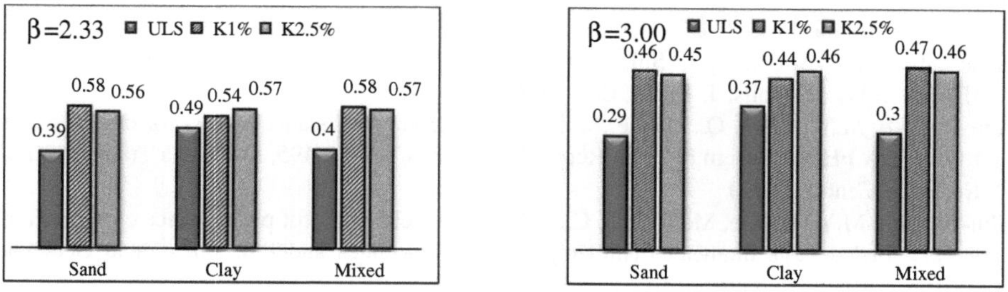

Fig. 7. ULS-factors versus K1% and K2.5% factors: (a) β = 2.33; and (b) β = 3.00

5 Summary and Conclusion

This study aimed at establishing the LRFD design recommendations for large-diameter bored piles to account for both, the strength and serviceability limit states, using measured load-displacement responses from EGYPT database. Following the typical LRFD reliability-based calibration framework, the resistance factors were developed for several pile groups, eight groups based on pile diameter, length, and soil profile. The load-transfer method was used to predict the load-displacement curves at the pile head. The limited settlement criterion used in the analysis was limited to 1% and 2.5% of the pile diameter, which satisfies most of the structural requirements. Summarized below are the major outcomes.

- Load-transfer analysis was conducted for the entire database that was upgraded to house new t-z information including separated skin-friction and end-bearing responses.
- LRFD resistance factors were calibrated for the load-transfer model using FOSM and MCS for redundant and non-redundant pile groups to account for the strength and serviceability limit states.
- A comparison between ULS and SLS-factors indicated that the resistance and efficiency factors are higher in case of SLS by around 28%.

Acknowledgements. The authors are grateful to the British University in Egypt (BUE) at which some of this research was conducted as part of final year graduation project.

References

AASHTO LRFD Bridge Design Specifications: Customary U.S. Units, 6th edn. (2013 Interim), Washington, DC (2007)

AbdelSalam, S.S., El-Naggar, M.E.: LRFD for large-diameter bored piles in Egypt. In: Proceeding of GeoCongress 2014, 23–26 Feb, Atlanta, GA (2014)

AbdelSalam, S.S., Baligh, F.A., El-Naggar, H.M.: Reliability and construction control of vibro piles. Ain Shams Eng. J. **7**(2), 885–893 (2016)

AbdelSalam, S.S., Baligh, F.A., El-Naggar, H.M.: A database to ensure reliability of bored pile design in Egypt. Proc. ICE J. Geotech. Eng. **168**(2), 131–143 (2015)

AbdelSalam, S.S., Sritharan, S., Suleiman, M.T.: LRFD resistance factors for design of driven H-piles in layered soils. J. Bridg. Eng. **16**(6), 739–748 (2011)

Abu-Farsakh, M.Y., Chen, Q., Haque, Md.N.: Calibration of resistance factors for drilled shafts for the new FHWA design method. Report No. FHWA/LA. 12/495, Louisiana Transportation Research Center (2013)

Abu-Farsakh, M.Y., Haque, MdN, Tsai, C.: A full scale field study for performance evaluation of axially loaded large diameter cylinder piles with pipe piles and PSC piles. Acta Geotech. (2016). doi:10.1007/s11440-016-0498-9

Alawneh, A.S.: Modeling load–displacement response of driven piles in cohesionless soils under tensile loading. Comput. Geotech. **32**, 578–586 (2006)

Australian Standard, A.S.: 5100: Bridge Design. Standards Australia International (SAI), Sydney (2004)

Becker, D.E.: Eighteenth Canadian geotechnical colloquium: limit states design for foundations, part I: an overview of the foundation design process. Can. Geotech. J. **33**(6), 956–983 (1996)

CEN: Eurocode 7: Geotechnical Design—Part 1: General Rules, EN 1997-1:2004. European Committee on Standardization, Brussels (2004)

Chin, F.V.: Estimation of the ultimate load of piles not carried to failure. In: Proceedings of the 2nd Southeast Asian Conference on Soil Engineering, Singapore, pp. 81–90 (1970)

Civil Tech Software: All pile program manual. Simultaneously published in the U.S. and Canada. Printed and bound in the United States of America (2014)

Davisson, M.: High capacity piles. In: Proceedings of Soil Mechanics Lecture Series on Innovations in Foundation Construction. American Society of Civil Engineers (ASCE), Illinois Section, Chicago, USA, pp. 81–112 (1972)

DIN: DIN 4014: Part 2: Large Bored Piles, Manufacture, Design and Permissible Loading. Beuth Verlag, Berlin (1990)

ECDF (Egyptian Code of Deep Foundations): Soil Mechanics and Foundation Engineering. ECDF Part 4, Deep Foundations, 6th edn. Housing and Building National Research Centre (HBRC), Cairo (2001)

El-Naggar, H.M.: Applying of the design approach of load and resistance factor in deep foundation design using pile load test database in Egypt. Master dissertation, Helwan University, Egypt (2016). doi:10.13140/RG.2.1.1229.2722

Hansen, J.B.: Discussion, hyperbolic stress–strain response: cohesive soils. J. Soil Mech. Found. Div. ASCE **89**(SM4), 241–242 (1963)

Haque, MdN, Abu-Farsakh, M.Y., Tsai, C., Zhang, Z.: Load testing program to evaluate pile set-up behavior for individual soil layers and correlation of set-up with soil properties. J. Geotech. Geoenviron. Eng. (ASCE) (2016). doi:10.1061/(ASCE)GT.1943-5606.0001617

Minitab 17 Statistical Software: [Computer software]. Minitab, Inc., State College, PA (2010) www.minitab.com

Misra, A., Roberts, L.A.: Probabilistic analysis of drilled shaft service limit state using 't–z' method. Can. Geotech. J. **43**(12), 1324–1332 (2006)

Misra, A., Chen, C.H.: Analytical solutions for micropile design under tension and compression. Geotech. Geol. Eng. **22**(2), 199–225 (2004)

Misra, A., Roberts, L.A.: Service limit state resistance factors for drilled shafts. Géotechnique **59**(1), 53–61 (2009)

Misra, A., Roberts, L.A., Levorson, S.M.: Reliability of drilled shaft behavior using finite difference method and Monte Carlo simulation. Geotech. Geol. Eng. **25**, 65–77 (2007)

Paikowsky, S.G., Tolosko, T.A.: Extrapolation of pile capacity from non-failed load tests. Federal Highway Administration (FHWA), Washington, D.C., USA, FHWA Report No. FHWA-RD-99-170 (1999)

Paikowsky, S.G., Birgisson, B., McVay, M., Nguyen, T., Kuo, C., Baecher, G. Ayyab, B., Stenersen, K., O'Malley, K., Chernauskas, L., O'Neill, M.: LRFD for deep foundations. NCHRP-507, TRB, Washington D.C. (2004)

Phoon, K.K., Kulhawy, F.H.: Serviceability limit state reliability-based design. In: Reliability-based Design in Geotechnical Engineering: Computations and Applications, pp. 344–383. Taylor and Francis, London (2008)

Roberts, L.A., Gardner, B.S., Misra, A.: Multiple resistance factor methodology for service limit state design of deep foundations using t–z model approach. In: Proceeding: Geo-Congress 2008, New Orleans, LA (2008)

Numerical Study of Tapered Piles in Sand

Mohsen Mohammadizadeh[1(✉)] and Moein Mohammadizadeh[2]

[1] Department of Civil Engineering, Sirjan Azad University, Sirjan, Iran
mohammadizadeh@iausirjan.ac.ir
[2] Young Researchers and Elite Club, Sirjan Branch,
Islamic Azad University, Sirjan, Iran
mohammadizadeh@yahoo.com

Abstract. So much research has been done showing that bearing capacity of tapered pile in sands is more than the cylindrical pile of the same volume and length. ABAQUS 6.10 Software, which is based on the finite element method, has been used for analyzing the models. In order to verify numerical results, a comparison between numerical and experimental results has been made. Parametric studies have been carried out to determine the influence of contributing factors such as tapered angle and relative density of sand on maximum shear strength and bending moment of piles. In all cases, tapered piles had better performances than the cylindrical ones under lateral loading. Under lateral loading, maximum bending moment occurred in the top one-third of the pile length because cross section of tapered piles were more than the cylindrical piles in this place. Furthermore, the maximum shear force occurred in the pile head, where tapered piles had better material distribution.

1 Introduction

Tapered piles are a type of piles with a variable cross-section, which is bigger on top of the pile than the bottom of that. In recent years, an increasing interest is observed for using tapered piles. It can be mentioned that tapered piles have more suitable distribution of materials for various loading transfer than their cylindrical piles of the same volume and length. Few studies have been done to the type of cylindrical piles and moreover, relationship theory have not been proposed yet. Evaluation of ultimate strength and deformation of vertical piles under lateral loading are a complex issue because it involves the interaction between a semi-rigid structural element, pile, and an elasto-plastic material, soil. One of the first studies on tapered pile was conducted in sand by Nordlund. He concluded that bearing capacity of tapered pile increases due to be the more compact the soil around pile at the time driven pile (Nordlund 1963). Mayerhof conducted some investigations for single tapered piles whose conclusion was that in sandy soils under vertical loadings, frictional capacity of tapered piles is about 1.5 times of that of prismatic piles (Mayerhof 1976). Zil'berberg and Sherstnev showed that axial bearing capacity of driven tapered piles increases by 2–2.5 times comparing to the capacity of cylindrical piles with the same volume and mean radius (Zil'berberg and Sherstnev 1990). El Naggar and Wei conducted a laboratory study on tapered piles under lateral loading and showed that confining pressure affects the load-deformation

M. Abu-Farsakh et al. (eds.), *Advances in Analysis and Design of Deep Foundations*,
Sustainable Civil Infrastructures, DOI 10.1007/978-3-319-61642-1_9

behavior of tapered pile (El Naggar and Wei 1999). Ismael represented that the use of tapered piles step leads to a significant increase in lateral bearing capacity and decrease in displacement. It is concluded that moment maximum occurs at the top of the pile where the cross-section of pile is larger than the other part (Ismael 2006). Zhan et al. analyzed two series of tapered piles under axial loading in sand with low dilatancy by finite element method. The conclusion was that the resistance of body increases as taper angle increased (Zhan et al. 2012). Hataf and Shafaghat showed that the optimized tapered angle of a single pile under axial loading in sand depends on the soil internal friction angle. It is inferred in this paper that friction bearing of pile increases and its end bearing capacity decreases as tapered angle increased, however, the percentage of increase in frictional bearing is more than the percentage of decrease in the end bearing capacity at the optimal angle (Hataf and Shafaghat 2015).

2 Numerical Investigation

2.1 Finite Element Model, Boundary Conditions and Material Properties

Numerical analysis is widely used to analyze piles subjected to lateral and axial loading due to their low cost and fairly good prediction of the behavior of different materials. ABAQUS software is a finite element modeling program which is designed for modeling a variety of material behavior in both static and dynamic situations. In this paper, the Mohr-Coulomb model (MCM) was used for two types of soil. The concrete pile material is assumed to be elastic because the rigidity of pile is much bigger than the soil. The model includes 8-node linear brick, reduced integration (C3D8R) elements for soil and pile. The efficiency and behavior of tapered pile have been investigated in two types of sand. In addition, the mechanical properties used for the pile and the soils are summarized in Table 1.

Table 1. Properties of concrete and soil

Material	Behavioral model	Elastic modulus (MPa)	Poisson's ratio	Cohesion (KPa)	Angle of internal friction (°)	Unit weight (KN/m^3)
Concrete Pile	Elastic	25000	0.2	–	–	24
Soft Sand	Mohr-Coulomb	20	0.3	1	28	14
Dense Sand	Mohr-Coulomb	55	0.35	1	35	19

The 3D model shown in Fig. 1 is asymmetric because the lateral loading is applied only along one horizontal direction (X direction). Pile-soil contact was applied in both tangential and normal directions, whereas in the normal direction hard contact was utilized and frictional behavior was chosen for the tangential direction. The coefficient of friction is constant value $\mu = \tan\left(\frac{2}{3}\varphi\right)$ between sandy soil and pile surface. Additionally, the effect of reducing the friction coefficient is ignored between the pile and

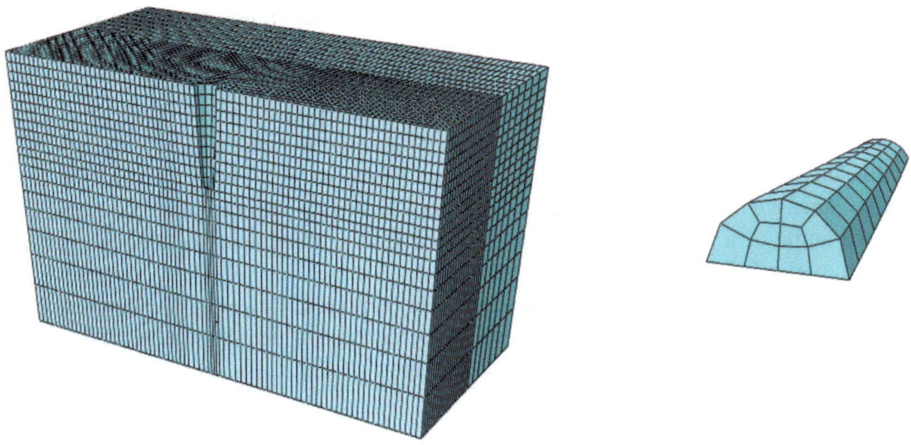

Fig. 1. 3D model of piles group foundation and soil

the adjacent soil. Boundary conditions are 30 times the pile diameter from the center of the pile in the horizontal direction and 2 times the pile length from the end of the pile in the vertical direction. Two types of piles were included, two tapered pile and two cylindrical pile, as shown in Table 2.

Table 2. The dimension of tapered piles used in this article

Index	Length (m)	Bottom radius (m)	Top radius (m)	Volume (m^3)	Taper angle (deg)
C 20	5	0.2	0.2	0.628	0
T 31	5	0.1	0.3	0.680	2.29
C 40	5	0.4	0.4	2.51	0
T 62	5	0.2	0.6	2.72	4.57

Pile installation causes a change in the stress fields around the pile, One method uses the K value defined by $p'_r = K \cdot \sigma'_{vo}$, where p'_r is the radial pressure applied on the cavity wall, σ'_{vo} is the initial effective overburden stress and K value depends on the soil nature, the diameter of the pile and the installation procedure. Based on this definition, the effects of pile installation is analysed with finite elements. K values for tapered piles and cylindrical piles were considered in soft sand and dense sand, as shown in Table 4 that these values have been obtained based on K-pressure method presented in the paper of Satibi et al. (2007).

2.2 Verification

Rybnikov carried out tests in the Irtysh Pavlodar region of the former Soviet Union and used bored-cast-in-place tapered piles (Rybnikov 1990). The behavior of five tapered piles and two cylindrical piles has been investigated. The vertical bearing capacity of pile was determined from the load-average settlements equal to 24 mm. The soil profile was made up of three layers. In addition, the properties used for the soil and pile are summarized in Tables 1 and 3, respectively. In this article, the load-settlement

Table 3. Soil properties used in Rybnikov's Test site

Index	Sandy loam	Ordinary loam	Sand
Layer thickness (m)	5.8	2.1	2.4
Soil density (g/cm^3)	1.74	1.68	1.81
Angle of internal friction (deg)	19	18	32
Cohesion (MPa)	0.012	0.026	0.004
Elastic modulus (MPa)	20	20	26
Angle of dilation (deg)	12	12	14

Table 4. Values of K used for simulation of pile installation

Index	Sand types	K
C 20	Soft Sand	2.1
	Dense Sand	2.7
T 31	Soft Sand	2.3
	Dense Sand	3.0
C 40	Soft Sand	2.4
	Dense Sand	3.2
T 62	Soft Sand	2.6
	Dense Sand	3.4

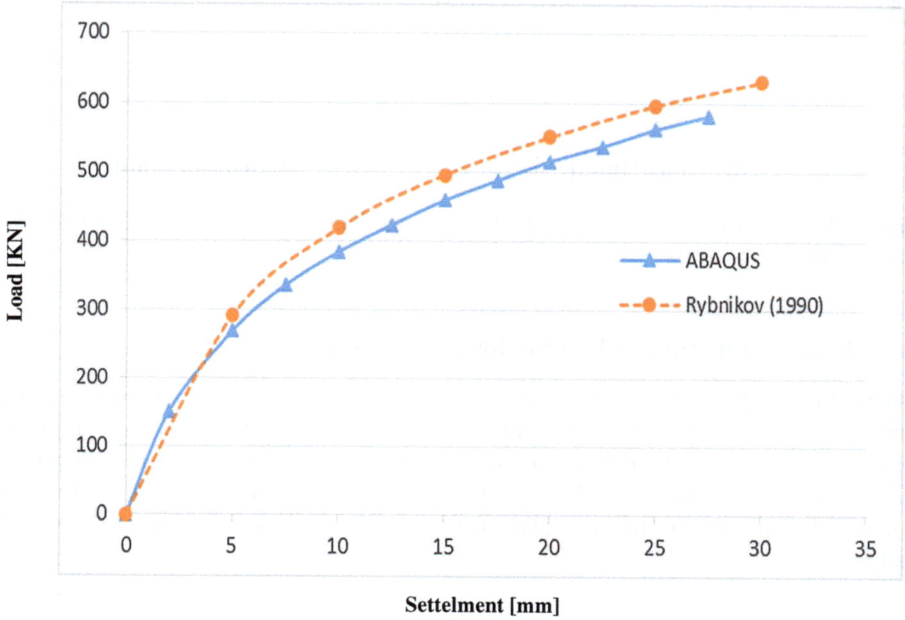

Fig. 2. Comparison of tapered pile analysis result with Rybnikov's Test Site

curve obtained from finite element were compared with the result obtained by experimental results for tapered pile with top and bottom radius of piles equal to 300 mm and 100 mm (2.66°) respectively and a length of 4.5 m, leading to a difference of less than 5% as shown in Fig. 2.

3 Results and Discussions

3.1 Effect of Vertical Load on the Response of Piles

Two types of concrete piles were compared with the same average diameter and close volume. The load-settlement curves of short piles (the length to diameter ratios of the piles are less than 10) are investigated under vertical loading as shown in Fig. 3. The results of loading of the two piles in dense and loose sandy soil showed that tapered piles had settlement less than uniform piles for a given same loading. Therefore, the performance of the tapered piles was better than cylindrical ones for places which have been more settlement restrictive. Increase in the bearing capacity of tapered piles were less than cylinder piles with the same mean radius in two sandy soil, although these changes were more for dense sand.

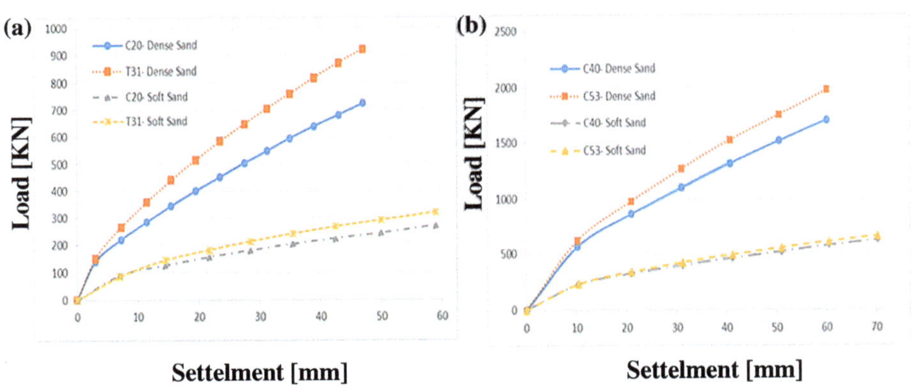

Fig. 3. The load-settlement curves of piles with average diameter 0.4 m (a) and 0.8 m (b) under vertical load

3.2 Effect of Lateral Load on the Response of Piles

Lateral load of piles was applied to the ground surface in two steps that piles equalize under vertical loads with a safety factor of 3 in the first step (one-third of the ultimate bearing capacity) and the horizontal lateral load was applied by lateral movements to pile head in the second step. The results of this modeling presented in the form of horizontal displacement, moment and shear force along piles are shown in Figs. 4, 5 and 6.

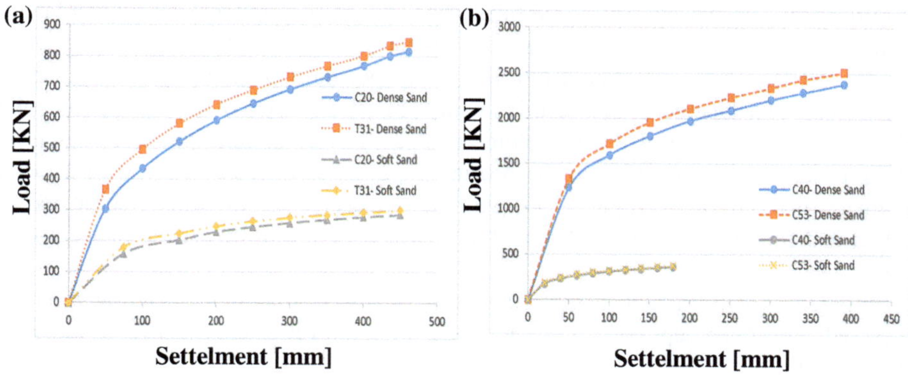

Fig. 4. The load-settlement curves of piles with average diameter 0.4 m (a) and 0.8 m (b) under lateral load

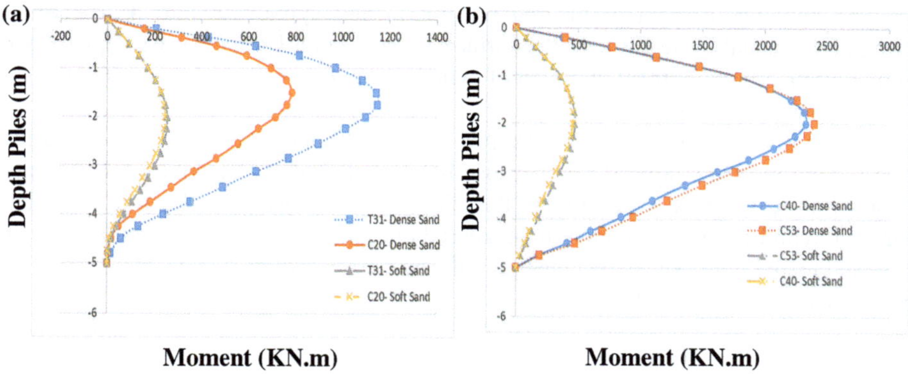

Fig. 5. Variation of bending moment for piles with average diameter 0.4 m (a) and 0.8 m (b)

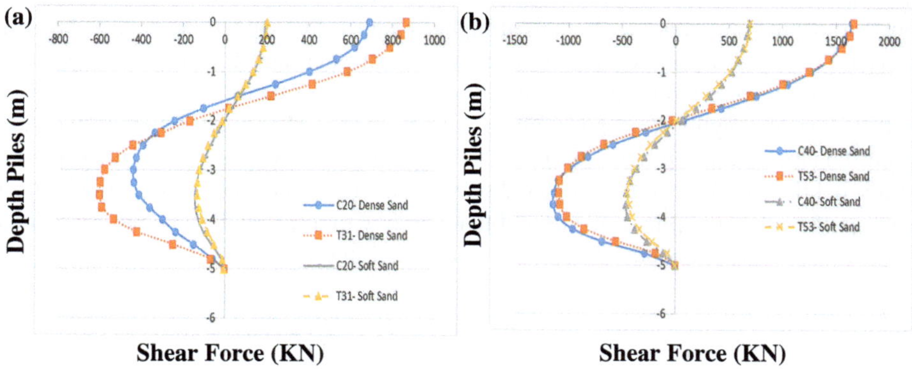

Fig. 6. Variation of shear force for piles with average diameter 0.4 m (a) and 0.8 m (b)

Figure 4, performance of tapered piles was better than cylindrical ones under lateral loads. Lateral bearing capacity of pile depends on the superstructure. In this paper, the bearing pile has been considered in a pile head specific deflection to easily compare cylindrical with tapered pile. For piles with a mean diameter of 0.2 m, tapered pile (T31) with an increase of 8% in volume compared to cylindrical pile (C20), can be observed an increase of 15% in the bearing capacity of pile. For piles with a mean diameter of 0.4 m, tapered pile (T53) with an increase of 2% in volume compared to cylindrical pile (C40), can be observed an increase of 7% in the bearing capacity of pile.

Figure 5, for piles with a mean diameter of 0.2 m, maximum bending moment occurred almost at a depth of 1.6 to 2 m above ground level where shear force was zero at this depth.

Tapered pile (T31) in comparison with the cylinder pile (C20), observed an increase of 5% in the bending moment of pile for dense sand. Tapered pile can provide a greater cross section area than to the cylindrical one, leading less amount of stress in cross section of the tapered pile than the same cylindrical pile.

The bending moment of tapered pile (T31) and cylindrical pile (C20) are approximately equal in height near the surface for soft sand. By increasing depth of soil and after maximum moment, moment of cylindrical piles are more than tapered piles in a specific depth.

Figure 6, the shear force of tapered pile is greater than the shear force of cylindrical pile at the top of the tapered pile because bearing capacity of tapered piles is greater than cylindrical one. At the bottom of the pile and by increasing the depth of soil, shear force of cylindrical piles are more than tapered piles. Forces are transmitted more to the soil at tapered pile, since the tapered pile has cross-sectional area more than the cylindrical pile at higher heights.

4 Conclusions

Tapered piles are more efficient and economical than cylindrical piles having the same material volume. Tapered piles, which have greater top cross-sections than bottom ones, transmit force more to the soil at higher heights. Percentage increase in bearing capacity of tapered piles were less than cylindrical piles under vertical loading, and these changes were more observed for dense sand. Performance of tapered piles was better than the cylindrical piles under lateral loads. Maximum bending moment occurred almost at a depth of 1.6 to 2 m above ground level where shear force was zero at this depth. The shear force of tapered pile is greater than the one of cylindrical pile at the top of the tapered pile. It is inferred that bending moment and shear force obtained after the maximum amount are more in cylindrical piles.

References

ABAQUS: ABAQUS: 6.10 Analysis User's Manual, ABAQUS/Standard Doc (2010)

El Naggar, M.H., Wei, J.Q.: Response of tapered piles subjected to lateral loading. Can. Geotech. J. **36**(1), 52–71 (1999). doi:10.1139/t98-094

Hataf, N., Shafaghat, A.: Optimizing the bearing capacity of tapered piles in realistic scale using 3D finite element method. Geotech. Geol. Eng. **33**(6), 1465–1473 (2015). doi:10.1007/s10706-015-9912-6

Ismael, N.F.: Analysis of Lateral Load Tests on Step Tapered Bored Piles in Calcareous Sands. In: GeoCongress 2006@ sGeotechnical Engineering in the Information Technology Age, pp. 1–6. ASCE (2006). doi:10.1061/40803(187)66

Mayerhof, G.G.: Bearing capacity and settlement of pile foundations. J. Geotech. Geoenviron. Eng. 102.ASCE# 11962 (1976). ISSN: 1090-0241

Nordlund, R.L.: Bearing capacity of piles in cohesionless soils. J. Soil Mech. Found. Div. **89**(3), 1–36 (1963)

Rybnikov, A.M.: Experimental investigations of bearing capacity of bored-cast-in-place tapered piles. Soil Mech. Found. Eng. **2**, 48–52 (1990). doi:10.1007/BF02306100

Satibi, S., Yu, C., Leoni, M.: On the numerical simulation of a tube-installed displacement pile. In: Proceedings of 1st International Conference of European Asian Civil Engineering Forum. Jakarta, Indonesia (2007)

Zhan, Y., Wang, H., Liu, F.: Numerical study on load capacity behavior of tapered pile foundations. Electron. J. Geotech. Eng. **17**, 1969–1980 (2012)

Zil'berberg, S.D., Sherstnev, A.D.: Construction of compaction tapered pile foundations (from the experience of the "Vladspetsstroi" trust). Soil Mech. Found. Eng. **27**(3), 96–101 (1990). doi:10.1007/BF02306664

A Full Scale Field Study: Evaluation of the Magnitude and Time Extent of Excess Pore Water Pressure During the Installation of Auger Pressure Grouted Displacement Piles in Downtown Orlando

Amr Sallam(✉) and Mohamed Alrowaimi

Terracon Consultants, Orlando, FL, USA
{amr.sallam,mohamed.alrowaimi}@terracon.com

Abstract. Central Florida's economy is growing strong and diverse, with that, comes taller structures, hotels, resorts, amusement structures, and public venues. Augered-Pressure-Grouted-Displacement (APGD) piles increasingly became the deep foundation of choice for heavily loaded structures in Downtown Orlando. The combination of favorable subsoil conditions, quick installation, superior skin friction, and minimal drilling spoil made APGD piles very attractive option especially where contaminated soils are encountered. The installation of the APGD piles causes instant increase in pore water pressure at the annulus of the drilled hole during drilling and grouting stages. The subsoil conditions in Downtown Orlando include a unique layer of silty to clayey sands and sandy clays, which exhibits high excess pore pressure during the installation of APGD piles. Piles installed in close proximity to a freshly grouted pile might cause squeezing and occasional grout outflow triggering potential necking. During the installation of 18-inch diameter 80-feet APGD piles for a 23-story mixed-use building in Downtown Orlando, grout outflow was observed when piles were installed closer than 15 ft of a freshly grouted pile. A field study was conducted to evaluate the magnitude and extent of excess pore water pressure during both drilling and grouting stages of APGD piles. Five piezometers were installed between two 60 ft apart piles to depths of 50 to 60 ft with screened section within the clayey soil. Water pressure probes were installed and data was continuously collected. The study showed sharp increase in pore water pressure during drilling and grouting stages. However, the increase in pore pressure was significantly lower than the estimated critical pore pressure needed to cause pile necking. A typical distance of 20 times the pile diameter should be considered for the next pile in the vicinity of a freshly installed pile to completely avoid the potential for pile communication problems in Downtown Orlando.

1 Introduction

During the last two decades, a record-high number of high-rise buildings may have been set in Central Florida. Due to high column loads associated with those towers, deep foundations are usually utilized. Popular deep foundations in Central Florida

M. Abu-Farsakh et al. (eds.), *Advances in Analysis and Design of Deep Foundations*,
Sustainable Civil Infrastructures, DOI 10.1007/978-3-319-61642-1_10

include, driven piles such as closed and open-ended pipe piles, H-piles, and precast concrete piles, and cast in place piles such as Augered Cast in Place (ACIP) piles and Augered Pressure Grouted Displacement (APGD) piles. Among the above systems, the APGD pile is becoming the pile of choice in Downtown Orlando. The APGD pile is similar to the conventional ACIP pile being a cast-in-place pile, however, the APGD pile is a displacement type, which does not result in drilling spoil, rather, laterally displaces the excavated soils. The APGD pile utilizes a special drilling tool that is capable of displacing the penetrated soil horizontally. The method is superior where the subsoil is environmentally contaminated. If constructed in materials that has the ability to densify in response to lateral displacement, the APGD piles may result in increasing shaft resistances higher than those developed during the installation of the conventional ACIP piles (NeSmith 2002). The horizontal displacement of the penetrated soils could be at its immediate horizontal position as in loose to medium dense soils or after being transported upward to the displacement element as in medium to dense soils, NeSmith (2003). Construction platforms are equipped with display unit that provides drilling tool depth, grout pressure, grout volume, and torque.

Once the displacement tool's tip reaches the desired depth, the downward travel of the tool is stopped and concrete pumping begins. Withdrawal of the tool starts when a target grout pressure is established. For a typical application in loose to medium dense granular materials, target installation pressures will generally be in the range of 10 to 20 psi for lift off and shaft construction (NeSmith 2004). The withdrawal rate depends on the grout pressure and is usually varied to maintain the target grout pressure. Although the grout volume is continuously checked to ensure that the delivered grout volume is greater than the theoretical pile volume, it should be emphasized that the pile is mainly cast based on the target grout pressure.

The ideal soil profile for the APGD piles is clean, granular, well-graded sand, loose near the surface, with a gradual uniform increase in density with depth. NeSmith (2002) stated that the existence of saturated fine-grained materials can impact pile constructability and quality in two ways: (i) the spacing between piles must be increased to preclude communications between piles, and (ii) the generation of pore water pressure during construction in the vicinity of the pile, which can have negative effect on pile integrity. The unique layer of silty to clayey sands and sandy clays typically encountered in Downtown Orlando exhibits high excess pore pressure during the installation of APGD piles. When a pile is installed in close proximity to a freshly grouted pile, occasional grout outflow of the fresh pile may occur, which might affect pile integrity.

During APGD piles installation in a Downtown Orlando project, few piles have communicated and grout overflow was observed triggering the field study presented in this paper. The goal of the field study was to evaluate the magnitude and extent of excess pore water pressure during drilling and grouting phases of APGD piles within the typical subsoil condition of Downtown Orlando.

2 Project Description

The project site is located in the Downtown area of the City of Orlando, Florida, USA and it consists of a 23-story mixed-use tower. The ground level will include retail and commercial spaces topped with 6-levels of parking, 15-stories of residential units, and 1-story of amenities with a pool.

3 The Geotechnical Study

3.1 General Geology

Orlando is located in Orange County which, is in the north-central part of peninsular Florida and to the east and southeast of the crest of the Ocala Uplift. The area is underlain by extensive deposits of Eocene age carbonates covered by younger dolomite, limestone, sand, clay, and shell beds. The dissolution of limestone and the marine processes are the dominant forces responsible for the development of the surface features observed in the County.

3.2 Field Explorations and Laboratory Testing

The geotechnical exploration for the project included three Standard Penetration Test (SPT) borings, and six Cone Penetration Test (CPT) soundings to depths of 60 to 100 ft below existing ground surface within the building area. Figure 1 presents a plot of the subsoil layers, average laboratory test results, and the SPT N-values, corrected to hammer type and overburden pressure, versus the depth.

3.3 Foundation Recommendations

Based on the encountered subsoil conditions and the variability in column loads, the authors recommended the following foundation options; (i) solid mat foundation on existing soil, (ii) solid mat foundation on improved soil, and (iii) deep foundations in the form of ACIP or APGD piles. The mat foundation option was evaluated, however, contact pressures of 2800 to 5050 psf, based on an average subgrade reaction modulus of 40 pci, were provided by the structural engineer. A settlement analysis was performed utilizing the Finite Element Analysis software (PLAXIS) for the highly loaded critical strip of the mat. The settlement results ranged from 3 to 5 in. for a solid mat, which was acceptable. The project civil engineer requested the solid mat be discontinued to allow for the installation of stormwater vaults, which resulted in high contact pressures triggering more total and differential settlement, therefore, the mat foundation was eliminated.

The final recommendation was for deep foundations in the form of AGPD or ACIP piles with a minimum pile to pile spacing of 3 times the pile diameter. The deep foundation system should minimize the expected total and differential settlements to ½ and ¼ in., respectively. Table 1 presents a summary of the recommended axial and

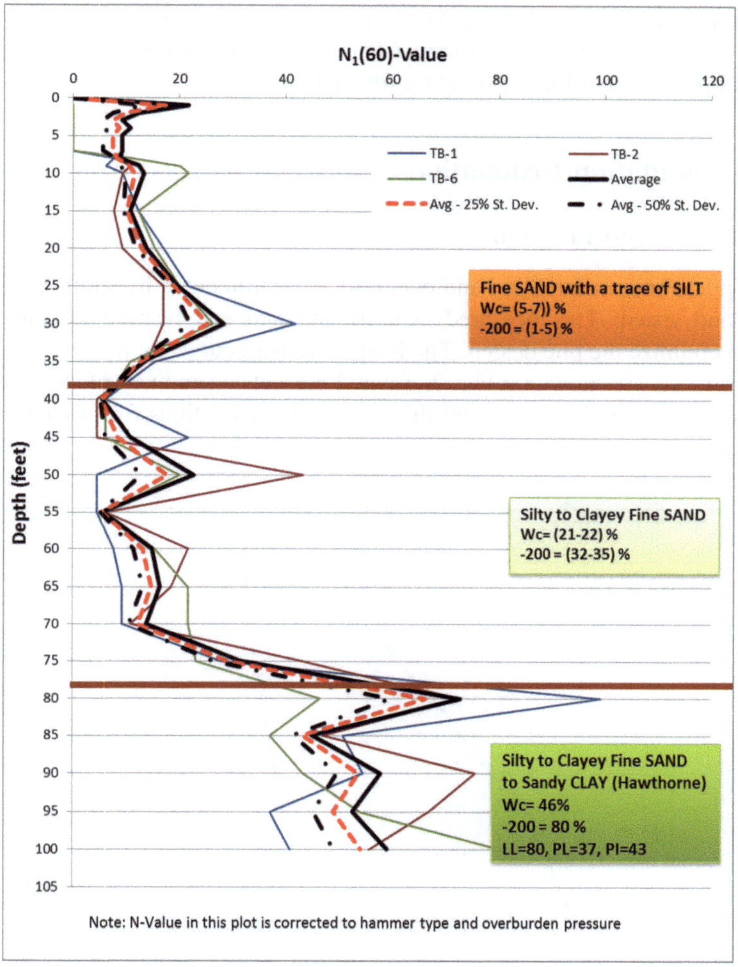

Fig. 1. Corrected SPT N-value versus Depth

Table 1. Axial and lateral capacities of ACIP and APGD piles

Pile type	Net pile length (ft)	Pile diameter (in.)	Allowable compression pile load (tons)	Allowable tension pile load (tons)	Lateral loads corresponding to target lateral deflections, tons			
					Fixed head support		Free head support	
					0.5″	1.0″	0.5″	1.0″
ACIP	85	18	150	80	14	19	6	9
APGD	75	18	150	80	14	19	6	9

lateral capacities of the ACIP and APGD piles. A deep foundation system that consisted of 18-inch diameter 80-feet long APGD piles was selected by the project contractor and design team for the building foundation.

4 Pile Foundation Installation

4.1 Pile Load Test Program

Two test piles, one compression and one tension, instrumented by strain gauges, were installed and tested. The measured skin friction and end bearing resistances were utilized to optimize the pile design. The load movement curves of the compression and tension load tests are shown on Fig. 2. Table 2 presents a summary of some of the key parameters of the pile load tests and the final recommendations for production piles.

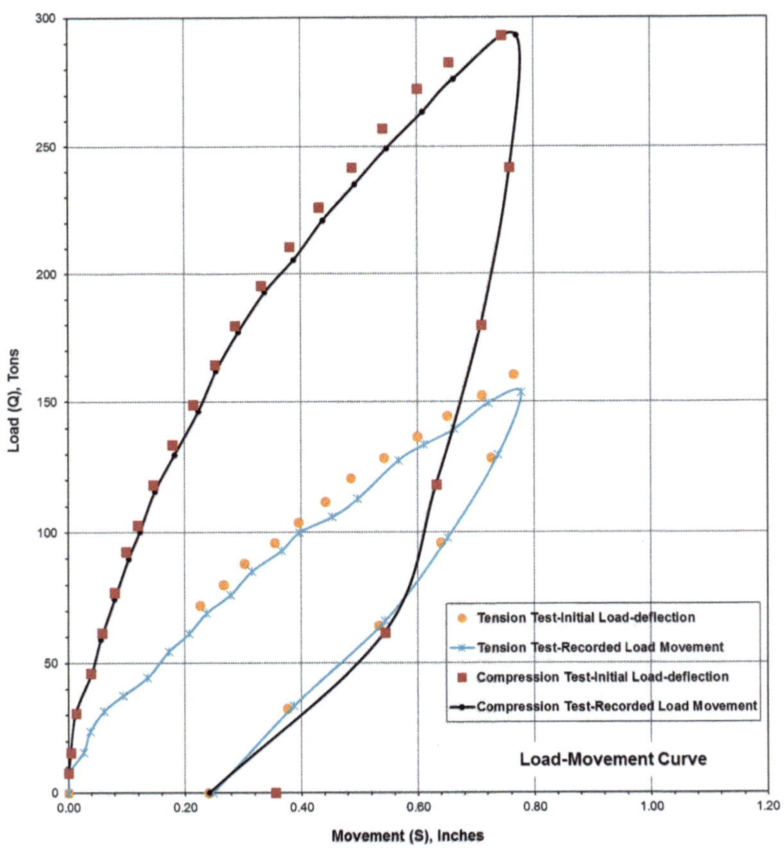

Fig. 2. Load-movement curves for tension and compression test piles

Table 2. Key test pile parameters and final recommendations

Test pile #	Test pile (tension)	Test pile (compression)
Load at 0.3 in. Per Florida Building Code, tons	85	180
Ultimate load per Florida Building Code, tons	160.38	292.91
Allowable Load (per field test), tons utilizing different methods stated above	80	150
Recommended Shortening of the pile to provide design allowable load with safety factor of 2.0, feet	5 ft	
Recommended Total Pile length measured from existing grades, feet	80 ft	
Recommended Net Pile length measured from bottom of pile cap (10 ft from current grades), feet	70 ft	

4.2 Piles Communication Issue

A week into pile installation, Terracon's site engineer observed grout was flowing out of a pile that was just installed during drilling and grouting of an adjacent pile. Communication between freshly installed piles is a serious problems for cast-in-place piles. Therefore, Terracon field engineer instructed the pile contractor to change the sequence of the pile installation by: (i) allowing larger spacing between fresh piles, (ii) using a thicker grout, (iii) slowing down during both drilling and grouting stages, and (iv) decrease the grout pressure during pumping. These temporary measure were recommended to maintain productivity while trying to really understand the root cause of the problem. Terracon approached both of the client and the pile contractor to be authorized to further investigate the reasons behind this problem. The pile contractor agreed to fund the installation of groundwater piezometers to evaluate the increase, dissipation, extend, and magnitude of excess pore water pressure during the drilling and grouting stages of APGD piles.

It should be noted that during pile installation, the foundation contractor was restrained from comfortably sequencing pile installation due to the fact that the site was significantly tight and the general contractor had occupied a portion of the site with portables, material, and machinery. This obstacle made it hard for the foundation contractor to move and install widely spaced piles freely within the site limits, therefore, the foundation contractor was trying to install piles as close as possible to meet the productivity target.

5 Field Investigation of Pile Communication Issue

5.1 Installation of Piezometers

Terracon team selected the south-east corner of the project site to utilize for the field work, which included the installation of five piezometers. Each piezometer consisted of screens, PVC pipes, seals, pore water pressure probes, and corresponding cables.

The typical depth of the piezometer was about 60 ft with screened section between 35 to 50 ft, within the clayey soils. At each piezometer location, a 2-inch diameter PVC pipe was installed to accommodate the probes. The annulus area between the outer PVC pipes and the soil, within the top 35–40 ft, were filled with grout/bentonite to seal the screened portion of the piezometer. Each piezometer was prepared with connections to be hocked up to a data logger to collect the pore water pressure data on a regular bases. The pore water pressure probes were submerged in the piezometer well for a period of 24 to 48 h to ensure full saturation prior to collecting the data. Figure 3 shows a sketch of the water pressure monitoring system.

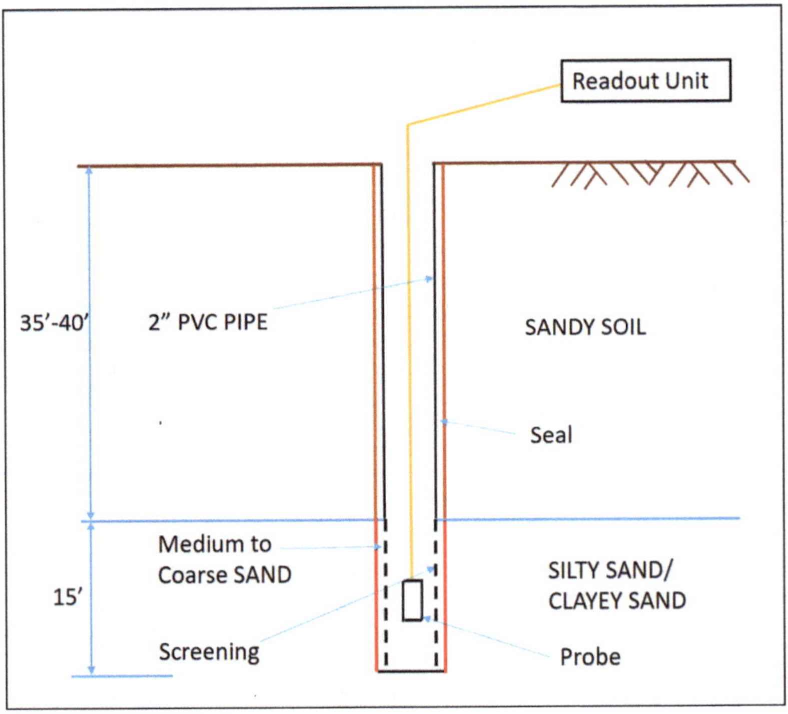

Fig. 3. Sketch for pore water pressure monitoring system

Two piles (Pile # 1397 and 1408), located on two adjacent pile caps, were selected for this study. The five piezometers were installed between two selected piles and were spaced about 10 to 12 ft apart. Pile # 1397 was installed first and the pore water pressure data was collected simultaneously for the five piezometers during both drilling and grouting stages. Similarly, during the installation of Pile # 1408, the pore water pressure data was monitored at all five piezometers. In addition to monitoring during installing the two consecutive piles, Pile # 1406 was later included in the scope since it was installed immediately after Pile # 1408. Pile # 1406 was about 12 ft from the centerline of the two piles. Figure 4 shows a plan view of location of the installed piezometers (PZ-1, PZ-2, PZ-3, PZ-4, & PZ-5).

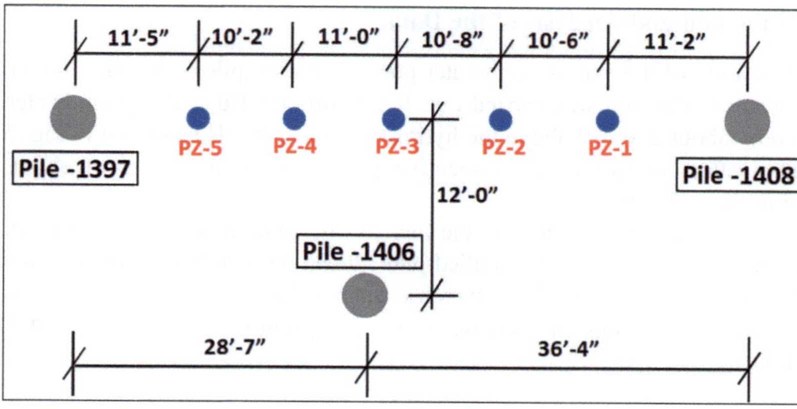

Fig. 4. A plan view of probes and piles locations

5.2 Collected Data

Figure 5 illustrates the changes of the groundwater levels in the five piezometers due to the installation of all three piles. As shown in Fig. 5, the groundwater readings were collected before pile installation in order to create a baseline for the measurements. The sequence of pile installation was Pile # 1397, Pile # 1408, then finally Pile # 1406.

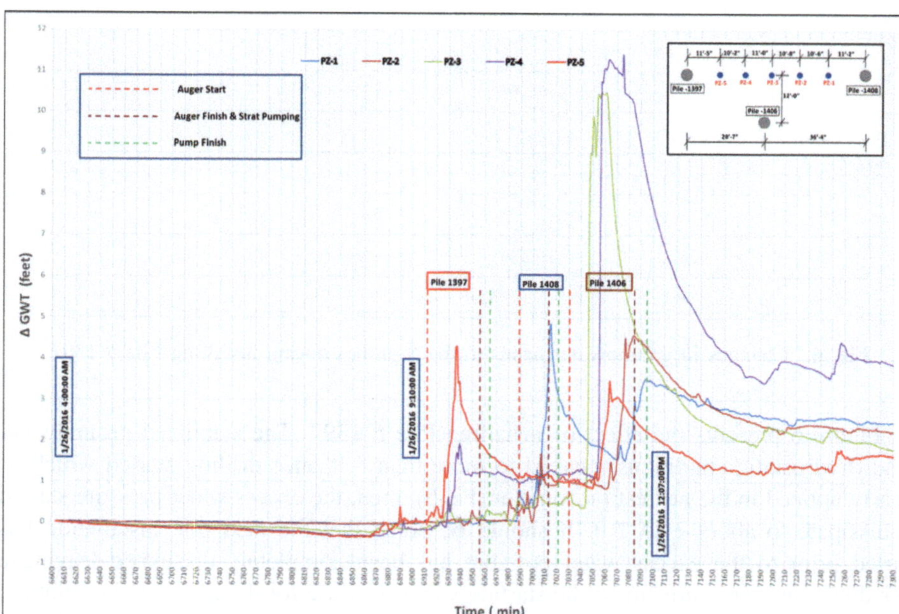

Fig. 5. Changes in groundwater heads in the 5 probes during pile installation

5.3 Evaluation and Analysis of the Data

A detailed study of the excess pore water pressure during pile installation, in terms of ground water fluctuation was carried out. It was observed that the groundwater head increased by about 2 to 4 ft above the hydrostatic water levels as shown in Fig. 5. The cumulative effect of installing consecutive piles was on the order of about 11 ft increase in the water head.

Figures 6, 7, and 8 demonstrate the increase in groundwater head during drilling and grouting of each of the three studied piles. It can be seen that all figures showed a general trend of groundwater head increase mainly during the drilling stage and followed with a less significant increase during the grouting stage of the APGD pile installation.

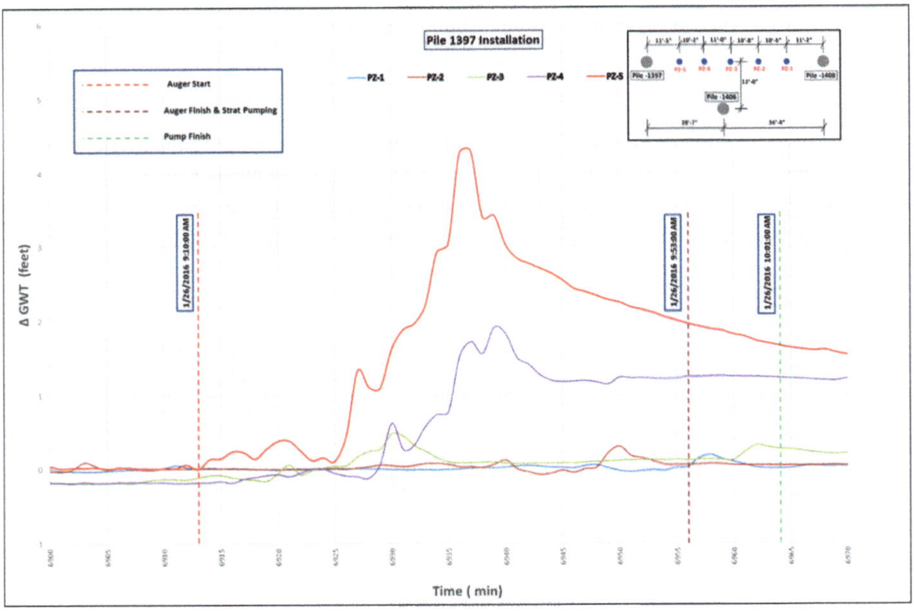

Fig. 6. Changes in groundwater heads in the 5 probes during installing Pile # 1397

Figure 6 presents the effect of installing Pile # 1397. The stabilized groundwater head (hydrostatic conditions) peaked to more than 4 ft once drilling started which, is clearly noticed in the nearest piezometer (PZ-5), then, the excess water pressure started to dissipate to about 1 to 2 ft. It should be noted that the next pile installation has started prior to the excess water pressure has been completely dissipated leaving a "residual water pressure" to be the starting water pressure for the next pile installation.

Similar behavior was observed during installation of Piles # 1408 and 1406 as shown in Figs. 7 and 8. However, both piles have started with a "residual water pressure" that was built up during the installation of the previous piles. Plots of the increase in groundwater head due to pile installation were developed to estimate the

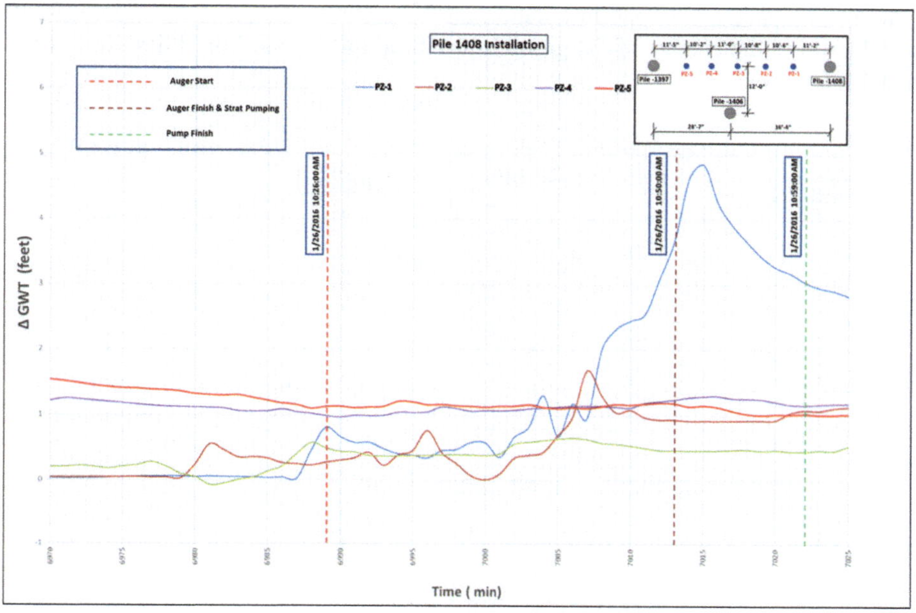

Fig. 7. Changes in groundwater heads in the 5 probes during installing Pile # 1408

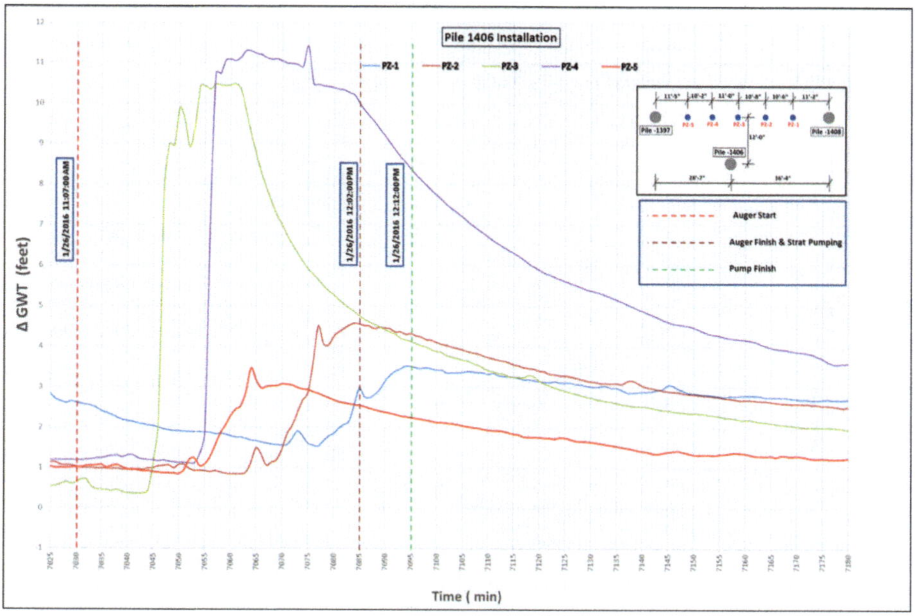

Fig. 8. Changes in groundwater heads in the 5 probes during installing Pile # 1406

zone of influence of the pile installation on the groundwater pressure. Figures 9, 10, and 11 shows the change of the groundwater head due to installing Piles 1397, 1408, and 1406 respectively.

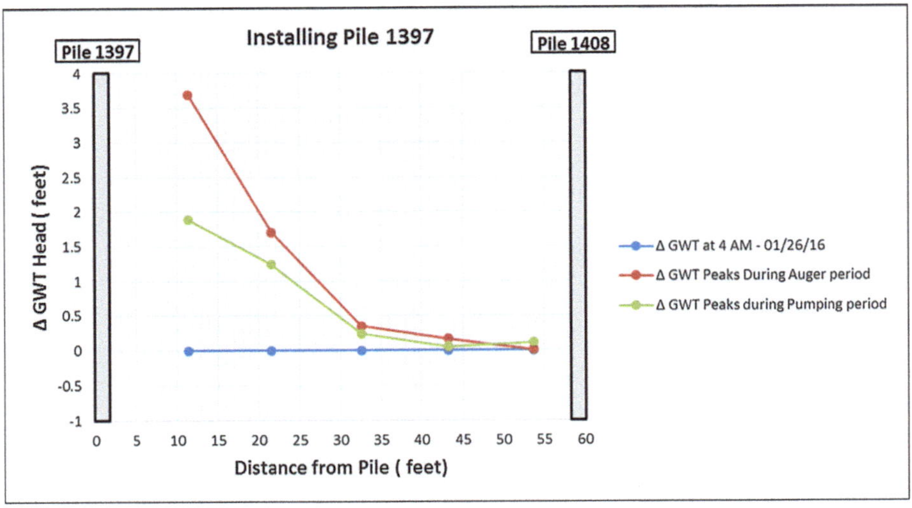

Fig. 9. Peaks of the groundwater changes during Pile # 1397 installation

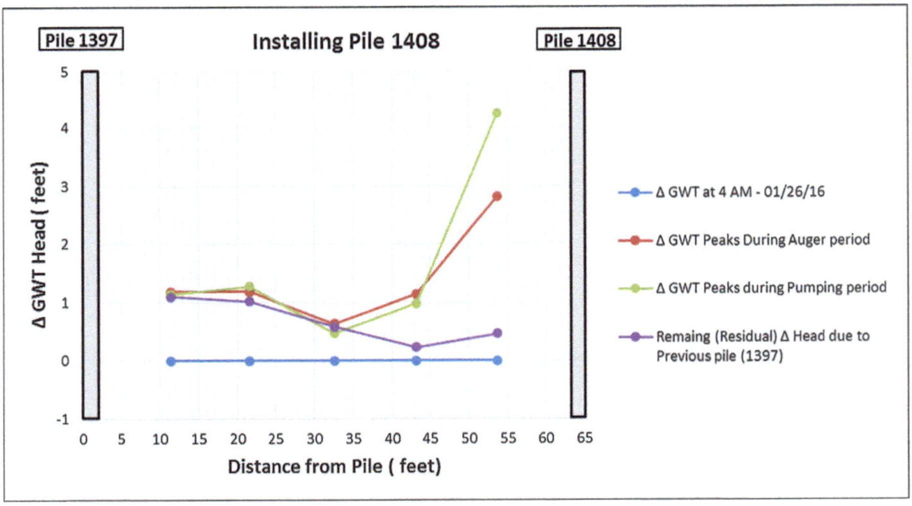

Fig. 10. Peaks of the groundwater changes during Pile # 1408 installation

All results were in general agreement on the extent of the influence zone of the increase in groundwater head during pile installation, which was estimated to be about 20 times the pile diameter. This distance was very clear in the first pile (Pile # 1397)

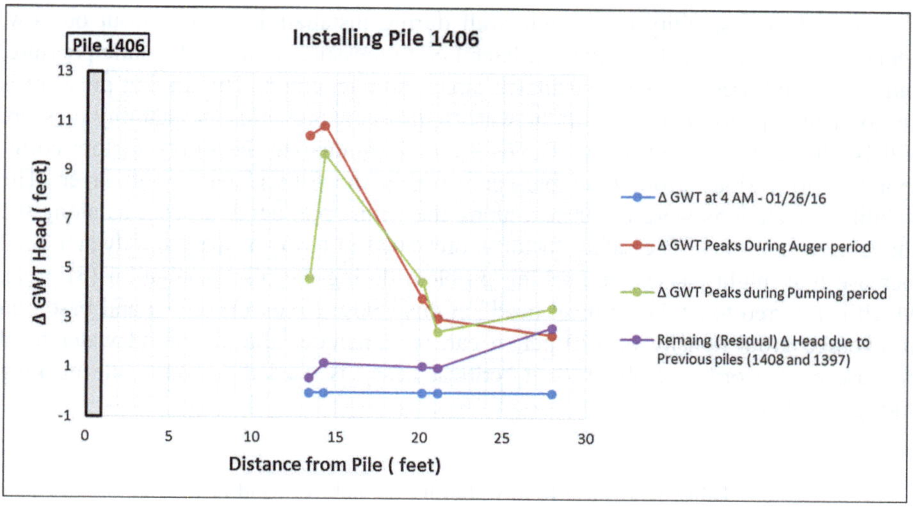

Fig. 11. Peaks of the groundwater changes during Pile # 1406 installation

since there was no "residual water pressure" from prior nearby pile installation (Fig. 9). It can be observed that no increase in the pore water pressure was observed at piezometers located beyond 30 ft away from the pile currently installed. However, Piles # 1408 and 1406 had residual excess pore water pressures, which were built-up during previous pile installations. Therefore, groundwater heads, at these piles, were assumed to be completely stabilized at those residual pressure heads as shown in Figs. 10 and 11. In Fig. (12), Pile # 1397 was chosen to estimate the increase in ground water head at the exact pile location by extrapolation. The increase in the groundwater head during drilling Pile # 1397 was about 6.5 ft based on polynomial equation correlation (R = 0.99) of the groundwater increase as shown in Fig. 12.

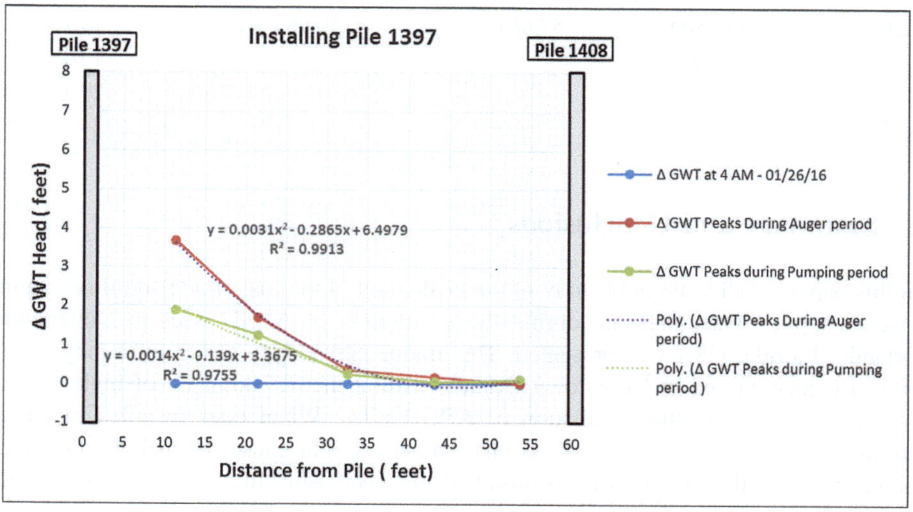

Fig. 12. Peaks of the groundwater changes and the trend lines during Pile # 1397 installation

To study the stability of the pile wall during installation against grout outflow, which might lead to necking, the soil lateral earth pressure, water hydrostatic pressure, and grout fluid pressure were calculated along the pile length. The driving pressure is the soil lateral pressure along with the water hydrostatic pressure. The stability pressure will be the grout fluid pressure. Tables 3 and 4 summarize the safety factors corresponding to various grout fluid pressures within the Silty/Clayey Sand layer. The stability calculations were performed along the whole pile depth as well as along the silty/clayey soil only. The safety factor is presented in terms of the groundwater head increase that might cause pile necking problem during the pile installation (SF = 1), which is referred to as the "critical head" in this study. Given the above and based on an average grout density of 140 pcf, it can be concluded that a ground water head increase on the order of about 24 ft "critical head" is needed to cause pile necking during installation.

Table 3. Critical head within the silty/clayey sand layer

Grout unit weight (pcf)	Grout fluid resultant (lb/ft)	Hydrostatic water resultant (lb/ft)	Lateral earth resultant at REST (lb/ft)	Critical head (ft)
110	70,125	33,228	32,697	4.5
120	76,500			11.3
130	82,875			18.1
140	89,250			24.9

Table 4. Critical head within whole pile length

Grout unit weight (pcf)	Grout fluid resultant (lb/ft)	Hydrostatic water resultant (lb/ft)	Lateral earth resultant at REST (lb/ft)	Critical head (ft)
110	137,500	57,689	50,969	10.7
120	150,000			15.4
130	162,500			20.1
140	175,000			24.7

6 Summary and Conclusions

In this paper, a full scale field study of the evaluation of the magnitude and time extent of excess pore water pressure during the installation of APGD piles in Downtown Orlando, Florida, USA was presented. The unique silty/clayey sands and sandy clays typically encountered in Downtown Orlando exhibit the development of high excess pore pressure during the installation of APGD piles. When piles installed in close proximity to a freshly grouted piles, this behavior might cause squeezing and occasional grout outflow of the freshly installed pile increasing the risk of potential pile necking.

The field study presented in this paper included the installation of five piezometers extended to the clayey soil layer. Two piles of two adjacent pile caps and an additional close-by pile were selected for this study, between which, the five piezometers were installed equally spaced and the pore water pressure data was collected during both drilling and grouting stages of the APGD pile installation.

The study showed sharp increase in pore water pressure during drilling and grouting stages. However, the increase in pore pressure was significantly lower than the estimated critical pore pressure needed for potential risk of pile necking. A minimum distance of about 20 times the APGD pile diameter should be considered for the next APGD pile in the vicinity of a freshly installed APGD pile in Downtown Orlando area in order to completely eliminate the risk of pile communication issues and potential for pile necking. Shorter spacing may be used with great care and continuous monitoring to avoid pile communications and potential necking.

Acknowledgments. The authors would like to extend their thanks to foundation contractor (MORETRENCH) for funding this field study as well as to the owner (SDG, LLC) for allowing us to use the project for the field experiment.

References

NeSmith, W.: Design and installation of pressure-grouted, drilled displacement piles. In: Proceedings of the Ninth International Conference on Piling and Deep Foundations, Nice, France (2002)

NeSmith, W.: A history of the use of ACIP piles for major projects in Atlanta. The deep Foundation Institute's Augered cast-In-Place Piles Committee Specialty Seminar, Atlanta, Georgia, pp. 99–104 (2003)

NeSmith, W.: Application of Augered, Cast-In-Place Displacement (ACIPD) Piles in New York City: The deep Foundation Institute's Augered cast-In-Place Piles Committee Specialty Seminar, McGraw-Hill Building, New York, NY, pp. 77–83 (2004)

NeSmith, M., NeSmith, W.: Anatomy of data acquisition system for drilled displacement piles. In: Proceedings of GeoCongress Geotechnical Engineering in the Information Technology Age, Atlanta, USA (2006)

Evaluating Pile Setup Using Numerical Simulation and Introducing an Elastoplastic Constitutive Model for Clays

Firouz Rosti[1(✉)], Murad Abu-Farsakh[2], and Carol J. Friedland[3]

[1] Department of Civil and Environmental Engineering, Louisiana State University, Baton Rouge, LA 70803, USA
frostil@lsu.edu

[2] Louisiana Transportation Research Center, Louisiana State University, Baton Rouge, LA 70808, USA

[3] Department of Construction Management, Louisiana State University, Baton Rouge, LA 70803, USA

Abstract. An elastoplastic constitutive model was developed to define cohesive soil behavior. During pile installation in saturated ground, the soil adjacent to the pile disturbs causing large displacements and numerous variation in the porewater pressure in the soil-pile interface zone. Therefore, the soil disturbance and the corresponding decline in the soil shear strength were included in the developed constitutive model. After end of pile driving (EOD), the surrounding disturbed soil tends to regain its strength over time due to both consolidation and thixotropic effects. In this paper, the soil thixotropy was simulated by applying a time-dependent reduction parameter, β, which affects both the interface friction and the soil shear strength parameters. In order to examine the proposed model, numerical simulation of pile installation and the following increase in the pile capacity over time (pile setup) was performed for a full-scale pile load test case study. Finite element (FE) software Abaqus utilized to simulate the pile installation and following pile load tests. Dissipation of the induced excess porewater pressure was modeled through applying conventional consolidation theory. The proposed model was developed based on disturbed state concept and application of the modified Cam-Clay model. Pile installation was modeled by combination of two phases in an axisymmetric FE model: creating a volumetric cavity expansion followed by applying a vertical shear displacement (penetration). The FE simulation results included: (1)-developed excess porewater pressure in the soil body during pile installation and its dissipation over time after EOD, (2)-increase in effective lateral stresses at the pile-soil interface, and (3)-the pile setup values attributed to both the soil consolidation and its thixotropic responses. Comparison of the FE simulation results with the measured values obtained from load tests conducted on a full-scale instrumented pile indicated that the developed constitutive model is able to appropriately predict pile installation and following setup.

1 Introduction

In Geotechnical Engineering cases such as deep penetration, soil disturbance and remolding of the soil particle occur during shear loading and it significantly affect the general soil behavior. For engineering problems that involve deep foundations, soil

M. Abu-Farsakh et al. (eds.), *Advances in Analysis and Design of Deep Foundations*, Sustainable Civil Infrastructures, DOI 10.1007/978-3-319-61642-1_11

type usually changes over depth due to differences in the stress history. Therefore, in such cases, incorporating an appropriate constitutive model that can capture the actual behavior for both normally consolidated (NC) and over consolidated (OC) soils is necessary. There are several elastoplastic constitutive model available in the literature attempted to model the soil response under different loading conditions. Most of the developed constitutive models for clays are based on the critical state soil mechanics (CSSM) concept (Pestana and Whittle 1999). The modified Cam-Clay (MCC) model proposed by Roscoe and Burland (1968) is the most well-known critical state model, which is able to appropriately describe the isotropic NC clay behavior. Since the MCC model indicates elastic response inside the yield surface, its prediction for OC clay is poor (Likitlersuang 2003). Series of bounding surface models have been developed to overcome this deficiency (e.g., Dafalias and Herrmann 1986). The bounding surface plasticity concept was used later to develop the MIT-E3 model by Whittle (1993). The bounding surface plasticity has been developed to provide smooth transition from elastic to fully plastic state for soils under general loading. Application of the critical sate models for heavily OC clays is limited, and it needs specific consideration. Yao et al. (2007) and (2012) introduced a unified hardening model using Hvorslev envelope to capture the heavily OC clay behavior. Linear and parabolic form of the Hvorslev envelope were adopted to adjust the conventional MCC model for heavily OC clay response under shear loads. Based on CSSM and bounding surface theory, Chakraborty et al. (2013a, b) developed a two surface elastoplastic constitutive model to define strain rate dependent behavior for clay. Chakraborty et al. (2013a, b) and Basu et al. (2014) used two-surface plasticity constitutive model for the clays, and it was implemented for analysis of shaft resistance in piles. Although, some of these models are able to describe both NC and OC clay behavior, but they usually include plenty of model parameters that requires performing several lab tests for their estimation. Furthermore, these models don't address directly the disturbance occurs in the soil body during shear loading in their elastoplastic formulation.

The disturbed state concept (DSC) developed by Desai and Ma (1992) is a powerful technique that is directly formulated based on the soil disturbance. In the DSC model, the soil response is obtained using two boundary (reference) state responses, which are named as relative intact (RI) state and fully adjusted (FA) or critical (c) state. The real or observed soil behavior is defined as a linear combination of RI and FA responses (Desai 2001).

Cohesive soils show certain degrees of thixotropic response under constant effective stress and constant void ratio. Thixotropy is defined as the "process of softening caused by remolding, followed by a time-dependent return to the original harder state" (Mitchell 1960). Thixotropy is a reversible process, which is mainly related to the rearrangement of the remolded soil particles, and it must be considered in constitutive models that deal with shear failure at the soil-structure interface such as driven piles, and the following increase in pile capacity with time after end of driving (or pile setup). Fakharian et al. (2013) described a reduction factor to incorporate soil remolding during pile installation in the numerical simulation. Barnes (1997) introduced a time-dependent exponential function to formulate inks thixotropic (strength regaining) response after remolding.

In this paper, a new constitutive model is developed and examined based on the combination of the critical state theory and the DSC [named as CSDSC model], which can describe behavior of both NC and OC soils as well as the soil disturbance caused by shear loading during deep penetration. In this study, thixotropic response of the clayey soils was formulated similar to the ink's thixotropy formulation presented in Barnes (1997). The proposed model requires only six model parameters, which it is less than the numbers of parameters of models that have been previously developed based on DSC. Furthermore, the proposed model predicts a smooth transition from the elastic state of the soil to the plastic state, which is usually observed during laboratory tests performed on soil samples.

2 Proposed Constitutive Model

The proposed model was developed in the framework of DSC formulation. Therefore, in this study, the relative intact (RI) state behavior was modeled using the MCC model. The fully adjusted (FA) response was defined based on CSSM concept, and it was assumed that the soil is in critical state when it becomes fully adjusted. Adopting critical state concept for both RI and FA behaviors imposes the proposed model to have two different values of critical state parameter (*M*) in order to describe the behaviors of RI and FA reference states: A value for intact material response (M_i) and a value for the fully adjusted response (M_c). The former one is not a real soil properties, and it can be specified based on the proposed model requirement (as will be described in the following sections). The latter one represents the critical state parameter of the soil, and it is obtained from laboratory test results. Since the proposed model was obtained by combining the DSC and the critical state MCC model, it is called the Critical State and Disturbed State Concept (CSDSC) model. In this model, the actual soil response in each load increment is related to the values of M_i and M_c using disturbance function, *D* with following equation:

$$M_a = (1 - D)M_i + DM_c \tag{1}$$

where M_a is the averaged (linearly combined) value for the *M* at each stage of loading process. The disturbance function *D*, which is related to the produced plastic strain in the soil body under shear is used to define the averaged response. The following exponential equation was proposed by Desai (2001), which relates soil disturbance *D* to the developed plastic strain in the soil body under applied load:

$$D = 1 - e^{-A*\xi_d^B} \tag{2}$$

where ξ_d is the trajectory of deviatoric plastic strain dE_{ij}^p, defined as $\xi_d = \int \left(dE_{ij}^p \cdot dE_{ij}^p\right)^{1/2}$; and *A* and *B* are material parameters, which are obtained from results of laboratory soil Triaxial tests.

At the initial stage of shear, the soil is assumed to be undisturbed ($D = 0$ and $\xi_d = 0$), which means Eq. 1 yields to $M_a = M_i$. However, with the proceeding of the applied load, the soil disturbs, the plastic strains develop in the soil body, the values of D and ξ_d increase, and eventually the D value approaches to 1. At this point, the soil reaches the critical state (i.e., $M_a = M_c$) condition.

M_c and M_i can be assumed to be constant, so the incremental form for Eq. 1 can be expressed as follows:

$$dM_a = (M_c - M_i)dD \tag{3}$$

From plasticity conceptual formulation and using the MCC model to represent the material response, the plastic multiplier λ^* can be defined as follows:

$$\lambda^* = \frac{\frac{\partial F}{\partial \sigma_{ij}} C^e_{ijkl} d\varepsilon^i_{kl}}{\frac{\partial F}{\partial \sigma_{mn}} C^e_{mnpq} \frac{\partial F}{\partial \sigma_{pq}} - \left[\frac{1+e}{\lambda-\kappa} p' p'_0 \frac{\partial F}{\partial p'}\right]} \tag{4}$$

where $d\varepsilon^i_{kl}$ is the incremental intact strain. On the other hand, for a specific yield surface F and in the case of associated flow rule used for plasticity formulation, the plastic strain increment is related to the deferential of the yield function F with respect to the stress tensor by the following equation:

$$d\varepsilon^p_{ij} = \lambda^* \frac{\partial F}{\partial \sigma_{ij}} \tag{5}$$

where λ^* is the plastic multiplier. The star sign in λ^* is used here for plastic multiplier to remove confusion. After mathematical manipulation incremental variation of D can be obtained as:

$$dD = \frac{\left(AB\xi_d^{B-1} e^{-A*\xi_d^B}\right) \frac{\partial F}{\partial \sigma_{ij}} \cdot C^e_{ijkl} \left[\frac{\partial F}{\partial \sigma_{rs}} \frac{\partial F}{\partial \sigma_{rs}} - \frac{1}{3} \frac{\partial F}{\partial \sigma_{uu}} \frac{\partial F}{\partial \sigma_{vv}}\right]^{1/2}}{\frac{\partial F}{\partial \sigma_{mn}} C^e_{mnpq} \frac{\partial F}{\partial \sigma_{pq}} - \left[\frac{1+e}{\lambda-\kappa} p' p'_0 \frac{\partial F}{\partial p'}\right]} \cdot d\varepsilon^i_{kl} \tag{6}$$

Equation 6 indicates that the incremental change in the disturbance function in related strain increment of the intact material.

In the proposed model, MCC formulations run in each increment (or sub-increment). However, while the soil shears, the critical state parameter M evolves gradually from M_i value and toward M_c value depending on the amount of developed plastic strain in each increment and obeying the DSC theory. Figure 1 presents the formulation of the proposed model in the $p' - q$ space. The point A represents stress state at the beginning of strain increment $d\varepsilon_n$. The MCC model is used to solve the governing equations for $d\varepsilon_n$ using averaged critical state parameter, M^n_a, and the new stress state is obtained at point B, which is located on the yield surface a_n. Then, updated value for the averaged critical state parameter M^{n+1}_a is obtained from the incremental value of dM_a by using Eqs. 3 and 6 for use in the next increment. The imaginary yield surface i_{n+1} will then be defined using the updated critical state parameter M^{n+1}_a and the hardening parameter p^{n+1}_c (the prime index in p'_c removed for

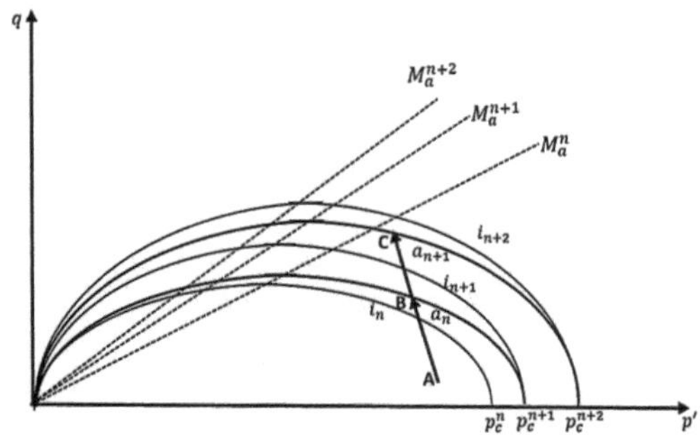

Fig. 1. Proposed (CSDSC) model representation in $p' - q$ space

simplicity). The current stress state (point B) is located inside the imaginary yield surface i_{n+1}, which causes the elastoplastic behavior for the material in the next steps until stress state reaches the critical state. The MCC model is then solved using the new strain increment $d\varepsilon_{n+1}$ to reach point C and so on. The main advantage of this approach is the possibility of specifying a small value close to zero for M_i since the observed behavior is captured by the disturbance parameters regardless of the chosen value for M_i. By choosing a very small value for M_i, the plastic behavior inside the yield surface is achieved; leading to a smooth transition between the elastic and plastic behavior.

3 CSDSC Model Parameters

The proposed CSDSC model has six parameters, including the following four critical state (MCC) model parameters: (1) The Poisson ratio ν, (2) the slope of the critical state line M, (3) slope of the normal compression line λ, and (4) slope of unloading-reloading line κ; and two parameters related to the disturbed state concept, defining the disturbance function D. The first four parameters can be obtained directly from laboratory tests (i.e., consolidation and Triaxial tests). The other two parameters namely, A and B define disturbance function D, and they can be obtained from Triaxial test results by some mathematical manipulation (Desai and Ma 1992).

4 Soil Thixotropy and the CSDSC Model

For cases like driven piles, which deal with change in the soil properties during different steps of installation and the following setup, it is necessary to adopt appropriate material properties at each step. Numerical simulation of pile setup using properties obtained from laboratory tests like Triaxial or consolidation tests on undisturbed soil samples yields unrealistic results. Therefore, the time-dependent reduction parameter $\beta(t)$ was applied in this study on the critical state parameter M and the soil-structure interface friction coefficient μ to incorporate the effect of soil remolding during pile installation and the following strength regaining with time after that:

$$M(t) = \beta(t)M, \quad \mu(t) = \beta(t)\mu, \tag{7}$$

Based on research performed by Barnes (1997) on the thixotropic strength regaining over time for inks, Abu-Farsakh et al. (2015) proposed the following definition for $\beta(t)$:

$$\beta(t) = \beta(\infty) - [\beta(\infty) - \beta(0)]e^{-\left(\frac{t}{\tau}\right)} \tag{8}$$

Where, the parameter t is time after soil remolding. $\beta(0)$ is the initial value for reduction parameter β immediately after soil shearing (t = *0*), which its value depends on the degree of remolding occurs in the soil during shear. $\beta(\infty)$ is the β value after long time from soil disturbance (t = ∞); and τ is a time constant that controls the rate of evolution of β. Abu-Farsakh et al. (2015) related τ to the soil t_{90}, which is the time for 90% dissipation of the excess pore water pressure at pile surface.

In this study, a similar formulation to the disturbance function D (i.e., Eq. 2) was also proposed, which relates the initial reduction parameter $\beta(0)$ to the deviatoric plastic strain trajectory using the following exponential function:

$$\beta(0) = \beta_R + (1 - \beta_R)e^{-A*\xi_d^B} \tag{9}$$

where β_R is the β value for the fully remolded soil, which indicates a maximum reduction of the soil strength during shearing, and its value is related to the soil sensitivity. In order to reduce complexity, the disturbed state parameters A and B were used to introduce a relation between $\beta(0)$ and ξ_d in Eq. 9. Figure 2a and b present the schematic representations of the variations of D and $\beta(0)$ versus the deviatoric plastic strain trajectory, respectively. These Figures show that while the soil disturbs, the D value approaches unity, and the $\beta(0)$ yields to β_R by proceeding the plastic strain.

5 Verification of the Proposed Model

In order to verify the developed CSDSC model, elastoplastic formulation of the model was coded in Fortran computer language and it the was implemented in FE software, Abaqus, using a user defiend material (UMAT) subroutine. Two case studies including a consolidated undrrained (CU) Triaxial test and a full-scale pile load test program were then simulated as discribed in following sections, and the obtained results compared with measured values from the experimental tests.

6 Case Study 1: Kaolin Clay

To verify the predictive capability of the proposed model, the results of laboratory Triaxial tests on Kaolin Clay performed by Yao et al. (2012) was simulated using the proposed CSDSC model. Three-dimensional model with a cubic porous element for soil specimen was used. The coupled porewater pressure analysis was used to define the multi-phase characteristic of the saturated soil. Triaixal stress state was applied

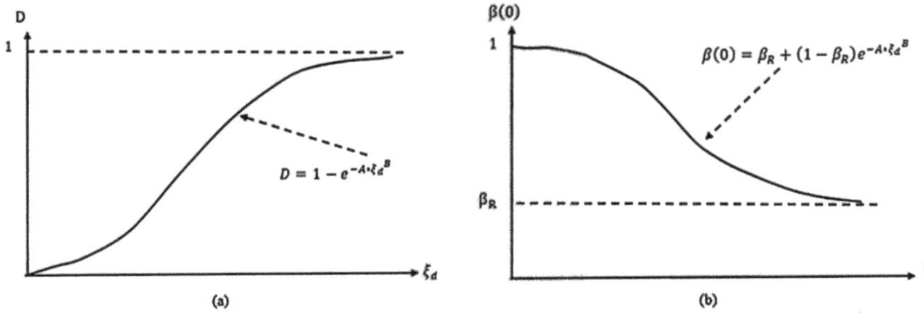

Fig. 2. Variation of soil characteristics during shear loading: (a) disturbance function D, and (b) soil strength reduction factor immediately after remolding, $\beta(0)$

using prescribed stresses for confining stress and using the prescribed displacement to apply deviatoric stress. The sample top surface was assumed to be free for drainage. The shear responses from underained Triaxial compression test for different stress history (OCR = 1, 1.20, 5,8,12) were simulated. The four model parameters that are related to the MCC model were obtained from Yao et al. (2012). The remained two model parameters that are related to the disturbed state concept (i.e. A and B) were obtained from the Triaxial test results following the procedure explained in a previous section. The calculated parameters are presented in Table 1.

Table 1. Model parameters for Kaolin Clay used for implementation (Yao et al. 2012)

M	λ	κ	ν	A	B
1.04	0.14	0.05	0.20	14.43	0.47

Using the model parameters presented in Table 1, the FE model was run with MCC model and the results for different stress paths in the undrained condition are presented in Fig. 3a. The figure shows that the MCC model is not able to capture appropriately the actual soil response under undrained shear loads, especially for OC clays. In the proposed model, the strong capability of the CSDSC in modeling the actual behavior of soils was demonstrated, and the results of numerical simulation for different stress paths using the proposed model are presented in Fig. 3b. The figure clearly indicates that the proposed model can predict the actual soil behavior for both the NC and OC soils with good agreement. The model is also able to capture the strain softening behavior of heavily OC soils. Figure 4 shows the results of proposed model for stress-strain relations and excess porewater presuure generated during traixial CU test at different over-consolidation ratios, which represents good agreement between the model prediction and meaured lab results. In this figure, the stress values are normalized with respect to the initial pre-consolidation pressure p'_0. Figure 4b shows that, for NC soil and lightly OC soil, the generated porewater pressure is positive, which is representation of soil contraction during undrained shearing. On the other hand, for heavily OC soils, the numerical simulation shows the generation of positive porewater pressure at the initial stage of the test followed by negative pore water pressure until

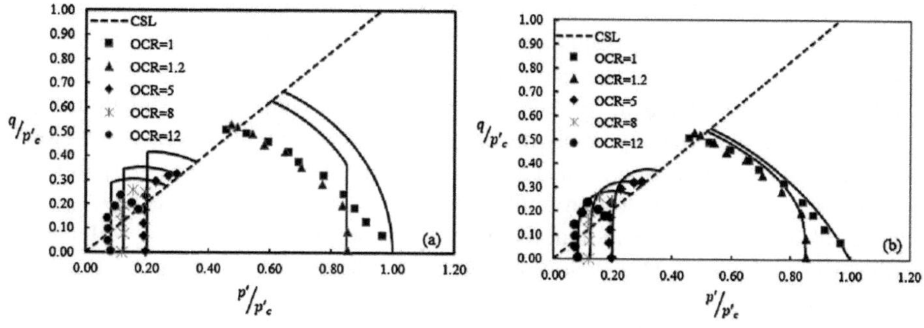

Fig. 3. Prediction of numerical simulation of undrained Triaxial test on Kaolin Clay (Yao et al. 2012) using (a) MCC model, and (b) proposed CSDSC model

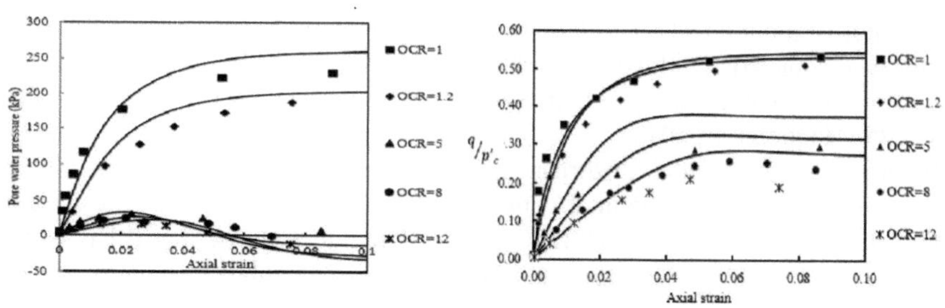

Fig. 4. Prediction of numerical simulation of undrained Triaxial test on Kaolin Clay (Yao et al. 2012) (a) excess porewater pressure, and (b) stress-strain curves

failure. This is an indication of soil dilative behavior, which is common in heavily OC soils. Based on the obtained results, the soil dilation in undrained condition increases by increasing OCR values.

7 Case Study 2: Full-Scale Pile Installation

The load test results for a full-scale instrumented test pile that was conducted at Bayou Laccasine Bridge site, Louisiana (Haque et al. 2014) was simulated using the proposed CSDSC model. The test pile was square concrete pile with 0.76 m width and a total length of 22.87 m. A 6.4 m long casing was installed and driven prior to pile installation to represent the scour effect at shallow depth. The test pile was fully instrumented with pressure cells, vibrating wire piezometers and sister bar strain gages that were installed at different depths of pile length, targeting specific soil layers. In addition, the surrounding soils were instrumented with nine multi-level piezometers located at the same depths as the pressure cells and piezometers installed at the pile's face. Both static load tests (SLTs) and dynamic load tests (DLTs) were conducted to obtain the pile resistance at different times after end of driving.

In this paper, the numerical simulation of the pile installation and following setup were performed using the Abaqus software and adopting the techniques described in Abu-Farsakh et al. (2015). The geometry of the soil and pile, the applied boundary conditions, and finite element mesh are shown in Fig. 5. Numerical simulation of pile installation was achieved by applying prescribed displacement to the soil nodes to create volumetric cavity expansion. The pile was then placed inside the cavity followed by applying a vertical penetration until the steady state condition is reached.

The subsurface soil condition at the pile site is mainly consists of clay soil, and the natural water table is 2.24 m below the ground surface. The subsurface soil domain was divided into eight layers based on the soil type and properties as presented in Table 2. In the table, w is the soil water content (%), S_u is the undrained shear strength (kPa), and K is the soil permeability (m/s).

The proposed CSDSC model was used to describe the elastoplastic behavior of the surrounding clay soil. The soil remolding during pile installation was incorporated in the constitutive model, and relating β_R to the soil sensitivity S_r with $\beta_R = (S_r)^{-0.3}$. This relation was depicted for Bayou Laccasine Bridge site based on available data for S_r and the pile resistance values obtained from field load tests, which yields a value of $\beta_R = 0.75$ (Haque et al. 2014).

Figure 6 represents the disturbance occurs in the soil immediately after pile installation for a typical horizontal path (path 1 in Fig. 5), which was obtained from numerical simulation using the CSDSC model. The figure shows that β has its maximum value $\beta_R = 0.75$ for soil adjacent to the pile face and approaches unity at a radial distance equal to eight times the pile size. At the same time, the disturbance function has a maximum value ($D = 1$) at the soil-pile interface, and it approaches to $D = 0$ at a radial distance equal to eight times the pile size along the same path.

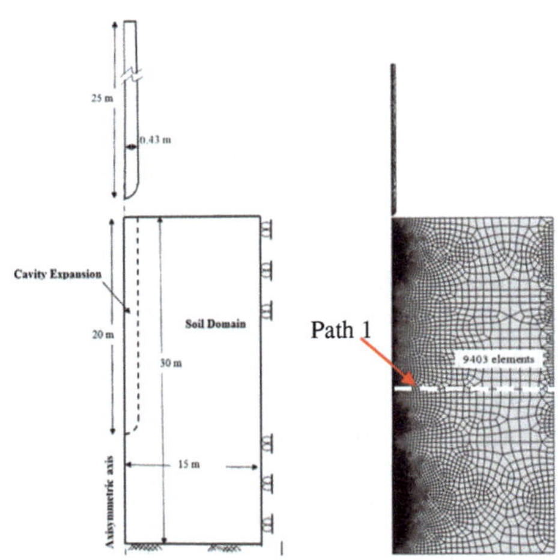

Fig. 5. Numerical simulation domain: (a) geometry and boundary conditions and (b) FE mesh (Abu-Farsakh et al. 2015)

Table 2. Soil material parameters for the test pile site (Abu-Farsakh et al. 2015)

Layer No.	Depth (m)	w (%)	S_u (kPa)	OCR	M	λ	κ	K (m/s) 10^{-9}
1	0–6.40	21	120	4	0.61	0.104	0.035	3.80
2	6.40–7.60	26	72	2.5	1.17	0.100	0.029	4.20
3	7.60–10	25	68	2	0.90	0.091	0.026	0.62
4	10–11.60	29	104	1.7	0.90	0.108	0.035	0.12
5	11.60–13	23	94	1.45	0.62	0.108	0.035	7.60
6	13–16	52	150	1.40	1.12	0.147	0.061	8.90
7	16–20	24	112	1.3	0.92	0.100	0.030	0.17
8	20–23	29	101	1	0.93	0.056	0.013	0.66

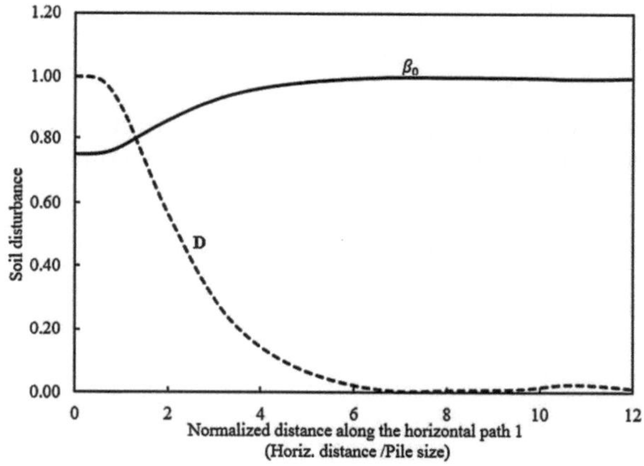

Fig. 6. Variation of β and *D* for a typical horizontal path in soil body immediately after pile installation

The numerical simulation using CSDSC model was compared with the measured field test results of the pile load tests. Figure 7a shows the comparison between the predictions of unit shaft resistance one hour after end of driving obtained using the CSDSC model and the measured values from the field load tests. The cumulative values of shaft resistance obtained from numerical simulation were also compared with the calculated values obtained from field tests, and the results are shown in Fig. 7. These figures clearly indicate that the CSDSC model is able to predict the pile resistance appropriately. For more verification, the increase in pile capacity after end of driving (or pile setup) was obtained from numerical modeling using the CSDSC model, and the model predictions were compared with the measured values from field load test results as shown in Fig. 8. The figure demonstrates that the proposed model is able to simulate pile setup with the model predictions of the pile resistance are slightly over-predicted the measured values.

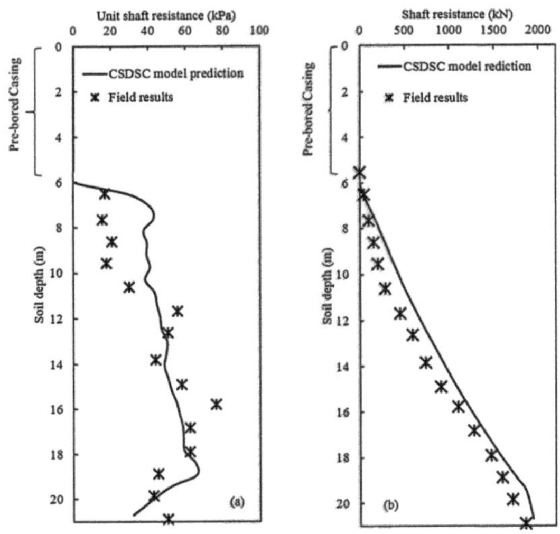

Fig. 7. Comparison between the proposed CSDSC model predictions with measured values from field test results (a) unit shaft resistance, and (b) total shaft resistance

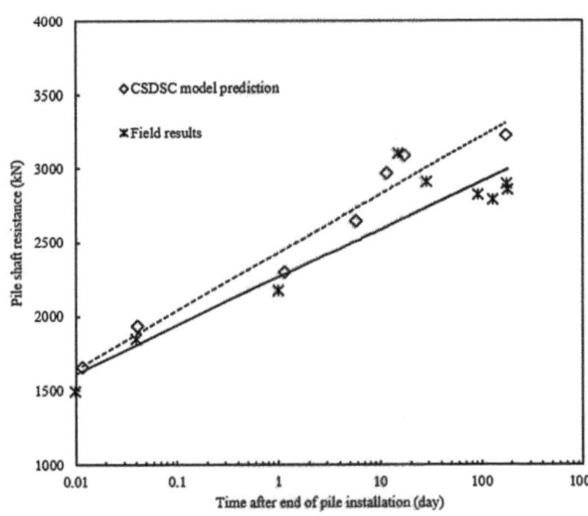

Fig. 8. Comparison between the proposed CSDSC model prediction and field measurement for pile setup

8 Summary and Conclusions

In this paper, an elastoplastic constitutive model for clay soil was developed and evaluated in application of pile installation and the following setup over time. The proposed model is based on the combination of Disturbed State Concept (DSC) and critical state concept of Modified Cam-Clay (MCC) model, which is referred as the CSDSC model. In the CSDSC model, the disturbance function D was applied to the critical state parameter M to adopt disturbed state concept. The soil remolding behavior was related to the state of deviatoric plastic strain developed in the soil body during shear loading for use to simulate deep penetration problems such as pile installation and the corresponding remolding of surrounding soil. The soil thixotropic response was incorporated in the pile setup phenomenon using a time-dependent function, which increases exponentially with time after end of pile driving. The proposed model was implemented in Abaqus software via a user defined subroutine UMAT. The responses of Kaolin Clay under undrained monotonic loads was simulated for verification. Furthermore, a full-scale instrumented pile driven in Louisiana clayey soil and the following setup were simulated using the CSDSC model, and the results obtained from FE model and those measured from field test results were compared. Based on the results of this study, the following conclusions can be made for the numerical simulation using the CSDSC model:

a) The developed CSDSC model has only six parameters, which is less than the previous elastoplastic models developed based on DSC, which makes it more effective in geotechnical engineering applications.
b) The proposed CSDSC model predictions were compared with laboratory Triaxial test results, which show that the model was able to appropriately capture the undrained shear responses for NC and OC clays.
c) The steep changes in stress paths inside yield surface, which is normally observed in MCC model, could be vanished in CSDSC model providing a smooth transition from elastic to plastic response. In other word, the proposed model is able to capture the smooth transitional behavior of the soil from elastic to elastoplastic and then to fully plastic states, which is usually observed during experimental tests performed on clayey soils.
d) Numerical simulation of a full-scale pile installation and following pile load tests after end of driving indicated that the proposed CSDSC model is able to simulate pile installation and capture the soil disturbance, soil thixotropy and pile setup appropriately.
e) The results demonstrated good agreement between the prediction of pile resistance and pile setup using CSDSC model and the measured values obtained from full-scale pile loads tests results.

Acknowledgements. This research is funded by the Louisiana Transportation Research Center (LTRC Project No. 11-2GT), Louisiana Department of Transportation and Development, LADOTD (State Project No. 736-99-1732), and The Louisiana Board of Regents Support Fund contract LEQSF (2012-2015)-RD-B-03. All these supports are greatly appreciated.

References

Abu-Farsakh, M., Rosti, F., Souri, A.: Evaluating pile installation and the following thixotropic and consolidation setup by numerical simulation for full scale pile load tests. Can. Geotech. J. **52**, 1–11 (2015)
Barnes, H.A.: Thixotropy-a review. Int. J. Non-Newton. Fluid Mech. **70**, 1–33 (1997)

Basu P., Prezzi M., Salgado R., Chakraborty T.: Shaft resistance and setup factors for piles jacked in clay. J. Geotech. Geoenviron. Eng. **140**(3) (2014)

Chakraborty, T., Salgado, R., Loukidis, D.: A two-surface plasticity model for clay. Comput. Geotech. **49**, 170–190 (2013a)

Chakraborty, T., Salgado, R., Basu, P., Prezzi, M.: Shaft resistance of drilled shafts in clay. J. Geotech. Geoenviron. Eng. **139**(4), 548–563 (2013b)

Dafalias, Y.F., Herrmann, L.R.: Bounding surface plasticity. II: application to isotropic cohesive soils. J. Eng. Mech. **112**(12), 1263–1291 (1986)

Desai, C.S., Ma, Y.: Modeling of joints and interface using disturbed state concept. Int. J. Num. Anal. Meth. Geomech. **16**, 623–653 (1992)

Desai, C.S.: Mechanics of Materials and Interface: The Disturbed State Concept. CRC Press, Boca Raton (2001)

Fakharian K., Attar I.H., Haddad H.: Contributing factors on setup and the effects on pile design parameter. In: Proceedings of the 18th International Conference Soil Mechanics and Geotechnical Engineering, Paris (2013)

Haque, M.N., Abu-Farsakh, M., Chen, Q., Zhang, Z.: A case study on instrumentation and testing full-scale test piles for evaluating set-up phenomenon. 93th Transportation Research Board Annual Meeting, vol. 2462, pp. 37–47 (2014)

Likitlersuang S.: A hyper plasticity model for clay behavior: an application to Bangkok clay. Ph. D. Dissertation, The University of Oxford (2003)

Mitchell, J.K.: Fundamental aspects of thixotropy in soils. J. Soil Mech. Found. Des. **86**, 19–52 (1960)

Pestana, J.M., Whittle, A.J.: Formulation of a unified constitutive model for clays and sands. Int. J. Num. Anal. Meth. Geomech. **23**, 1215–1243 (1999)

Roscoe, K.H., Burland, J.B.: On the Generalized Behavior of Wet Clay, Engineering Plasticity, pp. 535–610. Cambridge University Press, Cambridge (1968)

Whittle, A.J.: Evaluation of a constitutive model for over consolidated clays. Geotechnique **43**(2), 289–313 (1993)

Yao, Y.P., Sun, D.A., Matsuoka, H.: A unified constitutive model for both clay and sand with hardening parameter independent on stress path. J. Comput. Geotech. **35**, 210–222 (2007)

Yao, Y.P., Gao, Z., Zhao, J., Wan, Z.: Modified UH model: constitutive modeling of over consolidated clays based on a parabolic Hvorslev envelope. J. Geotech. Geoenviron. Eng. **138**, 860–868 (2012)

Analytical Models to Estimate the Time Dependent Increase in Pile Capacity (Pile Set-Up)

Murad Y. Abu-Farsakh[1(✉)] and Md. Nafiul Haque[2]

[1] Louisiana Transportation Research Center, Louisiana State University, Baton Rouge, LA 70808, USA
cefars@lsu.edu

[2] Louisiana Transportation Research Center, Baton Rouge, LA 70808, USA

Abstract. This paper presents the analyses of twelve prestressed concrete (PSC) instrumented test piles that were driven in different locations of Louisiana in order to develop analytical models to estimate the increase in pile capacity with time or pile set-up. The twelve test piles were driven mainly in cohesive soils. Detailed soil characterizations including laboratory and in-situ tests were conducted to determine the different soil properties. The test piles were instrumented with vibrating wire strain gauges, piezometers and pressure cells. Several static load tests (SLT) and dynamic load tests (DLT) were conducted on each test pile at different times after end of driving (EOD) to quantify the magnitude and rate of set-up. Measurements of load tests confirmed that pile capacity increases almost linearly with the logarithm of time elapsed after EOD. Case Pile Wave Analysis Program (CAPWAP) were performed on the restrikes data and were used along with the load distribution plots from the SLTs to evaluate the increase of skin friction capacity of individual soil layers along the length of the piles. The logarithmic set-up parameter "A" for unit skin friction was calculated of the 70 individual clayey soil layers, and were correlated with different soil properties. Nonlinear multivariable regression analyses were performed and three different empirical models are proposed to predict the pile set-up parameter "A" as a function of soil properties.

Keywords: Pile set-up · Static load test · Dynamic load test · Empirical model · Consolidation

1 Introduction

It is well-known that the axial capacity of piles usually increases with time after driving in cohesive soils. Many researchers (e.g., Komurka et al. 2003; Rausche et al. 2004; Fellenius 2008; Abu-Farsakh et al. 2016) have studied this increase in capacity, known as "set-up". Several empirical, analytical and numerical techniques have been proposed over the past few decades to predict the magnitude and rate of pile set-up with time. It has been well recognized that the magnitude of set-up is dependent upon the pile size, pile length, pile material, soil type and soil strength (Long et al. 1999; Chen et al. 2014).

M. Abu-Farsakh et al. (eds.), *Advances in Analysis and Design of Deep Foundations*, Sustainable Civil Infrastructures, DOI 10.1007/978-3-319-61642-1_12

Pile set-up phenomenon is mainly attributed to three main mechanisms: (1) Dissipation of excess pore water pressure (PWP) (or consolidation), (2) Thixotropic effect, and (3) Aging effect. During pile driving, the surrounding soil is displaced predominantly radially along the side and vertically and radially beneath the tip, thus generating a significant amount of excess PWP. In addition, the soil within the vicinity of pile face loses its strength due to an increase in excess PWP, disturbance of the soil structure and the soil remolding (McVay et al. 1999). As the excess PWP starts to dissipate, the effective stress of the disturbed soil starts to increase, and consequently set-up primarily occurs due to the increase in shear strength and the increase in lateral stresses against the pile (Rausche et al. 2004; Chen et al. 2014). Thixotropic effect (or regaining of soil strength of disturbed soil with time) also plays a significant role at the early stage of set-up (Ng et al. 2013; Haque et al. 2016a, b). Any set-up occurs after the completion of excess PWP dissipation is mainly due to "aging" effect (i.e., time dependent change in soil properties at a constant effective stress) (Schmertmann 1991; Wang and Gao 2013). Several empirical models (e.g., Skov and Denver 1988; Ng et al. 2013; Wang et al. 2015) have been proposed to estimate the pile set-up capacity with time. Of these models, the relationship developed by Skov and Denver (1988) is considered the most popular relationship due to its simplicity. They postulated that the pile capacity increases with the logarithm of time as follows:

$$\frac{R_t}{R_{to}} = A \log_{10}\frac{t}{t_o} + 1 \tag{1}$$

where: R_t = Total pile capacity at time, t; R_{to} = Total pile capacity at reference time, t_o; t = Time elapsed since end of initial pile driving; t_o = Initial reference time, a reference time before which there is no predictable R_{to} increase as a function of elapsed time; A = Set-up rate parameter (log-linear). The "A" parameter can be assumed, back-calculated from field data, or gleaned from empirical relationships available in the literature. However, most of the available models in literature (e.g., Skov and Denver 1988) did not consider the soil properties in their formulations and that the total capacity (R_t) was mostly used instead of the skin friction (R_s). However, very few models (e.g., Ng et al. 2013; Karlsrud et al. 2014) incorporated the soil properties in their models to predict pile set-up. Table 1 presents the most recent pile set-up developed models.

The construction of pile foundation usually becomes expensive. Each year, millions of dollars are spent in order to drive prestressed concrete (PSC) piles. Therefore, the incorporation of even a small percentage of pile setup into pile design, can result in significant cost savings. The accurate prediction/estimation of the increase in pile capacity with time can be incorporated into a rational design through (a) reducing the number of piles, (b) shortening pile lengths, (c) reducing pile cross-sectional area (using smaller-diameter piles), and/or (d) by reducing the size of driving equipment (using smaller hammers and/or cranes).

Table 1. Available pile set-up models

References	Models	Comments
Skov and Denver (1988)	$R_t = R_o\,[A \log(\frac{t}{t_o}) + 1]$	A = 0.2 for sand and 0.6 for clay t_0 = 0.5 for sand and 1.0 for clay
Bogard and Matlock (1990)	$R_t = R_u\,[0.2 + 0.8(\frac{\frac{t}{T_{50}}}{1+\frac{t}{T_{50}}})]$	$R_u = R_u$ occurs at 100% set-up $T_{50} = T_{50}$ is the time required to reach 50% set-up
Long et al. (1999)	$R_t = 1.1R_{EOD}t^{\alpha}$	α = 0.05 for lower bound α = 0.18 for upper bound bound
Svinkin and Skov (2000)	$R_u = R_{EOD}\,[B\{\log(t) + 1\} + 1]$	B = Empirical factor
Mesri and Smadi (2001)	$\tau = \tau_R\,(t/t_R)^{CDC\alpha/Cc}$	τ_R = Pile capacity at 1 day t_R = 1 day $C_\alpha/C_c = 0.02 \pm 0.01$ C_D = Empirical value
Karlsrud et al. (2005)	$R_t = R_{100}[A(\log(\frac{t}{t100}) + 1)]$ $A = 0.1 + 0.4(1 - \frac{PI}{50})OCR^{-0.8}$	R_{100} = Pile capacity at 100 days PI = Plasticity index OCR = Over consolidation ratio
Bullock et al. (2005)	$R_S = R_o\,[A\log(\frac{t}{t_o}) + 1]$	A = 0.1 (In absence of test) t_o = 1 day
Ng et al. (2013)	$R_t = R_{EOD}\,[A \times \log_{10}(\frac{t}{t_{EOD}}) + 1](L_t/L_{EOD})$ $A = \frac{f_c Cha}{N_a r^2 p} + f_r$	c_{ha} = Horizontal coefficient of consolidation N_a = SPT N value r_p = Equivalent pile radius f_c = Consolidation factor f_r = Remolding recovery factor
Haque et al. (2016a, b)	$f_S = f_{so}\,[A \log(\frac{t}{t_o}) + 1]$ $A = 0.57\,e^{-0.05qt}$	f_s = Unit skin friction f_{so} = Initial unit skin friction q_t = Corrected cone tip resistance t_o = 1 day

2 Objective

The objective of this study is to develop analytical models to estimate the set-up parameter "A" of individual soil layers from soil properties. The set-up parameter "A" of individual soil layers were back-calculated using the unit skin friction (f_s) rather than the total pile capacity (R_t) as proposed by Skov and Denver (1988) model. The soil properties of individual soil layers for each test pile location [i.e., undrained shear strength (S_u), Atterberg limits, sensitivity (S_t) and vertical coefficient of consolidation (c_v)] were obtained from the laboratory testing and/or interpreted from the piezocone penetration test (PCPT) or dissipation test. The back-calculated set-up parameters "A" were correlated with the selected soil properties and nonlinear analytical models were developed to estimate the skin friction set-up for individual soil layers along the pile length.

3 Test Site and Subsurface Geotechnical Condition

3.1 Test Location and Test Piles (TP)

Five different sites were selected in Louisiana to perform the pile set-up study. These sites include: Bayou Zourie, Bayou Lacassine, Bayou Teche, Bayou Bouef and LA-1. Detailed description of the Bayou Zourie, Bayou Lacassine and LA-1 project sites can be found in Chen et al. (2014), Haque et al. (2014), and Haque et al. (2016a, b), respectively. Figure 1 shows the approximate locations of the pile set-up project sites. This set-up study was performed only on 12 square PSC instrumented test piles driven in cohesive dominated subsurface soil conditions. The objective of this study is to develop analytical models that can predict pile set-up for individual soil layers along the pile lengths. In order to meet this criterion, the test piles in all sites were instrumented with strain gauges in order to calculate the increase in skin friction of individual soil layers after EOD. With the aim to understand the consolidation behavior with pile set-up, a combination of piezometers and pressure cells were also installed in selected test piles to measure the total and excess pore water pressure and hence effective lateral stresses on pile face. The pile ID, their width and length and the hammer type used for installation are presented in Table 2.

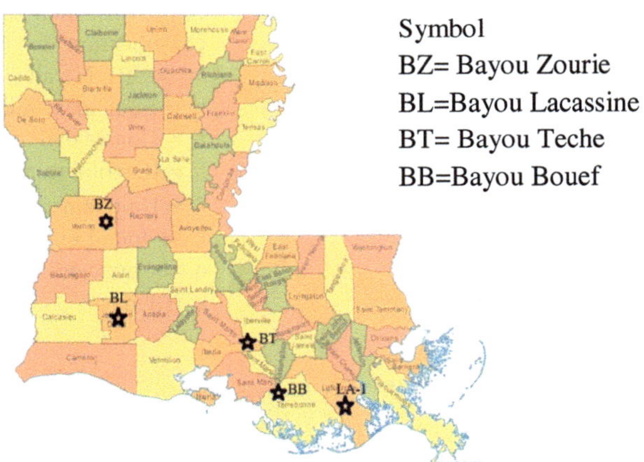

Fig. 1. Location of the performed projects result

3.2 Geotechnical Subsurface Characterization

Both laboratory and in-situ tests were conducted at each test pile location to evaluate the different soil properties. 7.6 cm Shelby tube samples were retrieved from boreholes drilled at different depths at each test pile location for comprehensive laboratory testing. Water content, unit weight, Atterberg limits, one-dimensional consolidation tests and unconsolidated undrained (UU) triaxial tests were performed on selected soil samples to

Table 2. Summary of test piles

Project Name	Pile ID	Pile Type	Width (mm)	Length (m)	Hammer Type
LA-1	LA-1-TP-2	PSC	406	39.6	Vulcan 010
	LA-1-TP-3		762	57.9	Vulcan 010
	LA-1-TP-4a		610	48.8	Vulcan 020
	LA-1-TP-4b		610	64.0	Vulcan 020
	LA-1-TP-5a		610	44.2	Vulcan 020
	LA-1-TP-5b		610	51.8	Vulcan 020
Bayou Lacassine (BL)	BL-TP-1		762	22.9	ICE I-46
	BL-TP-2		762	25.0	ICE I-46
	BL-TP-3		762	22.9	ICE I-46
Bayou Bouef (BB)	BB-TP-1		762	43.2	HPSI 2005
Bayou Zourie (BZ)	BZ-TP-1		610	16.8	ICE I-62 V-2
Bayou Teche (BT)	BT-TP-1		610	19.5	ICE-I-36

Table 3. Summary of the soil properties and set-up parameter "A"

Pile ID	Testing period (Days)	Nos and types of tests performed		Nos and types of soil layers		Soil properties		Set-Up parameter "A"		Set-Up ratio	
		DLT	SLT	Clayey	Sandy	S_u (kPa)	PI (%)	Clayey soil layer	Sandy soil layer	R_t/R_{to}	R_s/R_{so}
LA-1-TP-2	7	6	1	7	1	7–35	4–25	0.35–0.53	0.15	4.9	7.5
LA-1-TP-3	13	3	1	6	3	38–49	16–37	0.31–0.43	0.07–0.13	2.4	3.4
LA-1-TP-4a	6	6	1	8	3	8–45	46–77	0.38–0.51	0.13–0.24	5.0	9.9
LA-1-TP-4b	6	7	1	11	3	8–78	26–77	0.22–0.47	0.13–0.24	2.3	2.9
LA-1-TP-5a	6	5	1	4	2	23–44	20–50	0.24–0.33	0.23–0.24	4.2	5.2
LA-1-TP-5b	6	6	1	6	2	23–51	20–50	0.20–0.28	0.15	2.0	2.4
BL-TP-1	217	3	5	6	1	72–123	4–25	0.13–0.26	0.10	2.1	2.3
BL-TP-2	23	3	1	6	1	85–145	16–37	0.16–0.27	0.08	1.7	2.0
BL-TP-3	181	3	5	6	1	79–124	17–35	0.14–0.26	0.05	1.6	1.9
BB-TP-1	28	1	3	4	2	51–59	16–35	0.29–0.48	0.05	2.0	3.3
BZ-TP-1	76	3	2	3	2	116–157	37–75	0.15–0.29	0.17–0.26	1.5	1.8
BT-TP-1	32	4	1	3	3	10–20	36–52	0.28–0.40	0.02–0.09	1.2	1.2
Total Soil Layers = 94				**70**	**24**	**Average "A"**		**0.31**	**0.15**		

characterize the subsurface soil conditions. The in-situ testing program included both piezocone penetration tests (PCPT), piezocone dissipation tests and standard penetration test (SPT). One-dimensional consolidation tests were also performed to calculate the coefficient of consolidation (c_v) of soil in the absence of PCPT dissipation tests. Table 3 presents the information of the soil properties that have been used in this study.

4 Load Testing Program

The load testing program was designed to measure the increase in pile capacity with time (or pile set-up). Table 3 summarizes the total number of tests, testing period, set-up ratio (i.e., capacity during a load test over initial driving capacity) and the back-calculated set-up factor "A" for the individual soil layers along the 12 instrumented test piles. The details of load test results for Bayou Zourie, Bayou Lacassine and LA-1 project can be found in Chen et al. (2014), Haque et al. (2014) and Haque et al. (2016a, b), respectively. The dynamic measurements were acquired with Pile Driving Analyzer (PDA) during initial driving and subsequent restrike events on each test pile. The DLTs were performed in accordance with the ASTM D 4945-89. In addition, the CAPWAP was also used to evaluate the skin friction of individual soil layers.

SLTs were performed after 6 to 14 days from EOD to evaluate the increase in pile capacity with time in each test pile as compared to DLT restrikes. The compression SLTs were performed in accordance with the ASTM D-1143 quick test loading option in all the test piles. The ultimate load capacity from the pile load test was determined based on the modified Davisson interpolation method (1972). Figure 2 presents the result of a SLT that was conducted at TP-2 location of LA-1 site. Embedded strain gauges were used to calculate the R_s, R_{tip} and R_t during each SLT, and to estimate the distribution of R_s along the pile length. In order to capture the strain gauge measurements for every incremental load during SLTs, the data acquisition system was set to collect the data at two minute intervals during each SLT. The axial load transfer can be determined from the strain measurements, the cross-sectional area and the Young's modulus of the pile. Figure 2b depicts an example of the load distribution plot obtained during the SLT at TP-2 location of LA-1 site.

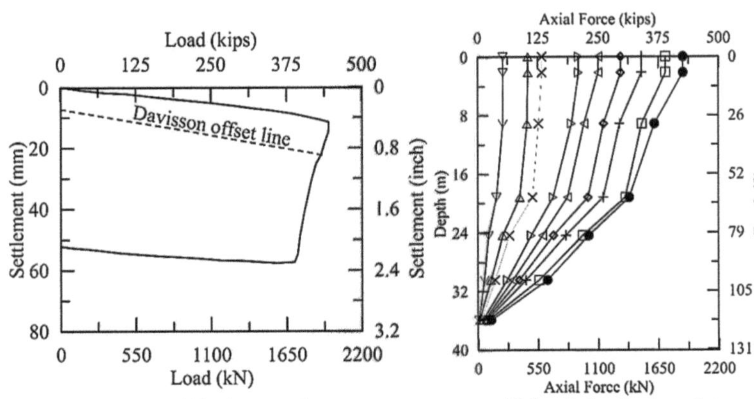

(a) Load-Settlement plot (b) Load distribution plot

Fig. 2. Results of static load test result

5 Results

5.1 Set-up in Terms of Total (R_t) and Skin Friction (R_s) Capacity

The total pile capacities estimated from the DLTs as well as the capacities measured by the SLTs are analyzed to study the set-up behavior for all test piles. However, performing repeated load tests numerous times can significantly affect the soil-pile interface and affect the true setup behavior. Performing frequent static and dynamic load tests on the same pile was a limitation of this study and can affect the result on some content. All the test piles exhibited significant amount of set-up as shown in Table 3. The results of skin friction set-up of all test piles are presented in Fig. 3. Figure 3 shows that the test piles of LA-1 site exhibited higher amount and rate of set-up compared to the test piles of other sites. The presence of very soft soil at the project location (i.e., near Gulf of Mexico) as compared to the other test pile location contribute to this behavior. The figure demonstrated that the skin friction capacities are best fitted to linear logarithmic of time with high coefficients of correlation (R^2). As seen in Table 3 and Fig. 3, the pile set-up was mainly due to increase in R_s. The R_{tip} was almost constant over time for all the test piles.

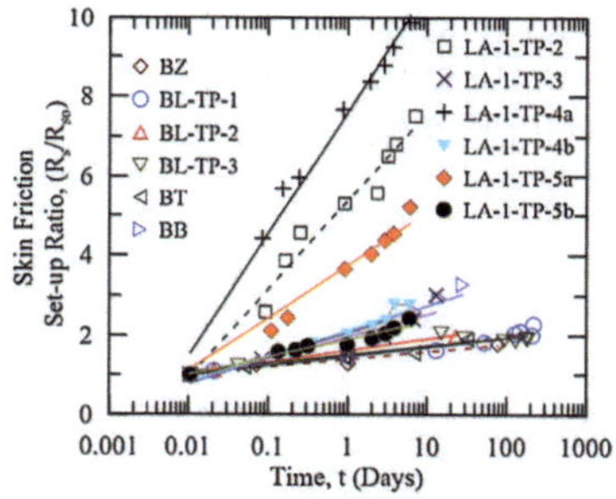

Fig. 3. Skin friction set-up for all test piles

5.2 Set-up of Individual Soil Layers

Most of the models available in literature to predict pile set-up consider either the total pile capacity set-up or the skin friction capacity set-up of entire pile. As a result, the soil properties of different soil layers along the pile length were not incorporated into those models, which results on difficult implementation of set-up models on different soil conditions. In this study, the unit skin friction (f_s) (i.e., skin friction /contact area) was used to analyze the set-up behavior for individual soil layers along the pile length (Eq. 2). The set-up behavior for individual soil layers along the pile length were calculated in this study with the aid of vibrating wire strain gauges measurements during the SLTs and from the CAPWAP analyses during the DLTs.

$$\frac{f_s}{f_{so}} = 1 + A \log\frac{t}{t_o} \tag{2}$$

Examples of analyses of set-up for individual soil layers are presented in Tables 4 for the TP-3 of Bayou Lacassine. The published literature (e.g., Paikowsky et al. 2005; Ng et al. 2013) documented that set-up is mainly dominant in clayey soil layers, and that small amount of set-up was observed in the sandy soil layers. In this study, the sandy soil layers exhibited smaller amount of set-up compared to the clayey soil layers due to quick dissipation of excess PWP after EOD. The tabulated data in Table 4 show that the clayey soil layers exhibited an average increase of 70% to 260% of set-up while sandy soil layers exhibited insignificant amount of set-up during the testing period compared to the EOD skin frictions.

Table 4. Example of set-up for individual soil layers (Test Pile-3 of Bayou Lacassine)

Layer ID	EOD	1st Res (60 min)	2nd Res (24 h)	1st SLT (15 days)	2nd SLT (29 days)	3rd SLT (93 days)	4th SLT (129 days)	5th SLT (175 days)	3rd Res (181 days)
3-1	Casing								
3-2	133/1.0	271/2.0	339/2.6	547/4.1	556/4.2	559/4.2	564/4.2	577/4.3	479/3.6
3-3	128/1.0	162/1.3	226/1.8	360/2.8	378/2.9	380/2.9	384/3.0	393/3.1	425/3.3
3-4	179/1.0	180/1.0	221/1.2	322/1.8	338/1.9	345/1.9	347/1.9	346/1.9	310/1.7
3-5	144/1.0	205/1.4	241/1.7	384/2.7	382/2.6	389/2.7	391/2.7	418/2.9	380/2.6
3-6*	637/1.0	673/1.1	716/1.1	840/1.3	690/1.1	576/0.9	525/0.8	578/0.9	661/1.0
3-7	229/1.0	302/1.3	363/1.6	527/2.3	444/1.9	447/1.9	449/1.9	454/1.9	464/2.0
3-8	45/1.0	58/1.3	65/1.4	120/2.7	121/2.7	125/2.8	126/2.8	130/2.9	137/3.0
Total skin friction (kN)	1495/1.0	1851/1.2	2171/1.4	3100/2.1	2909/1.9	2821/1.9	2786/1.9	2896/1.9	2856/1.9
End bearing capacity (kN)	765/1.0	791/1.0	720/0.9	681/0.9	667/0.9	693/0.9	657/0.9	645/0.8	765/1.0
Total Capacity	2260/1.0	2642/1.2	2891/1.3	3781/1.7	3576/1.6	3514/1.5	3443/1.5	3541/1.6	3621/1.6

*Sandy soil layer

5.3 Correlations Between Soil Properties and Set-up Parameter "A"

The set-up rate in this study as measured by "A" parameter is calculated using the unit skin friction for each soil layers along the pile length. 94 soil layers from 12 PSC test piles driven in five different project sites were included in the analyses. Clayey soil behavior was dominant in 70 clayey soil layers and the remaining 24 soil layers exhibited sandy soil behavior. The average value of "A" parameter for clayey and sandy soil layers are 0.31 and 0.15, respectively. The effects of soil properties on these back-calculated "A" parameters are investigated here in order to develop correlations between "A" parameter and the different soil properties. The soil properties that have significant influence on the set-up parameter "A" can be identified as the S_u, PI, c_h or c_v and S_t.

5.4 Effects of Undrained Shear Strength (S_u)

The S_u in this study is correlated with the rate of set-up parameter "A" for the individual clayey soil layers. S_u was experimentally measured for 70 clayey soil layers and it ranges from 7 kPa to 157 kPa. The clayey soil layers of LA-1 project with the lowest S_u values generally exhibited higher rate and magnitude of set-up compared to the clayey soil layers of the other sites. The correlation between S_u and set-up parameter "A" is presented in Fig. 4a. The figure clearly demonstrates that there is an inverse-power relationship between the "A" parameter and S_u value.

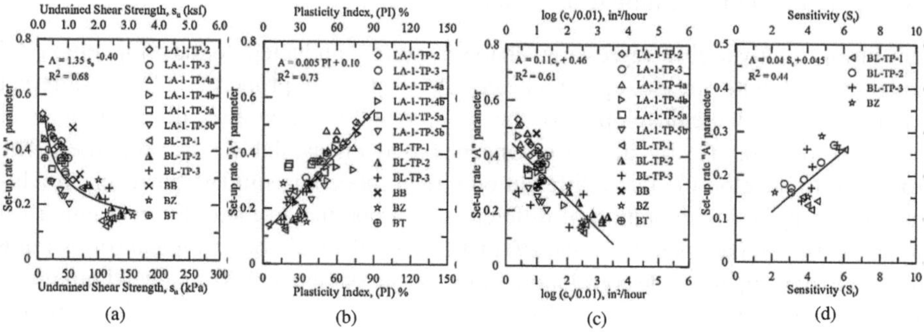

Fig. 4. Correlation of set-up parameter "A" with different soil properties.

5.5 Effects of Plasticity Index (PI)

Atterberg limit tests were performed on all the 70 clayey soil layers and the value of PI ranges from 4% to 77%. The correlation between the PI of clayey soil layers and the set-up parameter "A" is presented in Fig. 4b. The figure shows that a linear proportional relationship do exists between the PI and the "A" parameter with a relatively high coefficient of correlation (R^2 = 0.73) for this correlation.

5.6 Effects of Coefficient of Consolidation (C_v)

The coefficient of consolidation (c_v or c_h) is believed to be one of the most important factor to influence the set-up behavior of clayey soils. In-situ piezocone dissipation tests were performed and the c_h values were calculated using Teh and Houlsby (1991) interpretation method. Laboratory consolidation tests were also performed on soil samples collected at LA-1 and Bayou Teche pile sites. The correlation between the c_v and set-up parameter "A" for this study is depicted in Fig. 4c. In order to better represent the relationship, the normalized logarithmic value of c_v is considered in this analyses. The figure shows that there exists an inverse linear proportional relationship between the set-up rate parameter "A" and the log c_v values.

5.7 Effects of Sensitivity (S_t)

Due to thixotropic property of the soil, the subsequent remolding and reconsolidation of the disturbed soil at the soil-pile interface zone will also be associated with long-term increase in soil strength, depending on S_t values of the soil. The correlation between the

S_t and the set-up parameter “A” is presented in Fig. 4d. The figure shows that there exists a linear proportional relationship between the S_t and set-up parameter “A”, similar to the PI-set-up parameter “A” relationship.

5.8 Development of Empirical Models for “A”

Non-linear multivariable regression analyses were conducted to develop empirical models to estimate the set-up parameter “A” from soil properties. Three different empirical models were developed for the set-up rate “A” using three different levels of soil properties for use by design engineers based on available soil properties. The procedure to develop the three different empirical models is similar; however, the difference is only in incorporation of different soil properties in three different empirical models. Two soil parameters (S_u and *PI*) that are usually available in typical soil borelog are used to develop a simple correlation for the set-up parameter “A” in level-1 empirical model. Three soil parameters (S_u, *PI* and c_v) are incorporated in level-2 empirical model. c_h or c_v parameters are usually not available in typical soil borelog; however, it is believed to be the most important parameter that can incorporate the effect of consolidation on set-up model. The developed model of level-3 is complex to implement, but it incorporated the effect of S_t in this level. The correlation results between the set-up parameter “A” and the selected soil properties (Fig. 4) such as S_u, PI, S_t, c_v were used to develop the analytical models. All possible regressions procedures were examined to select the best subset of predictor. R-Square, adjusted R-Square, sum of square error (SSE) and mean square error (MSE) were used as criteria to assess best predictors. Once preliminary models were selected, detail statistical analysis such as significance of the model as whole (F test) and significance of the partial multiple regression coefficient (t test) was carried out. The following three analytical models were finally selected amongst all models after examining all of the statistical analyses.

$$\text{Level-1}: \ \mathrm{A} = \mathrm{f}\ (S_u, PI) = \frac{0.79 * \left(\frac{\mathrm{PI}}{100}\right) + 0.49}{\left(\frac{\mathrm{S_u}}{1\ \mathrm{tsf}}\right)^{2.03} + 2.27} \tag{3}$$

$$\text{Level-2}: \mathrm{A} = \mathrm{f}\ (S_u, PI, c_v) = \frac{1.12 * \left(\frac{\mathrm{PI}}{100}\right) + 0.69}{\left[\left(\frac{\mathrm{S_u}}{1\,\mathrm{tsf}}\right)1.44\right] * \left[\log\left(\frac{\mathrm{C_v}}{0.01\frac{\mathrm{in}^2}{\mathrm{hour}}}\right)\right]^{0.54} + 3.19} \tag{4}$$

$$\text{Level-3}: \mathrm{A} = \mathrm{f}\ (S_u, PI, c_v, S_t) = \frac{0.44 * \left(\frac{\mathrm{PI}}{100}\right)(\mathrm{S_t}) + 2.20}{\left[\left(\frac{\mathrm{S_u}}{1\,\mathrm{tsf}}\right)^{1.94}\right] * \left[\log\left(\frac{\mathrm{C_v}}{0.01\frac{\mathrm{in}^2}{\mathrm{hour}}}\right)\right]^{1.06} + 10.65} \tag{5}$$

Figures 5a, b and c present the comparison of measured versus predicted set-up parameter "A" of the 70 individual clayey soil layers. The models that had been developed to predict the set-up parameter "A" from soil properties for three different levels need to be incorporated in Eq. 2 in order to predict the set-up for f_s as:

$$\frac{f_s}{f_{so}} = 1 + \frac{0.79 * \left(\frac{PI}{100}\right) + 0.49}{\left(\frac{S_u}{1\,tsf}\right)^{2.03} + 2.27} \log \frac{t}{t_o} \tag{6}$$

$$\frac{f_s}{f_{so}} = 1 + \frac{1.12 * \left(\frac{PI}{100}\right) + 0.69}{\left[\left(\frac{S_u}{1\,tsf}\right)1.44\right] * \left[\log\left(\frac{C_v}{0.01\frac{in^2}{hour}}\right)\right]^{0.54} + 3.19} \log \frac{t}{t_o} \tag{7}$$

$$\frac{f_s}{f_{so}} = 1 + \frac{0.44 * \left(\frac{PI}{100}\right)(S_t) + 2.20}{\left[\left(\frac{S_u}{1\,tsf}\right)1.94\right] * \left[\log\left(\frac{C_v}{0.01\frac{in^2}{hour}}\right)\right]^{1.06} + 10.65} \log \frac{t}{t_o} \tag{8}$$

Where, t_o = 1 day and f_{so} = unit skin friction at 1 day restrike for individual soil layer.

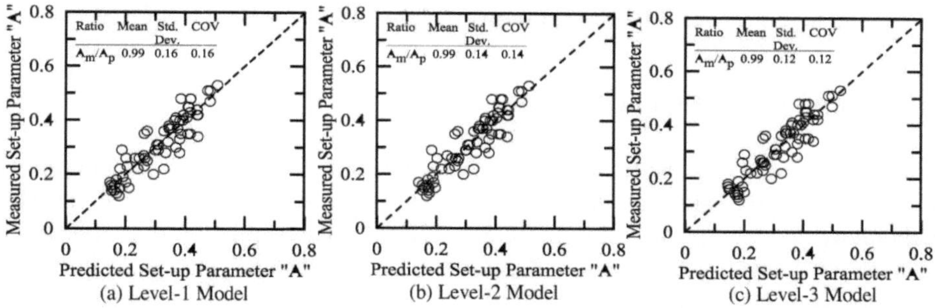

Fig. 5. Comparison of measured versus predicted "A" for development of models

Equations 6, 7 and 8 can be implemented to estimate the increase of f_s with time of individual clayey soil layer after EOD. The f_s value will be multiplied with the contact area of that layer to calculate the skin friction (R_{si}) of that layer. In the absence of sandy soil layers, the skin friction of all clayey soil layers along the pile length can be added to evaluate the total skin friction of the piles, R_s of the pile. A constant value of A = 0.15 (*Average value of "A" parameter of all sandy soil layers in this study*) is proposed here to estimate the f_s value for the sandy soil layers and hence to calculate the skin friction of the sandy soil layers. Since no set-up was observed for the end-bearing capacity (R_{tip}) in this study as well as reported in the literature (e.g., Ng et al. 2013), no set-up is considered in calculating R_{tip} in the proposed model to estimate the R_t.

5.9 Model Verification

Other available pile set-up data from LADOTD were analyzed here to verify the developed analytical set-up models in Eqs. 6, 7 and 8. The three developed models in Eqs. 6, 7 and 8 were used to predict the “A” for the individual soil layers of these 18 test piles, followed by calculating the R_t using the methodology described earlier. Figure 6 presents the comparison between the measured and predicted set-up parameter “A” for verification of these 18 test piles.

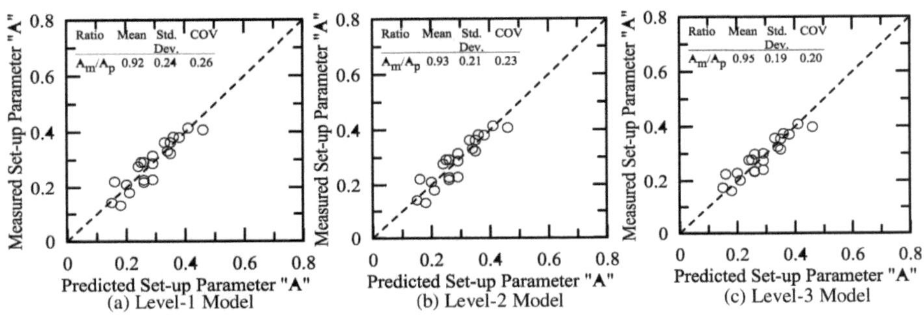

Fig. 6. Comparison of measured versus predicted “A” for verification of models

6 Summary and Conclusions

Pile set-up study was conducted on 12 instrumented test piles of five different projects in mainly cohesive soils with the presence of interlayers of sand and silt in Louisiana. Laboratory and in-situ soil testing were conducted at the test pile location in order to characterize the subsurface soil profile. Based on field measurements of load tests on PSC driven piles and the statistical regression analyses, the following conclusions can be drawn:

1. The total capacities measured by the static and dynamic load tests demonstrated that the set-up behavior follows a linear logarithmic rate of time after EOD. The end-bearing capacity was almost constant, and the majority of set-up was mainly attributed to increase in skin friction.
2. The CAPWAP analyses from the DLTs and the load-distribution plots from the SLTs were used to calculate the skin friction capacity for individual soil layers along the piles. Almost all the clayey soil layers exhibited significant amount of set-up compared to sandy soil layers.

3. The logarithmic rate of set-up parameter "A" was back-calculated for individual soil layers. A total 94 pile segments were considered in this study with the clayey soil behavior was dominant in 70 soil layers. The corresponding average values of the set-up rate "A" for clayey and sandy soil layers were 0.31 and 0.15, respectively, for this study.
4. The magnitude and rate of set-up "A" exhibited a good correlation with the different soil properties. The undrained shear strength (S_u), plasticity index (*PI*), coefficient of consolidation (c_v *or* c_h) and sensitivity (S_t) have shown significant influence on the set-up parameter "A". The set-up parameter "A" decreases with increasing S_u and c_v, and increases with increasing PI and S_t.
5. Three different multivariable non-linear regression models were developed to estimate the "A" parameter and the increase of unit skin friction capacity (f_s) with time for the clayey soil layers. The models incorporate different soil properties in three different levels with similar implementation procedure. The comparison between measured and predicted "A" parameter and "f_s" value are in good agreement.

Acknowledgements. This research is funded by the Louisiana Transportation Research Center (LTRC Project No. 11-2GT) and LADOTD (State Project No. 736-99-1732).

References

Abu-Farsakh, M.Y., Haque, M.N., Tsai, C.: A full scale field study for performance evaluation of axially loaded large diameter cylinder with pipe piles and PSC piles. Acta Geotech. (2016). doi:10.1007/s11440-016-0498-9

Bogard, J.D., Matlock, H.: In-situ pile segment model experiments at Empire, Louisiana. In: Proceedings of the Conference of Offshore Technology, Houston, TX, May 7–10, pp. 459–467 (1990)

Bullock, P.J., Schmertmann, J.H., McVay, M.C., Townsend, F.C.: Side shear setup. I: Test piles driven in Florida. J. Geotech. Geoenviron. Engng. **131**(3), 292–300 (2005)

Chen, Q., Haque, MdN, Abu-Farsakh, M., Fernandez, B.A.: Field investigation of pile setup in mixed soil. Geotech. Test. J. **37**(2), 268–281 (2014)

Davisson, M.T.: High capacity piles. In: Proceedings of the Soil Mechanics Lecture Series on Innovations in Foundation Construction, ASCE, Reston, VA, pp. 81–112 (1972)

Fellenius, B.H.: Effective stress analysis and set-up for shaft capacity of piles in clay. From Research to Practice in Geotechnical Engineering, Canada, pp. 384–406 (2008)

Haque, MdN, Abu-Farsakh, M., Chen, Q., Zhang, Z.: A case study on instrumenting and testing full-scale test piles for evaluating set-up phenomenon. J. Transp. Res. Rec. **2462**, 37–47 (2014)

Haque, MdN, Abu-Farsakh, M.Y., Zhang, Z., Okeil, A.: Developing a model to estimate pile set-up for individual soil layers on the basis of PCPT data. J. Transp. Res. Board **2579**, 17–31 (2016a)

Haque, M.N., Abu-Farsakh, M.Y., Tsai, C., Zhang, Z.: Load testing program to evaluate pile set-up behavior for individual soil layers and correlation of set-up with soil properties. J. Geotech. Geoenviron. Eng. (ASCE) (2016b). doi:10.1061/(ASCE)GT.1943-5606.0001617

Karlsrud, K., Clausen, C.J.F., Aas, P.M.: Bearing capacity of driven piles in clay, the NGI approach. In: Proceedings of the 1st International Symposium on Frontiers in Offshore Geotechnics, Perth, Sep 19–21, pp. 775–782 (2005)
Karlsrud, K.: Ultimate shaft friction and load displacement response of axially loaded piles in clay based on instrumented pile tests. J. Geotech. Geoenviron. Engng. **140**(2), 481–495 (2014)
Komurka, V.E., Wagner, A.B., Edil, T.B.: Estimating soil/pile set-up. Wisconsin Highway Research Program #0092-00-14, Wisconsin Department of Transportation (2003)
Long, J.H., Kerrigan, J.A., Wysockey, M.H.: Measured time effects for axial capacity of driven piling. J. Transp. Res. **1663**, 57–63 (1999)
McVay, M.C., Schertmann, J., Townsend, F., Bullock, P.: Pile friction freeze: a field investigation study. Research Report No. WPI 0510632, Vol. 1, Report submitted to Florida Department of Transportation (1999)
Mesri, G., Smadi, M.M.: Discussion: time-related increases in the shaft capacities of driven piles in sand. Geotechnique **51**(5), 475–476 (2001)
Ng, K.W., Suleiman, T.M., Sritharan, S.: Pile setup in cohesive soil. II: analytical quantifications and design recommendations. J. Geotech. Geoenviron. Eng. **139**(2), 210–222 (2013)
Paikowsky, S.G., Hajduk, E.L., Hart, L.J.: Comparison between Model and Full Scale Time Dependent Pile Capacity Gain in the Boston Area. ASCE Geotechnical Special Publication 132 Advances in Deep Foundations, pp. 1–16 (2005)
Rausche, M.F., Robinson, B., Likins, G.: On the prediction of long term pile capacity from End-Of-Driving information. In: Proceedings of the Current Practices and Future Trends in Deep Foundation, Los Angeles, CA, July 27–31, pp. 77–95 (2004)
Schmertmann, J.H.: The mechanical aging of soils. J. Geotech. Eng. **117**(9), 1288–1330 (1991)
Svinkin, M.R., Skov, R.: Set-up effect of cohesive soils in pile capacity. In: Proceedings of the 6th International Conference on Application of Stress Waves to Piles, S. Niyama and J. Beim, (Ed.), Sao Paulo, Brazil, Sep 11–13, pp. 107–111 (2000)
Skov, R., Denver, H.: Time-dependence of bearing capacity of piles. In: Fellenius, B.G. (ed.) Proceedings of the 3rd international conference on the application of stress-wave theory to piles, pp. 879–888. BiTech Publishers, Canada (1988)
Wang, Y.H., Gao, Y.: Mechanisms of Aging-Induced modulus changes in sand with inherent fabric anisotropy. J. Geotech. Geoenviron. Eng. **139**(9), 1590–1603 (2013)
Wang, J.X., Verma, N., Steward, E.J.: Estimating Pile Setup using 24-h restrike resistance and computed static capacity for PPC piles driven in soft Louisiana Coastal deposits. Geotech. Geol. Eng. **1**, 1–17 (2015)

The Effect of the L/D Ratio of Pile Group Using the Equivalent Pier Method Including Interaction

Pallavi Badry(✉) and D. Neelima Satyam

Geotechnical Engineering Laboratory,
International Institute of Information Technology, Hyderabad, India
pallavi.ravishankar@research.iiit.ac.in, neelima.satyam@iiit.ac.in

Abstract. The behavior of the pile under the load is governed number of piles in the group, pile spacing, pile length, pile diameter and type of the soil and its interaction with pile. In order to implement the interaction effect, the interfaces between soil and pile is needed to model which found to be very tedious, complex and numerically costly. To overcome this complication, the attempt is made to provide an easy to model the interaction problem. In this study, the equivalent pier method has been introduced to model the pile soil structure interaction system, where the pile group has been replaced with the single equivalent pier. In this study the L-shape asymmetrical pile group is modeled with a direct method using finite element procedure. The C++ program is developed using finite element procedures to achieve the dynamic analysis of the soil-pile structure interaction system. The SSI system is modeled for different L/D ratio including 10, 20, 30, 40 and 50. The responses of the system are obtained from the different L/D ratio of the pile group by carrying out the dynamic nonlinear soil structure interaction analysis of the 2001 Bhuj ground motion (M = 7.0). The responses of the SSI system have been checked for the reduced EPM configuration for the one of the mentioned L/D ratios. The results in terms of displacement and numerical statistics have been compared to the general pile layout (i.e. L/D = 20). The study concludes that the reduced EPM model offers ease to model complicated SSI system. Also, it has been observed that the EPM model has an acceptable accuracy in the responses and also numerically efficient with interaction effect under the dynamic loading condition.

1 Introduction

The pile behavior depends on the characteristic length of the pile (L) its cross section (Diameter D) (Basack 2009). Considering the length and depth ratio the pile is divided into two categories which alters its behavior under lateral load. When ration of the length of the pile and diameter exceeds 30 (L/D > 30), the pile is categorized as long pile. Whereas this ration limits 20 (L/D < 20), the pile is classified as short pile. Lateral loads and moments as an effect of wind, earthquakes and tides may act on piles in addition to the axial loads. Failure of a short rigid pile occurs when the lateral resistance of the soil has been exceeded. And supported structures have to withstand lateral loads and

M. Abu-Farsakh et al. (eds.), *Advances in Analysis and Design of Deep Foundations*, Sustainable Civil Infrastructures, DOI 10.1007/978-3-319-61642-1_13

overturning moments, but relatively small vertical forces which are often neglected. The behavior of the pile is governed by the interaction between the pile and the surrounding soil and precisely can be captured by the nonlinear 3-D soil structure interaction problem. The rigid pile behavior is explicitly captured by Roger (2006) and investigated that the short rigid piles experiences about a 58% more deflection surrounded by loose sands than the dry the state to submerged state (Chaudhary 2007). Thus the interaction between soil and pile is the crucial aspects which depends upon the type of soil and force transfer mechanism of soil to the pile. In case of a long, flexible pile the failure is associated with the moment at one or more points exceeds the moment of resistance and the failure takes place by the formation of one or two plastic hinges along the pile length. Various experiments conducted to study ultimate lateral bearing capacity of piles with the increase of soil stiffness the ultimate lateral resistance of a pile increases. Pile diameter would not have a significant effect on ultimate bearing capacity.

The effect of the piles on each other in pile group is very complicated and important. Therefore, it is necessary to estimate the effect of pile group's character on their behavior before design. In order to study the effects of piles on each other in pile group L/D ratio is used mostly instead of only diameter or only the length of the pile individually (Mohamad 2013). The suitable range of L/D ratio can be considered from 10 to 50 to capture the maximum variation in the response of the superstructure (Basack 2009).

2 Finite Element Analysis

In dynamic analysis, the total interaction response is the combination of the two parts namely kinematic and inertial interaction. Wolf (1985) has given an understandable shape to the SSI analysis by giving the detailed numerical methods. The soil structure interaction problem can be analyzed using the direct method and substructure methods.

In the present study the finite element method (FEM) has been incorporated by developing a program in C++, with object oriented methodology to analyze the interaction effect for pile supported buildings subjected to the transient loading condition.

The generalized equation of motion is used to get the response of the SSI system. When the system is subjected to the earthquake the spectral acceleration, combined with the system mass has been taken as external forces. Equation 1 shows the various components of the SSI system when it is subjected to the ground motion. Refer for derivation of the Eq. 1.

$$[M]\{\ddot{U}\} + [C]\{\dot{U}\} + [K]\{U\} = -[M]\{\ddot{U}_g\} \tag{1}$$

where, $[M]$, $[C]$ and $[K]$ are the mass, damping and stiffness matrices of the integrated system which includes the structure and foundation system.

$\{\ddot{U}\}$, $\{\dot{U}\}$ and $\{U\}$ are the acceleration, velocity and displacement of the system and Üg is the ground motion acceleration. [*C*], a damping matrix and can be given by Rayleigh damping given as $[C] = \alpha[M] + \beta[K]$, where coefficients α and β are Rayleigh damping coefficients. Rayleigh damping is more appropriate for transient dynamic analysis. The behavior of the material under the application of the load can be well understood by observing the stress strain curve. The behavior of soil material is very complex. It is often necessary to develop simple mathematical constitutive laws for practical applications. Several models are available in the literature deals with the constitutive behavior of granular and frictional material like Drucker-Prager, Mohr-coulomb, Duncan and Chan, Cam Clay, etc. In the present study, Draucker-Prager nonlinear material model is used for soil. The behavior of all the structural elements, including beams, columns, raft and piles are assumed to be linear elastic under the applied loads.

The nonlinear analysis has been carried out by using Initial Stiffness Method is adopted because its simplicity in the implementation. The Initial Stiffness Method advances the solution to the next load (time) step by satisfying the global equilibrium through iterations.

3 Validation of the Program

3.1 Geometry of the Model

In order to validate the finite element program developed to perform the dynamic nonlinear soil structure interaction analysis for pile supported building. In order to check the accuracy of the program developed in the present study, the 45.0 m tall square building supported by the pile foundation system (Hokmabadi et al. 2014) is model and the results are validated with the existing results.

In the study carried out by Hokmabadi et al. (2014) the prototype structure model on the shaking table, the model structure has been designed employing SAP2000 software considering the required characteristics of the model structure. The 3D numerical model consists of fifteen horizontal steel plates as the floors and four vertical steel plates as the columns. Steel plate grade 250, according to Australian standards, with the minimum yield stress of 280 MPa and the minimum tensile strength of 410 MPa, has been adopted in the design. The thickness of the steel plates has been determined in during the design process after several cycles of trial and error in order to fit the required natural frequency and mass of the model structure. The finalized base plate is a 500 × 500 × 10 mm steel plate while the floors consist of 400 × 400 × 5 mm plates and four 500 × 40 × 2 mm steel plates are used for the columns. The connections between the columns and floors are provided using stainless steel metal screws with 2.5 mm diameter and 15 mm length.

Similar to the model structure, the model pile should be subjected to the competing scale model criteria. In order to achieve a successful model pile design, the principal governing factors of pile response such as slenderness ratio L/d, moment curvature relationship, the flexural stiffness EI, relative soil/pile stiffness, yielding behavior/mechanism, and natural frequency of vibration should be adopted.

The details of the model including building geometry and the piles have been explained in the Fig. 1.

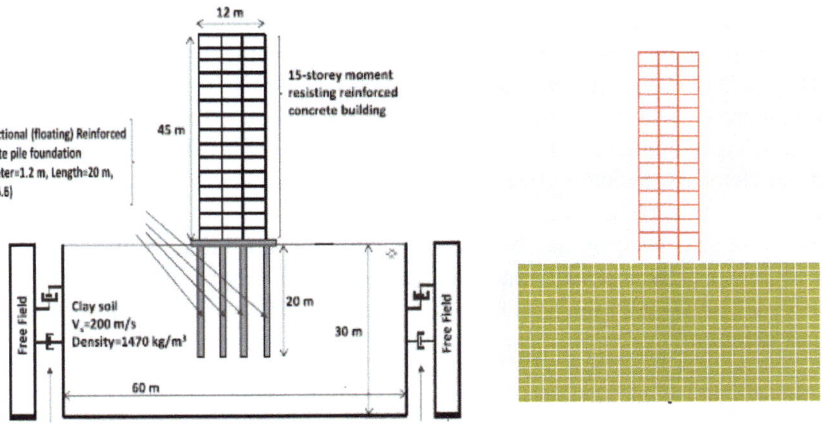

a. Model suggested by Hokmabadi (2014) b. Model developed in present study

Fig. 1. Details of the Finite Element model for validation.

3.2 Result Validation

The model is subjected to the El Centro earthquake which is applied at the bottom of the SSI model and the responses at each storey of the superstructure has been studied and compared with the existing research. The details of the earthquake used are provided in Table 1

Table 1. Earthquake details used for validation of the present study.

Earthquake	Country	Year	PGA (g)	Mw (R)	Duration	Type	Hypocentre Distance (km)
El Centro	USA	1940	0.349	6.9	56.5	Far Field	15.69

The dynamic nonlinear analysis has been carried out for the 45.0 m symmetrical building supported by the piles with the aspect ratio 16.6. The responses of the superstructure observed in the present study has been compared with the numerical model and the experimental prototype responses developed by the Hokmabadi et al. (2014).

The time history analysis responses for each storey including the interaction effect have been observed. Figure 2 shows the time history of the displacement observed at bottom and top of the superstructure.

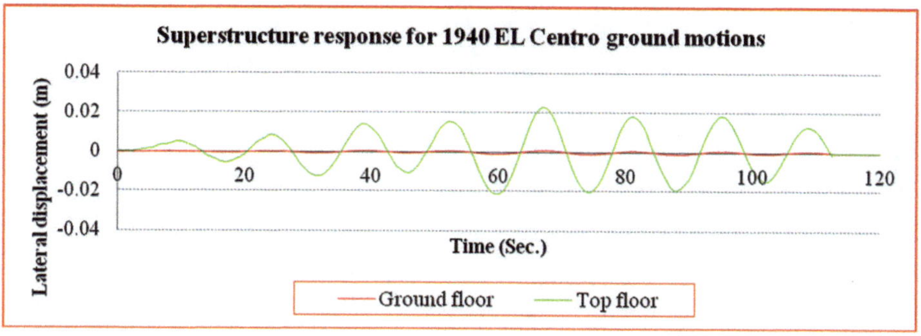

Fig. 2. Time history displacement of the superstructure at top and bottom location.

The responses obtained from the nonlinear dynamic analysis carried out in the present study have been compared with the experimental and the numerical model developed by Hokmabadi et al. (2014) for SSI effect study. The Fig. 3 shows the comparative responses at the different storey obtained from the present study and the experimental model.

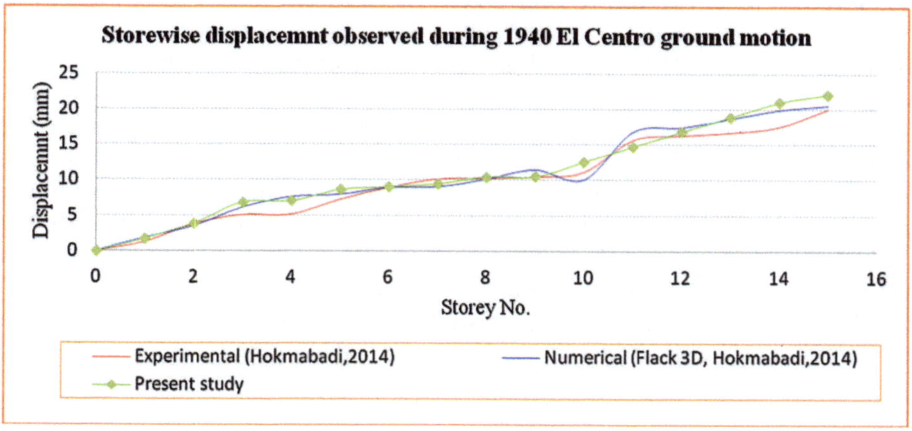

Fig. 3. Validation of Storeywise displacement observed during 1940 El Centro ground motion.

The story drifts also compared to understand the performance of the superstructure in terms of safety with the experimental and the numerical study (Fig. 4).

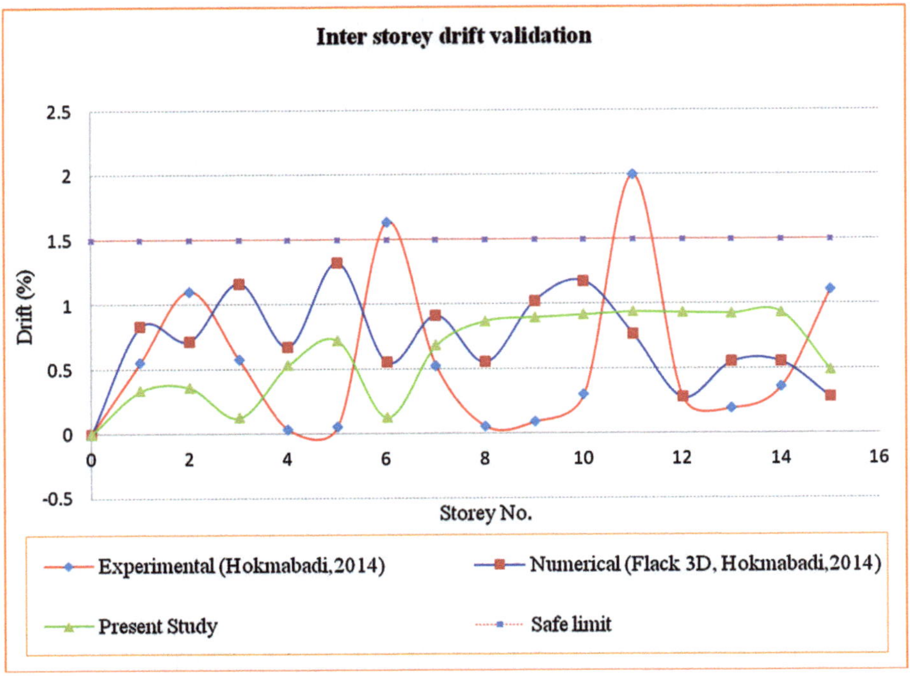

Fig. 4. Validation of Inter storey drift during 1940 El Centro ground motion.

4 Theory of Equivalent Pier Method (EPM)

Poulos and Davis (1980) proposed an equivalent pier method for heavy and large superstructures where a large pile group needs to analyze. Rajashekhar Swamy et al. (2011) adopted this method to find out the settlement analysis for the huge pile group without considering the superstructure and its interaction. In this method the pile groups as a whole pier to simplify the procedure for estimating the settlement of pile groups which equals that of single pile by means of load transfer functions. In this method, the pile group is replaced by a pier of similar length to the piles in the group and with an equivalent diameter (Deq.), estimated as follows (Poulos 1993).

The diameter of the equivalent pier is given by the following equation

$$D_{eq} = 2\sqrt{\frac{A_g}{\pi}} \quad \text{or} \quad 1.13 \ \text{to} \ 1.27\sqrt{A_g} \tag{2}$$

where, A_g plan area of pile group, including the soil between the piles.

The lower value in Eq. 2 is more relevant to predominantly end bearing piles, while the larger value is more applicable to predominantly friction or floating piles. Equivalent pier includes the soil entrapped in the pile spacing it is needed to modify the

Young's modulus in the analysis. The Young's modulus of the equivalent pier is given by the following formula

$$E_{eq} = \frac{(E_P - E_S)A_{np}}{A_g} + E_S \tag{3}$$

where E_p is the Young's modulus of the pile,

E_s is the Young's modulus of the soil penetrated by the piles

A_{np} is the total cross sectional area of the piles in a group

A_g is the plan area of pile group, including the soil between the piles.

Poulos (1993) and Randolph (1994) have examined the accuracy of the equivalent pier method for predicting group settlements, and have concluded that it gives good results. Poulos (1993) has examined group settlement as a function of the number of piles, for a group of end bearing piles. Thus, the applicability of EPM has been validated for the symmetric pile group, but there is no attempt has been made for the asymmetrical pile group (Fig. 5).

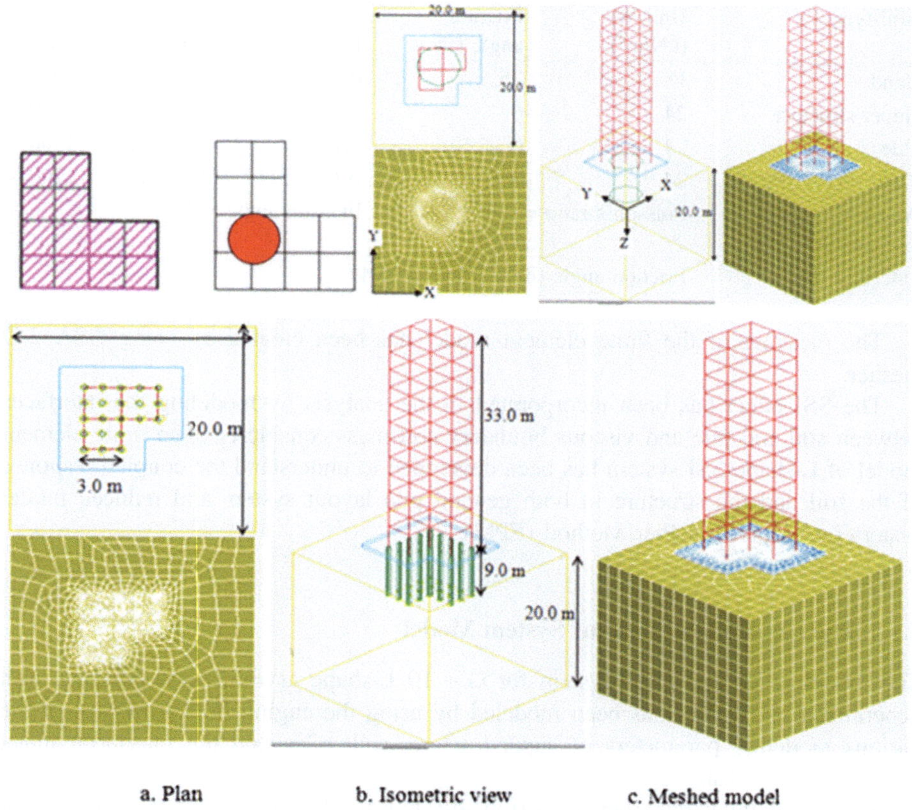

Fig. 5. Finite element model for different EPM configuration

5 Numerical Modeling

In this study, the finite element program has been developed in C++ to understand the numerical formulation of each aspect of SSI modeling, including element formulation, interface implementation, soil nonlinear solvers and connections between soil, foundation and superstructure. The 11 storey L-shape building has been modeled by the program developed in this study and the building, views have been created by sending the output file to the LS-PP freeware tool.

The G + 10 superstructure components, including beams and column have been modeled with 2 noded 3-D beam elements. The joints between beam and column are considered to be rigid. The connection between the raft and first storey column is modeled as the rigid connections. The half space of size 20 × 20 × 20 m is modeled using as sandy silt and the engineering properties of the soil domain has been explained in detail in Table 2. The nonlinear behavior of the supporting soil is captured using associative Drucker–Prager material model.

Table 2. Engineering properties of soil and structure considered

Soil type	Unit Wt. (kN/m^3)	Friction angle (°)	Poisson's Ratio	E (kN/m^2)	Vs (m/s)
Sand	18	35	0.35	445,872	300
Super structure	24	0	0.15	2.0×10^7	1200
Pile	24	0	0.15	2.0×10^7	1200
Raft	24	0	0.15	2.0×10^7	1200
Material model parameters	Poisson's ratio = 0.35		Friction angle = 35°		
Interface data	Friction angle (δ) = 1/3 ϕ′ = 11.4°				

The meshing of the finite element model has been created by using GSA 2-D mesher.

The SSI effect has been incorporated in the analysis by modeling the interfaces between soil and pile and viscous boundary soil mass considered. The finite element model of L-shape SSI system has been developed to understand the coupled response of the soil and the structure in both general pile layout system and reduced model system i.e. Equivalent Pier Method (EPM)

5.1 The General Pile Layout System Model

The modeling of the DSSI system for G + 10 L-shape asymmetrical building with generalized pile layout has been modeled by using the engineering properties of the various modeling parameters of superstructure, soil, piles and the interface/contact explained in the Table 2.

The 0.5 m thick raft with the 1.0 m offset from all the sides of the base of the superstructure have been modeled with the 3-D brick elements. The circular piles with 0.45 m and 9.0 m length have been modeled with the 3-D brick elements. The L-shape layout of piles accommodates the 23 piles spaced at 1.5 m c/c. The joints between the raft and pile have been modeled with the rigid connections.

5.2 Equivalent Pier Model (EPM)

In the present study the existing pile group is replaced by the equivalent pier with modified diameter and the modulus of elasticity (Table 3). The method is good enough for the symmetric pile group but need to extend its applicability in an asymmetrical pile group and the equivalent pier model has been developed. In this study, the attempt has been made to understand its applicability to the asymmetrical pile layout.

Table 3. Details of an equivalent pier model adopted for numerical models.

Area no.	L (m)	B (m)	No. of Piles participating	A_g (m^2)	D_{eq} (m)	Es (kN/m^2)	Ep (kN/m^2)	E_{eq} (kN/m^2)	Location (x,y) (m,m)
1	6	3	21	27	5.8	445872	2.93×10^7	3.67×10^6	(2.5,2.5)

6 Seismic Analysis of SSI System

Both the model general pile layouts and reduced EPM model have been analyzed for static and dynamic loading conditions layouts. Initially the SSI system is analyzed for static load in order to get the initial stress condition which includes the self weight of the superstructure and the foundation system. The static analysis has been carried out by applying the fixed boundary condition in normal direction, i.e. constraining the displacements only in the normal direction to surface to the nodes of the extreme element of the soil volume considered.

The displacement so obtained at the end of static analysis has been considered as the initial response for the dynamic analysis. The 2001 Bhuj ground motion (PGA = 0.31G, E-W) has been applied at the bottom nodes of the soil domain and the analysis has been carried out for the peak response which lies in the 15 s (Fig. 6).

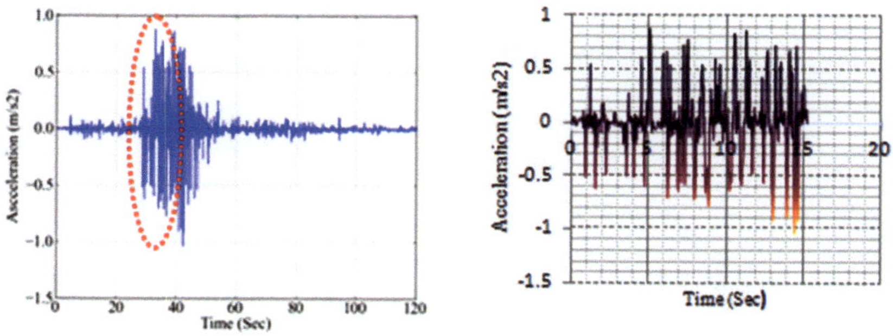

a. Bhuj ground motion b. Part of ground motion considered for study

Fig. 6. Bhuj ground motion and Part of ground motion considered for study

6.1 Different L/D Ratios Considered

In the present study various L/D ratios including 10, 20, 30, 40 and 50 are considered to understand its effect on the superstructure response under dynamic condition. The details of the combination of pile diameter and the pile length of the general pile layout of the L shape building have been given in the Table 4. The effect of the different L/D ratio of the foundation system has been noted by estimating the critical relative stiffness ($Kcr = EpIp/EsL^4$) of the pile in analytical solution, but in Finite element method the element stiffness of piles are taken care of the stiffness of the piles with the combination of its aspect ratio.

Table 4. Details of L/D ratio considered for the study.

Sr. no.	Pile Length (m)	Pile Dia. (m)	L/D Ratio
1	5	0.5	10
2	9	0.45	20
3	8	0.28	30
4	10	0.25	40
5	6	0.12	50

The effect of diameter and the length of the pile of the superstructure response have been studied by carrying out the dynamic soil structure interaction analysis for the different L/D ratio combination varying from 10 to 50 and analysis as been carried out for 2001 Bhuj (M = 7.7, 15 s duration) ground motion. In each case the spacing of the piles has maintained about 3.5D, accordingly the number of piles gets vary with the each asymmetrical pile group. The variation in the number, diameter and spacing contributes in the total pile stiffness which ultimately affects the superstructure response.

7 Results and Discussion

The exact impact of these parameters on superstructure response has been studied. Figure 7 shows the floor wise displacement experienced for different L/D ratios considered for analysis in the direction of applied ground motion for L shape building.

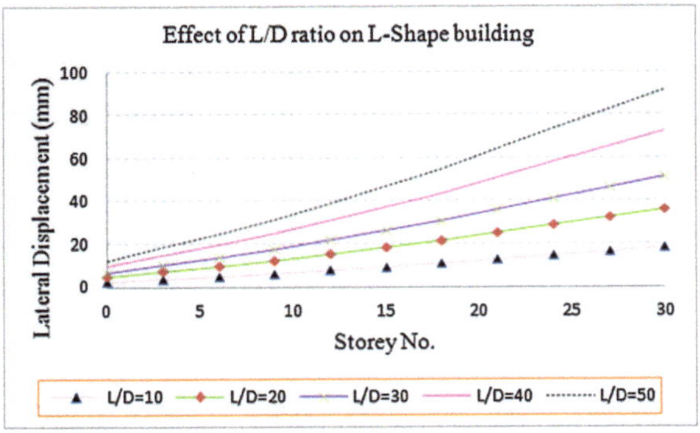

Fig. 7. Storey wise displacement at different L/D ratio

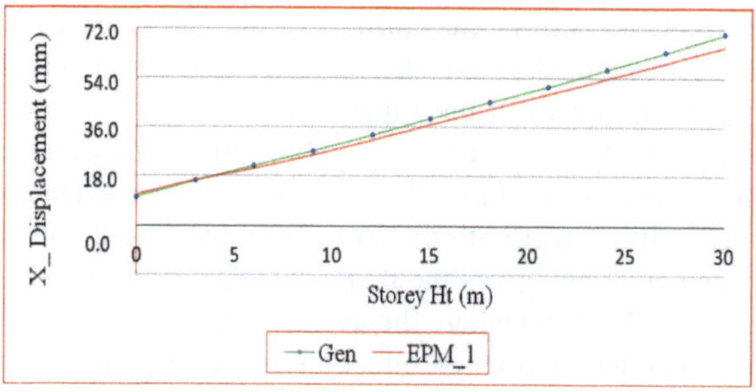

a. X-Direction response

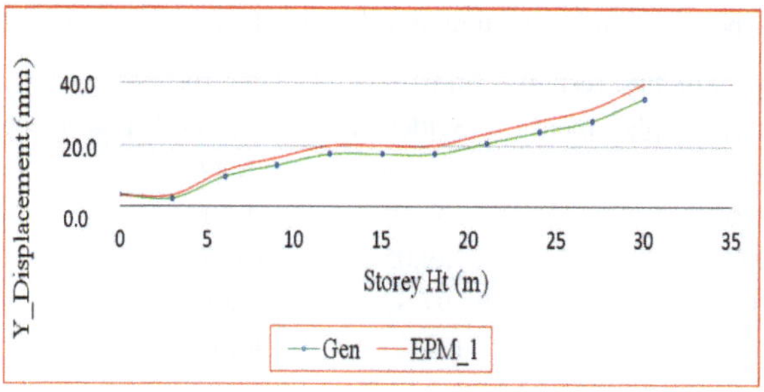

b. Y-Direction response

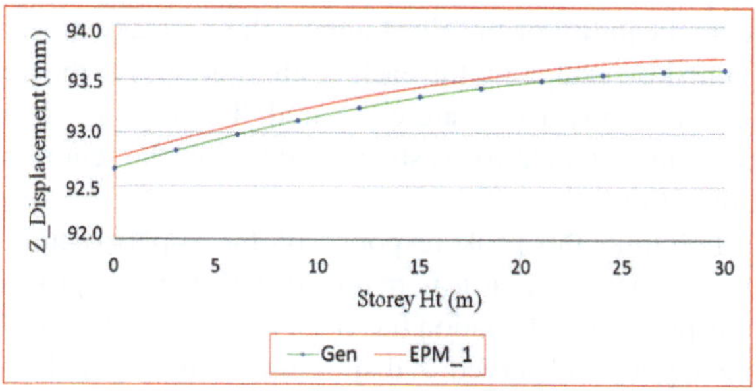

c. Z-Direction response

Fig. 8. Response comparison for various configurations in Z direction for C-Shape building

The displacement observed in lateral direction where the earthquake has been applied to the model (Fig. 8a, b).

Also the settlement in the pile which is located at the center of the geometry asymmetrical shape has been observed to understand the soil failure under seismic soil structure interaction analysis (Fig. 8c). The Table 4 shows the peak lateral displacement and settlement at superstructure and foundation system observed under dynamic analysis.

From Fig. 7 and Table 5 it has been found that the displacement values for the L/D ratio 10, 20 have, more closure than the L/D 30, 40, 50. The diameter of L/D ratio 10 and 20 is found to be 0.5 m average, while other L/D ratios show the average 0.2 m diameter. Thus the diameter is more the stiffness of the pile is found to more and offers comparatively less displacement of the structure. But for lesser pile diameter, though the length is more, the responses of the superstructure are found to 3 times more than the higher diameter case (Avg. Pile diameter = 0.45 m). Thus more area is in contact with the soil with respect to interfaces for this L/D ratio experiences more displacements. As more area is in contact with the soil contributes more contact displacements in the nodal displacements thus gives the displacements in higher sides than the other L/D ratio ranging from 30, 40, 50 where the average diameter and length is found to be 0.25 m and 8.0 m. Thus, the interaction effect alters the response of the superstructure for the different L/D configuration and the area of foundation in contact with the soil.

Table 5. Peak response of superstructure and the foundation system for different L/D ratio.

L/D Ratio	Superstructure response (mm)		Pile response at top (mm)	
	Lateral displacement	Settlement	Lateral displacement	Settlement
10.00	81.59	12.19	3.47	12.39
20.00	39.26	34.51	9.91	35.41
30.00	55.97	59.05	16.97	60.60
40.00	78.09	97.82	28.06	100.21
50.00	98.72	139.53	40.05	143.05

7.1 Applicability of EPM Approach

To understand the applicability of the EPM approach to the asymmetric pile group the displacements obtained by analyzing the each configuration, i.e. the general pile layout and EPM which consists only 1 pier are compared. The Fig. 8(a–c) shows the storey wise peak displacements obtained for L shape building in the tenure of earthquakes in each X, Y and Z directions.

In case of X direction, the peak response of the superstructure is found to be deviated with 4 to 8% (Avg.) when it is modeled with EPM approach.

In Y direction response, the deviation has been found with the EPM model at −16% than the general pile layout is observed. Z displacement shows the deviation −1 to −2% for EPM configurations which shows the system attains more stiffness when the EPM configuration has been model which leads to the underestimating the responses as compared to the general pile layout.

7.2 Numerical Statistics

The numerical expense of the general pile layout and EPM have been estimated from the finite element model. The numerical expense has been expressed in terms of DOFs, elements obtained after meshing the model, no. nodes and the least element size for each model. Table 5 shows the quantitative metric manifested with each configuration of the L shape building plan.

In SSI analysis foundation system modeling is very much complex as needed to provide an interface for the elements in contact. Thus, complexity increases with increasing the no. of such element. As per the conclusion draw in applicability of EPM for asymmetrical pile groups, including interaction effects is the applicable with acceptable deviation in the response. It has been observed that in EPM mechanism, it is needed to model only one single pier, thus modeling complexity reduced at the countable extent as the location of interfaces application can be reduced to the greater range.

It has been observed that the no. of contact nodes has been greatly reduced from 1900 to 300 (round off) which gives the measure of reducing the complexity and time reduction in iterating the contact displacements. This is found to be the countable advantage to EPM approach.

In EPM approach the no. of elements reduces to 9,000 (average) from 22,000 and DOF s reduces 27,000 from 72,000 (round off) which proves the numerical efficiency of the approach (Table 6).

Table 6. Quantitative metric for each EPM configuration.

Numerical attributes	Dof s	Elements	Nodes	Contact nodes	Least element	A Critical time step (s)	Analysis time (h)
General	72,792	22,695	24,264	1,974	0.11	5×10^{-5}	51.17
EPM	27,372	8,376	9,124	309	0.9	8×10^{-5}	15.35

In this study the solution has been obtained by the explicit solver where the stability of the solver depends upon the time step taken in the analysis, which in turn the function of the least element size in the finite element model.

In general pile layout the critical time step is needed to be taken as 5×10^{-5} s. corresponding to the 0.11 m element size of the pile (with pile dia 0.45 m) which is average least element size in models. When the dynamic load (duration 15 s.) applied to the system the solution obtained is in 51 h for the general pile configuration. But when the EPM model concerned the average element size is obtained as 0.9 m which allow to take the critical time step 8×10^{-5} s for analysis and gives the converged responses in the 15 h.

It has been observed that the time required to get the solution is reduced to 72% for EPM configuration than the general pile layout for L shape building layout, the EPM approach is satisfactory for the SSI problems where the numerical cost and CPU memory is required very high.

8 Conclusions

The following are the conclusions drawn from the present study.

1. Based on the experimental average values of maximum lateral deflections of pile supported building has been checked and it has been found that the top storey response of the present study and the experimental study is f story found to be nearly same with the average 6% deviation. Thus it is valid and qualified method of simulation with sufficient accuracy which can be employed for further numerical dynamic soil-structure interaction investigations.

2. The interaction effect alters the response of the superstructure for the different L/D configuration and the area of foundation in contact with the soil. The study noticed that the bigger diameter with the minimum length provides a good stability in the foundation system, but alters the displacement depending upon the asymmetry manifested with the superstructure.
3. This shows that the kinematic interaction is same for both the configuration, but when the earthquake wave reaches to the bottom of the superstructure due to its asymmetrical stiffness distribution, the deviation in the responses of the superstructure is found to be more for EPM model w.r.t. the general pile layout. Thus, it can be concluded that the torsion at the base plays an important role to respond the asymmetrical building under the applied dynamic loading.
4. EPM mechanism can be adapted well in huge SSI problems where the analysis time is one of the critical issues in getting the responses of the system. Thus, the study concludes that the EPM approach is numerically efficient with the acceptable accuracy including interaction effect.

References

Basack, B.: Influence of l/d and e/d ratios on the response of cyclically loaded single pile in uniform clay. Asian J. Civ. Eng. (Build. Hous.) **10**(3), 307–320 (2009)

Chaudhary, M.T.A.: FEM modelling of a large piled raft for settlement control in weak rock. Eng. Struct. **29**(11), 2901–2907 (2007)

Mohamad, F.N.: Effect of L/D ratio on pile group behaviour in different soil types using 3-D finite element method. Glob. J. Sci. Eng. Technol. **10**(2), 8–13 (2013)

Roger, F.: Some aspects of soil–structure interaction according to Eurocode 7 Geotechnical design. CERMES (Soil Mechanics Teaching and Research Centre, ENPC-LCPC) (2006). doi:10.1080/17499510902788843

Rajashekhar Swamy, H.M., Krishnamoorthy, Prabakhara, D.L., Bhavikatti, S.S.: Relevance of interface elements in soil structure interaction analysis of three dimensional and multiscale structure on raft foundation. Electron. J. Geotech. Eng. **16**, 199–218 (2011)

Poulos, H.G.: Settlement prediction for bored pile groups. In: van Impe, W.F. (ed.) Deep foundations on bored and auger piles. A.A. Balkema, Rotterdam (1993). doi:10.1061/40865(197)1

Poulos, H.G., Davis, E.H.: Pile Foundation Analysis and Design. Wiley, New York (1980). doi:10.1061/40865(197)1

Randolph, M.F.: Design methods for pile groups and piled rafts. In: Thirteenth International Conference on Soil Mechanics and Foundation Engineering, India (1994)

Hokmabadi, A.S., Fatahi, B., Samali, B.: Assessment of soil–pile–structure interaction influencing seismic response of mid-rise buildings sitting on floating pile foundations. Comput. Geotech. **55**, 172–186 (2014). doi:10.1016/j.compgeo.2013.08.011

Wolf, J.: Dynamic soil-structure interation (No. LCH-Book-2008-039), Prentice hall, Inc. (1985)

Deep Foundations Case Histories in the East Coast of United States

Raymond R. Mabnkbadi(✉) and Aravinda M. Ramakrishna

Hardesty & Hanover LLC, New York, NY, USA
{rmankbadi, aramakrishna}@hardesty-hanover.com

Abstract. The East Coast of the United States is the region that borders the North Atlantic Ocean. The East Coast is the most urbanized region of the United States, which creates numerous challenges in the construction of civil engineering works. These challenges often necessitate the use of deep foundations. This paper presents a discussion on the aspects related to design and construction of deep foundations unique to the East Coast of United States. Challenges such as limited accessibility; maintenance and protection of adjacent structures; and increased risk management are discussed with the aid of demonstrative case histories.

1 Introduction

The East Coast of the United States (herein referred as East Coast) is the most populated coastal area in the United States and accounts for approximately 36% of the country's total population (see Fig. 1). Cities, such as New York, Boston, and Washington, DC which drive country's economy are located in this region.

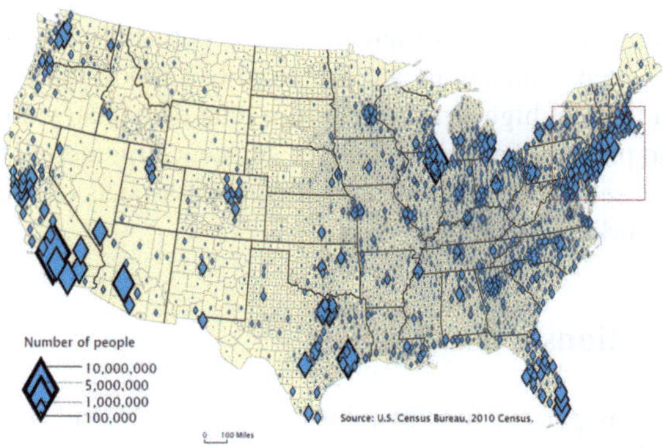

Fig. 1. Population Density (Source: U.S. Census Bureau)

The East Coast region is classified as having low seismic hazard, but the actual seismic risk is higher because of population density, concentration of buildings and economic importance. Recent studies suggest seismic hazards appear to be larger than previously considered.

M. Abu-Farsakh et al. (eds.), *Advances in Analysis and Design of Deep Foundations*, Sustainable Civil Infrastructures, DOI 10.1007/978-3-319-61642-1_14

The East Coast seaports from Eastport, Maine (ME), through Key West, Florida FL facilitate freight flow and international trade. All months of the year, the freights from these seaports travel under the bridges that cross the essential waterway.

Sinkholes and karst related geological hazards are also common in the eastern coast. The interface between the rock and soil in the karst terrain is highly irregular. Damage caused by sinkholes is the highest in Florida and Pennsylvania, due to various water-soluble rocks found in the East Coast (Fig. 2).

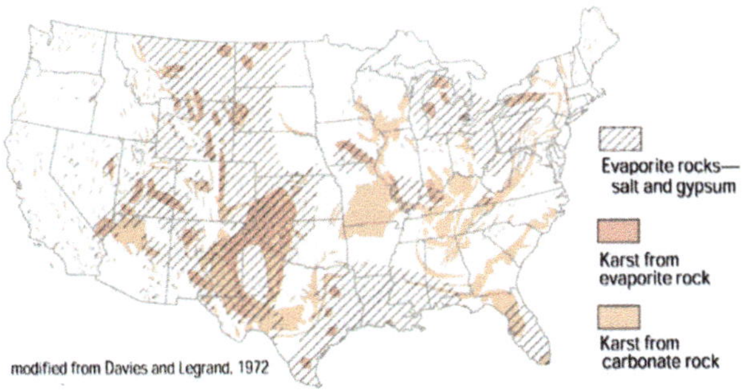

Fig. 2. Water-Soluble Rock Map (Source: U.S. Geological Survey)

As mentioned, population density, concentration of buildings and higher risk management associated with the East Coast makes the construction of civil engineering works in this region to be challenging. The design and construction of foundations for such civil engineering works are often complicated tasks due to these challenges. Furthermore, civil engineering structures are required to be designed for higher loads due to higher risk (vulnerability of people and property that are exposed to seismic hazards) in the event of earthquake and/or cargos collisions for structures crossing waterways. Hence, deep foundations are preferred to support civil engineering structures in this region to meet high structural design load demands while limiting the size of the foundation to prevent the undermining of adjacent structures. This paper presents case histories on the selection, design, and construction aspects of deep foundations by the authors while working in the East Coast.

2 Deep Foundations Case Studies

The following sections present four case histories of design and construction of deep foundations in the East Coast of United States.

2.1 The Willis Avenue Swing Bridge Replacement, New York

Introduction. The $612 million Willis Avenue Swing Bridge replacement project over the Harlem River (see Fig. 3) was the largest ever undertaken by the Movable Bridge

Fig. 3. Project Aerial View

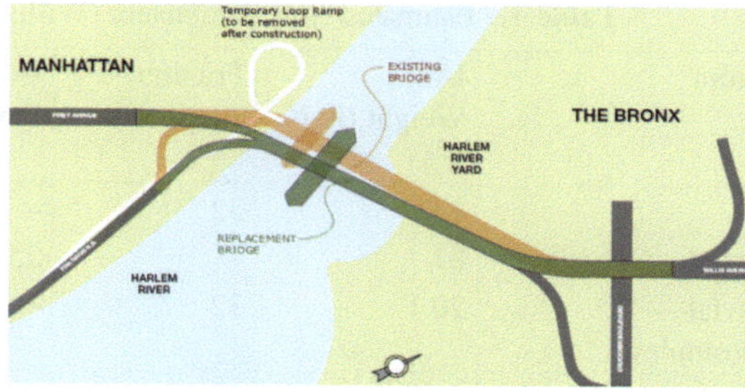

Fig. 4. Replacement Structure Alignment

Group of the New York City Department of Transportation (NYCDOT). This project involved the replacement of the 100-year-old Willis Avenue Swing Bridge with a new off-line swing bridge that incorporated a new 345-foot-long through-truss-type swing span and 3,000 feet of approach viaducts, including the ramps connecting FDR Drive and Bruckner Boulevard. The original Willis Avenue Bridge, constructed in 1901, exhibited the effects of age, weather, and wear due to the continual daily usage by motor vehicles. The NYCDOT asked that a new bridge be constructed adjacent to and just south of the existing bridge (see Fig. 4). Thus traffic could continue to use the old structure until the new bridge opened.

Subsurface Conditions. The subsurface conditions at the project site can be divided into four different strata: fill, alluvial deposits, glacial deposits, and bedrock. The fill consists of a heterogeneous mixture of sand, silt, clay, peat, gravel, and building material. Underlying the fill material is a layer of fluvial deposits. The fluvial deposits are subdivided into soft cohesive soils and medium dense cohesionless soils. Underlying the fluvial deposits is a layer of glacial deposit; the layer is approximately 70 feet. The upper portion of the glacial stratum consists of varved silt and clay and sand, whereas the lower portion of this glacial stratum is comprised of sand and gravel. The bedrock underlying this project site generally consisted of Inwood Calcitic and Dolomitic Marbles overlying Fordham Gneiss within the Harlem Lowland. The rock formations at the project site are separated by a major tectonic thrust fault referred to as Cameron's Line resulting in complex arrangement of faults and shear zones with low RQD. Table 1 summarizes estimated design parameters.

Foundation Design and Recommendation. This structure was defined as a critical bridge according to NYCDOT's Seismic Design Guidelines; therefore, the substructure design are designed for higher seismic risk. The substructures are designed such that bridge must not collapse during seismic event.

The replacement structure's new abutments on the Manhattan side of the project and for the retaining walls were founded on H-piles. H-piles were chosen because structural load demand was moderate and construction accessibility was not an issue. Closer to the waterfront and in water, substructure units consist of clusters of drilled shafts. This selection was influenced by the lateral deformation of the structure, which can be caused by scour event; or design earthquake levels. The substructure units consist of

Table 1. Estimated Soil Parameters (Willis Avenue Swing Bridge)

Stratum	Unit Weight (lb./ft^3)	Friction Angle (°)	Undrained Shear Strength (psf)	Compressive Strength (psi)
Fill	128	32	–	–
Alluvial—Cohesive	87	–	250–375	–
Alluvial—Cohesionless	20.1	32	–	–
Glacial—Varved	117	30	1000–2000	–
Glacial—Sand and Gravel	132	36	–	–
Bedrock—Faults/Shear zones	135	32	–	530–3000
Bedrock—No Faults/Shear zones	175	41	–	>7000

clusters of micro piles at both the Manhattan and the Bronx sides of the project, where working in limited overhead conditions was required. The new abutments at the Bronx side of the project are supported on spread footings which are founded on competent bearing stratum.

Presence of Cameron's Line created unique design challenges since majority of rock-quality designation (RQD) values were less than 25% and rock resistance estimation as given by O'Neill and Reese (1999) method, and adopted in the American Association of State Highway and Transportation Officials (AASHTO) design specification is sensitive to RQD values as shown below:

$$Skin\ Friction = 0.65 P_a \alpha_E \sqrt{\frac{q_u}{p_a}}$$

The empirical reduction factor α_E is determined as function of RQD. O'Neill and Reese (1999) applied an empirical reduction factor α_E to account for the degree of fracturing.

Theoretical estimation suggested rock socket length in excess of 20 ft, therefore, project specification called for Osterberg Cell (O-cell) load on four drilled shafts for rational design approach. O-cell testing was conducted by LOADTEST, Inc. in accordance with the procedure outlined in ASTM D1143 (Standard Test Method for Piles under Static Axial Load) for Quick Load Test Method. The O-cell test results indicated higher axial capacity than theoretical estimation as shown in Fig. 5. Based on the load test results production drilled shafts socket length was optimized.

2.2 The Sarah Mildred Long Bridge Replacement, Maine and New Hampshire

The $158.5 million Sarah Mildred Long (SML) Bridge replacement project is a lift bridge located on U.S Route 1 Bypass between Kittery, Maine, and Portsmouth, New

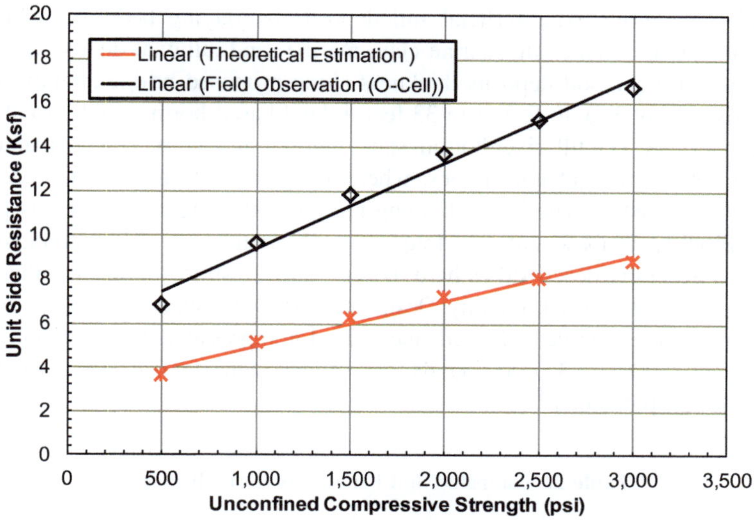

Fig. 5. Theoretical Estimation versus Field Observation

Hampshire. The existing Sarah Mildred Long Bridge is structurally deficient and its current posted load of 20 tons for highway traffic is obsolete. The project involves construction of a new 2,631-foot two-level bridge (road and rail) over the Piscataqua River with a major lift span system and 1,554-foot off approaches. The span above the navigation channel is a tower-driven vertical lift span providing navigation channel that is approximately 200 feet wide. The south end of replacement is proposed to begin immediately west of the current Portsmouth abutment. The bridge will arc as much as 300 feet upstream from the existing bridge before arcing back toward the existing alignment (See Fig. 6).

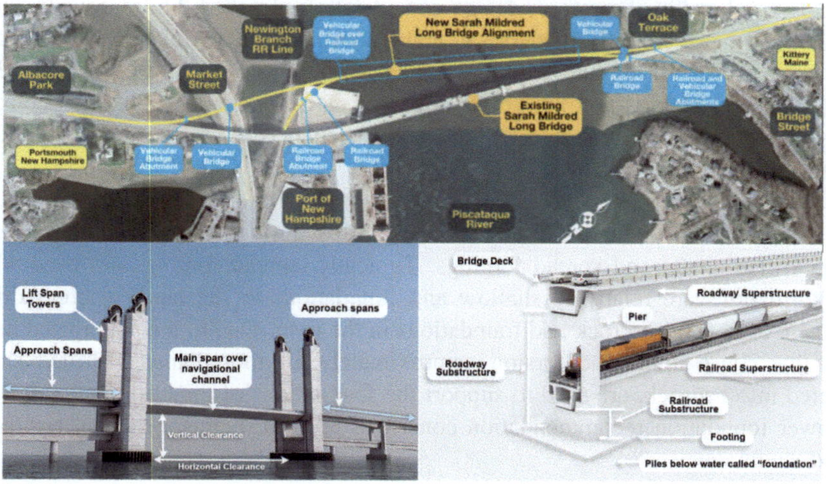

Fig. 6. Project Features

Subsurface Conditions. The surficial soil deposits overlaying bedrock includes fill, river bottom sand, organic silt, marine clay, marine sand, alluvial deposit and glacial till. The total surficial soil deposits thickness along proposed bridge alignment varies from approximately 3 to 40 feet: 3 to 33 feet in land-based borings and 2 to 40 feet in over-water borings. The fill, river bottom sand, marine sand, alluvial deposit and glacial till exhibit cohesionless characteristics, whereas organic silt and marine clay exhibit cohesive characteristics. The bedrock along the proposed alignment consists predominately of Phyllite. Diabase and (less frequently) Quartzite were also encountered. The Phyllite was generally described as hard to very hard, fresh to slightly weathered, fine grained and gray. The Rock Quality Designation (RQD) ranged from 0 to 100. The testing of rock core sample indicated unconfined compressive strength ranging from approximately 5,900 to 37,000 pounds per square inch (psi). Table 2 summarizes estimated design parameters.

Table 2. Estimated Soil Parameters (SML Bridge)

Stratum	Model	Unit Weight (lb./ft^3)	Soil Modulus (pci)	Friction Angle (°)	Unconfined Compressive Strength (psi)	Initial Modulus, E_{ir} (psi)
River Bottom Sand	Reese Sand	118	25	30	–	–
Alluvial Deposit/Glacial Till	Reese Sand	125	60	35	–	–
Bedrock	Strong Rock	171	–	38	8600	5000

Foundation Design and Recommendation. The SML Bridge Replacement footing elements and foundations are located within the river and subject to access by commercial vessel traffic. Therefore, design consideration included the improved channel orientation which currently hinders safe navigation and the vessel impact resisted by pier. A river user survey has determined the controlling vessel impact load as 12000 kips. It was determined that the vessel impact resisted by the pier was favorable; the design vessel size was large and an independent fender system would add significant cost to the project.

Two preferred foundation types were identified through the geotechnical design study: drilled shafts and spread footings. Typically, spread footings are used where rock formations are relatively shallow and foundations in the land, whereas drilled shafts are used for deep rock and foundations in the river. Due to the anticipated heavy loads that cause uplift and torsion, approximately 10 foot diameter drilled shafts, socketed into bedrock, are used to support the foundations of the bridge in water. The lift tower foundation recommendation consist of 8, 10-ft diameter heavily reinforced drilled shafts.

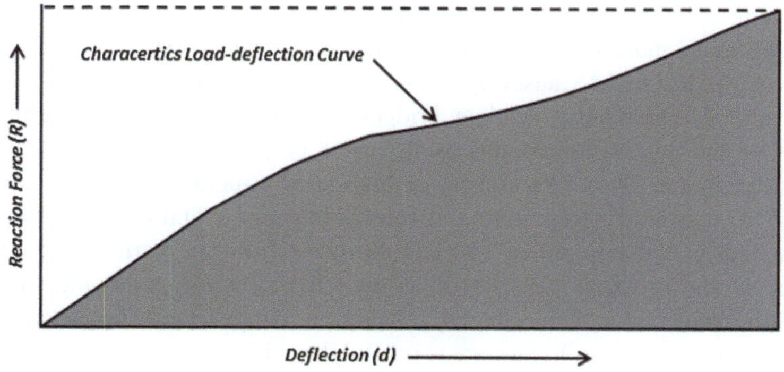

Fig. 7. Energy Absorption Capacity of Tower Foundation

Design included a detailed understanding of the dynamic relationship foundations play on movable structures thus, enabling the bridge to operate effectively and allowing easier movement of large vessels. Finite element program FB-MultiPier that incorporate soil/rock resistance using P-Y curves and Drilled Shaft foundation properties was used to generate a characteristic load-deflection curve as shown in Fig. 7. This curve is then used to determine a tower foundation system's potential energy absorption capacity to withstand impact from errant vessel.

2.3 Route 52 Causeway Replacement Project, New Jersey

Introduction. Route 52 Causeway, a $400 million design and reconstruction project, was one of the largest projects ever let out for bid by the New Jersey Department of Transportation (NJDOT). The old Route 52 Causeway was approximately a two-mile

Fig. 8. Aerial View

long crossing of the Great Egg Harbor Bay between the City of Somers Point on the New Jersey mainland and Ocean City on the barrier island (see Fig. 8). The reconstruction of the Route 52 Causeway included replacing the existing four structurally deficient and geometrically obsolete bridges and construction of the embankment roadways on the tidal marsh islands using staged construction. The crossings at Ship Channel and Beach Thorofare consist of high-level structures (vertical clearance of around 55 ft) that hold northbound and southbound traffic and are constructed on an alignment that is a maximum of 100-feet east/west from the current alignment. The remainder of the crossing consists of a low-level structure following the existing alignment but widened at the north end. In addition, ramps were constructed to link the causeway to the pedestrian access areas on Rainbow Island and the Visitor Center on Garrets Island which accommodate fishing and recreation.

Subsurface Conditions. In general, the stratigraphy may be divided into two layers consisting of loose to medium dense sand and silt and clay material on top of deeper medium dense to very dense Cohansey sand with interbedded clay lenses. Based on the Standard Penetration Test (SPT) results, the overlaying loose to medium dense sand and silt and clay material seemed unsuitable to withstand the load required from the new structures. The deeper non-cohesive soils consists of poorly graded fine and fine-to-medium sand, typically containing less than 5% fines. The relative density of deeper Cohansey sand generally ranged from medium-dense to very dense based on the field SPT data. Many of the field-measured SPT blow counts for this layer were recorded as exceeding 100 blows per foot. For design purposes, the following soil parameters were established based on field and laboratory test program results (Table 3).

Table 3. Estimated Soil Parameters (Route 52 Causeway)

Stratum	Unit weight (pcf)	Friction ϕ (°)	Cohesion (psf)	Soil Modulus (pci)	ε_{50} (%)	Shear Modulus G (ksi)
Sand	105–115	26–31	N/A	20–60	N/A	2.5–6.5
Silt and Clay	90–105	N/A	150–450	N/A	2	1.0–2.0
Cohansey Sand	118–130	33–40	N/A	60–125	N/A	11.5–30
Clay Lenses	110–120	N/A	500–1700	N/A–200	1–0.7	2.0–5.5

Foundation Design and Recommendations. Foundation design recommendation called for 30-inch. square prestressed concrete piles to support the piers and 24-inch. square prestressed concrete piles to support the abutments.

The new alignment required pile driving in proximity to the existing bridge in service. Therefore earth-borne vibrations generated during pile driving were assessed during design phase to mitigate structural damages to the adjacent structure in service.

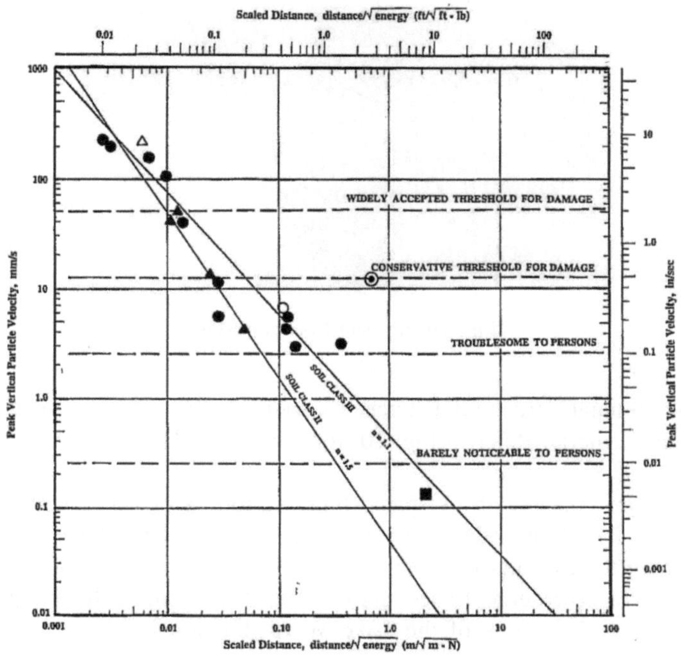

Fig. 9. Peak Vertical Particle Velocity versus Scale Distance

The AASHTO guidance suggest limiting pile driving induced earth-borne vibrations measured as Peak Particle Velocity (PPV) within 1.0 to 1.5 in./s to prevent structural damage to adjacent transportation facilities. The anticipated PPV from proposed pile driving operation was estimated using Woods and Jedele (1985). Woods and Jedele (1985) developed a chart that can predict anticipated PPV resulting from a source of known energy magnitude, distance and soil class (refer to Fig. 9). For a source of known energy magnitude, the PPV can be predicted at any distance by entering Fig. 9 with "scaled distance" information. PPV value can be estimated for the representative soil class through which the energy is being transmitted.

Using Fig. 9, the pile driving energy that induce PPV within 1.0 to 1.5 in./s was determined and project specific pile installation procedures were developed to mitigate vibration related structural damages by specifying the amount energy contractor should use while driving piles in proximity to the existing structure. All the new foundation piles were installed successfully without incident.

2.4 Flagler Memorial Bridge Replacement Project, Florida

Introduction. The project described herein is the replacement of the existing Flagler Memorial Bridge on SR A1A in Palm Beach County, Florida. The overall condition of the existing structure was poor and exhibits substantial deterioration. The replacement

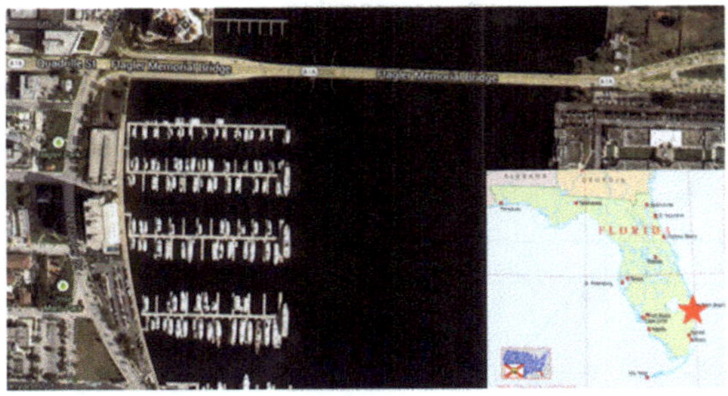

Fig. 10. Project Location Map

is designed to eliminate structural deficiencies, address substandard geometry, and meet bridge clearance criteria. Refer to Fig. 10, for the Project Location Map.

The new structure consists of 11 spans: four, 12-foot wide lanes with 8-foot wide shoulders and a 15.5-foot wide center median. Portions of the Lake Worth Lagoon were filled along the east and west approaches of the new bridge structure in order to meet bridge clearance criteria (12-feet above Mean High water). Retaining walls were be utilized along splash zone of the Lake Worth Lagoon in order to minimize the project footprint.

Subsurface Conditions. The subsurface conditions at the project site consist of loose to medium dense fine sand overlaying deeper loose to very dense fine sand with cementitious sand. The density of the deeper sand layer was not anticipated to improve with an increase in depth. A layer of dense to very dense limestone/coquina rock formation with soil seams found between the upper and deeper sand layer. Field observation indicate limestone/coquina rock primarily consist of carbonate minerals that are soluble in slightly acidic waters with high porosity, permeability and voids. The nature of this rock formation is anticipated to be unpredictable. The degree of rock decomposition in several borings was severe and; the rock is anticipated to behave similar to that of gravel. Table 4 summarizes estimated design parameters established based on the results of SPT and/or lab test data.

Table 4. Estimated Soil Parameters (Flagler Memorial Bridge)

Stratum	Unit Weight (pcf)	Friction Angle ϕ (°)	Soil Modulus (pci)	Unconfined Compressive Strength (psi))	Shear Modulus G (ksi)
Upper Fine Sand	105–120	26–33	20–60	N/A	2.5–6.5
Limestone/Coquina rock	135–145	35–45	N/A	315–3118	70–170
Deep Fine Sand	118–130	32–36	60–125	N/A	11.5–20

Foundation Design and Recommendation. Based on the erratic subsurface condition and anticipated design loads, a foundation system consisting of five foot diameter drilled shafts with redundant shaft layout was recommended to support the new structure.

The laboratory test results of rock cores reviewed for uniformity using histogram indicated significant site variabilities. Refer to Fig. 11 for details. The highly variable properties of Limestone/Coquina rock prompted drilled resistance estimation by relying on both compression strength and split tensile strength of rock cores following method proposed by Prof. McVay (McVay et al. 1992) with data reduction to obtain reliable side shear resistance as follows.

$$Skin = \left[\frac{1}{2} * \sqrt{q_u} * \sqrt{q_t}\right] * RECOVERY$$

where: q_u is the unconfined compression Strength of rock core and q_t is the splitting Tensile Strength.

Figure 12 present the estimated calculated ultimate skin resistance of Limestone/Coquina rock together with lower and upper bound. The design considered use of pressure grouting the tips of the drilled shaft to improve the shaft's load-carrying capacity and possibly avoid construction of longer shafts in highly unpredictable and susceptible-to-caving soils. Pressure grouting shaft tip improves stiffness and load carrying capacity for soil beneath the shaft tip. The technique involves injecting cement grout into the soil beneath the shaft tip through the Cross-hole Sonic Logging (CSL) test tubes. The post grouting operation is performed after drilled shaft concrete has attained a minimum compressive strength of 4000 psi. See Fig. 13 for schematic.

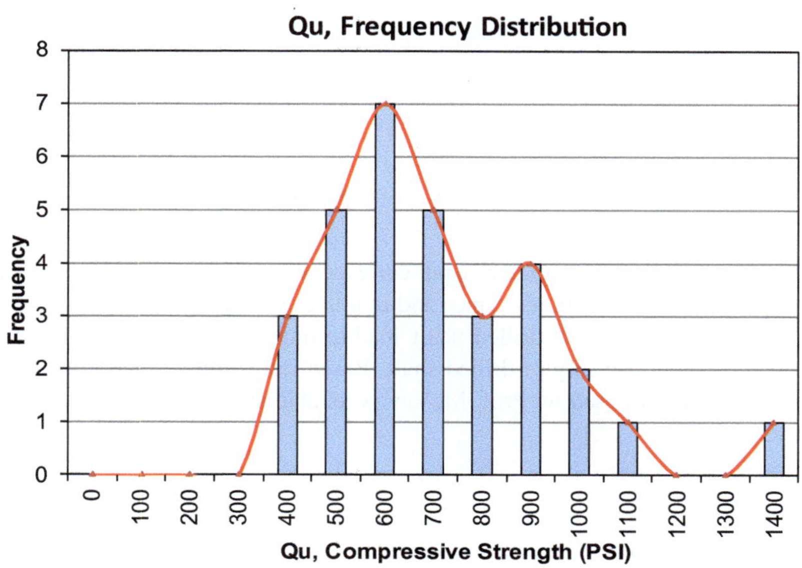

Fig. 11. Frequency Distribution

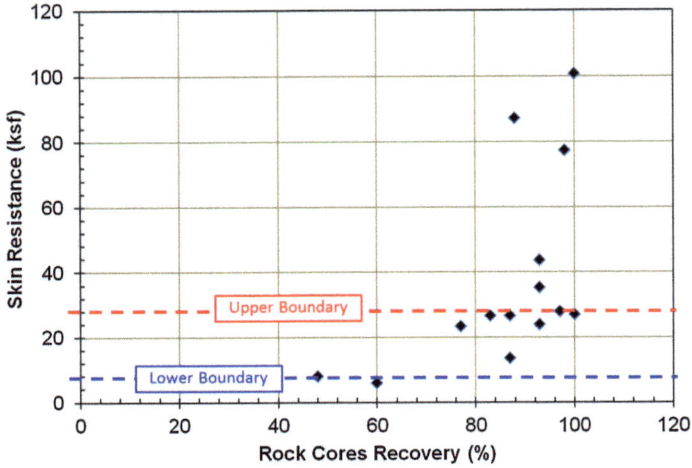

Fig. 12. Estimated skin resistance of Limestone/Coquina rock

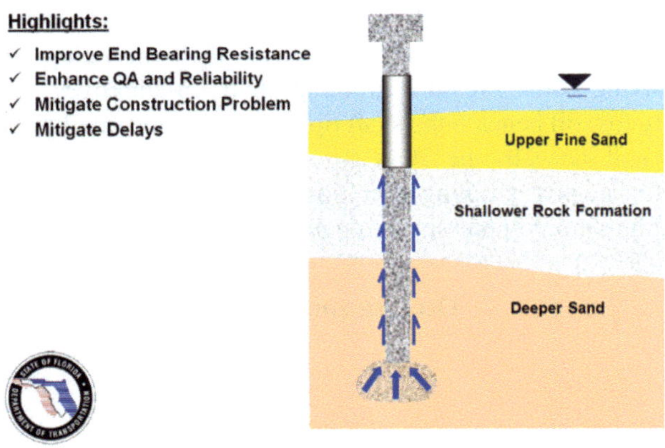

Fig. 13. Post Grouted Shaft Tip

The estimated shaft embedment depth corresponds to 50% of the ultimate capacity is being provided by the skin resistance and reminder from grout injected stiffened soil mass beneath shaft tip. The drilled shaft load-testing program was implemented to verify load carrying capacity of drilled shaft. Based test results, it was concluded that empirical method proposed by Prof. McVay is reliable and post grouting operation as effective.

3 Conclusion and Summary

The regional features of the East Coast often make the design and construction of foundations for civil engineering works a challenging task. In addition to subsurface conditions and structural design loads, the foundation recommendations in urban areas like this are commonly dictated by the project's construction staging, proximity to adjacent existing structures and site accessibility, and risk management. Additional difficulties arise from the complicated geology, seismic risk, and risk of cargo collisions structures crossing waterways.

As exemplified through the case studies presented in this paper, deep foundation options such as micropiles, drilled shafts, and driven piles can be utilized successfully to support civil engineering structures in urban settings.

Acknowledgements. The authors are grateful to their employer Hardesty & Hanover, LLC (H&H) for the support and commitment during the preparation of this manuscript.

References

McVay, M.C., Townsend, F.C., Williams, R.C.: Design of socketed drilled shafts in limestone. ASCE J. Geotech. Eng. **118**(10), 1626–1637 (1992)

O'Neill, M.W., Reese, L.C.: Drilled Shafts: Construction Procedures and Design Methods. Publication No. FHWA-IF-99-025. Federal Highway Administration, Washington (1999)

Woods, R.D., Jedele, L.P.: Energy-attenuation relationships from construction vibrations. In: Vibration Problems in Geotechnical Engineering, Proceedings of a Symposium Sponsored by the Geotechnical Engineering Division, ASCE, Detroit, Michigan, October 1985, pp. 229–246 (1985)

Numerical Modeling of a Caisson Foundation Retrofitted with Helical Piles

Serhan Guner(✉)

University of Toledo, Toledo, OH, USA
serhan.guner@utoledo.edu

Abstract. While the nonlinear finite element analysis methods have been commonly used for the performance assessment of existing structures, their use for the retrofit design of concrete foundations has remained limited. One reason for this is the sophisticated modeling process which requires knowledge, experience, and caution. This study demonstrates the applicability and benefit of the nonlinear finite element modeling for the performance-based structural retrofit design of caisson foundations. The foundation system investigated supports a self-supporting telecommunication tower located in Canada. The addition of new antennas and the change in the design standards requires the caisson foundations of this tower to be retrofitted with new cap beams and helical piles to resist significant additional tensile forces. A two-stage analysis and design process is conducted with the help of a continuum-type finite element analysis method, treating reinforced concrete as an orthotropic material and employing the constitutive relations of the Disturbed Stress Field Model. General modeling guidelines and the points for caution are discussed for the retrofit design of caisson foundations using nonlinear analysis methods.

1 Introduction

Nonlinear finite element analysis methods have seen significant advancements in the past decade. Various constitutive models and element formulations have been proposed. While these methods have been widely used by researchers, their practical application for the strengthening of reinforced concrete foundations has remained limited. One reason for this is the sophisticated modeling process which requires knowledge, experience, and caution. The objective of this study is to demonstrate a modeling methodology which can be employed when conducting a retrofit design in a design office environment. This methodology was developed during an actual design project to strengthen the caisson foundations of a number of existing telecommunication towers.

Self-supporting towers are commonly constructed using a triangular plan layout with three caisson foundations. Each caisson resists significant amounts of axial compression and tension loads in addition to a small shear force. Due to the changing wind direction, the axial load fluctuates between tension and compression, creating reversed-cyclic loading conditions. This makes the design of caissons for self-supporting towers more challenging than that of other types of caissons subjected to compression loads only. Caissons typically develop their tensile resistance by skin

M. Abu-Farsakh et al. (eds.), *Advances in Analysis and Design of Deep Foundations*,
Sustainable Civil Infrastructures, DOI 10.1007/978-3-319-61642-1_15

friction only, as opposed to skin friction and tip bearing for the compressive resistance. As such, retrofitting existing caissons to increase their tensile load resistance presents significant challenges.

A number of retrofit designs are used in industry to increase the axial load capacities of existing caissons. For example, weight blocks connected to caissons with epoxied dowel bars are commonly used to provide a small amount of additional tensile resistance. For larger overloads, new anchors consisting of new caissons, micro piles, or helical piles are commonly used. One challenge in designing these retrofit solutions is to ensure that the retrofitted system indeed works as a whole to carry the additional loads. The connections between the existing caissons and the new elements are one critical aspect that requires special attention due to the brittle nature of concrete which does not permit simple bolted or dowelled connections.

The literature investigating the structural behaviour of retrofitted caisson foundations remain very limited. One study was performed by Abdalla (2002) who presented a case study involving self-supporting and guyed tower foundations, and proposed repair and strengthening solutions. However, no numerical analysis and verification studies were presented. Another study was published by Guner and Carrière (2016), which forms the basis of this paper.

2 Proposed Analysis Methodology

2.1 Structure Definition

The tower examined has a height of 90 m with a face-width of 12.2 m at the base, as shown in Fig. 1. The tower is located in a residential area of Toronto, Ontario. It was designed and constructed in the early 1970's. Due to the high demand to add antennas on this tower, the tower mast has been reinforced several times in recent years. The tower has three caisson foundations; one caisson is shown in Fig. 2. There is an existing equipment building located at the centroid of the tower, which further limits the available area and the head clearance for the retrofit design. Each caisson has a diameter of 1067 mm, and a length of 10.7 m. The reinforcement includes 30-#9 longitudinal reinforcing bars and #3 circular hoops spaced at 300 mm, as indicated on the original design drawings. These drawings also specify a concrete compressive strength of 27.6 MPa, a reinforcing steel yield strength of 414 MPa, and a concrete cover of 76 mm. The soil profile includes by up to 2.7 m loose to compact sand and silt fill, 1.9 m compact silty sand, and glacial till of clayed silt and some sand, with a water table at about 11 m, as indicated in the geotechnical investigation report.

The structural analysis results indicated the maximum factored uplift and compression reactions to be 1530 kN and 1740 kN, respectively, at each caisson, considering the increased antenna loading and the latest versions of the design standards. The factored uplift capacity was calculated to be 675 kN using the geotechnical resistance factor of 0.375 in the Canadian CSA S37 standard (2001). Considering the 136 kN self-weight of the caisson, an overload factor of 2.1 was obtained. Consequently, an additional uplift capacity of 800 kN was required per caisson. Due to the limited space available on the tower site, two helical piles, each with 400 kN factored

Fig. 1. The self-supporting tower examined

Fig. 2. One of the caisson foundations to be retrofitted

tensile capacity, was employed in the proposed design (see Figs. 3, 4, and 5). The design of the helical piles was conducted in a separate geotechnical study.

The main challenge in using helical piles is the design of an effective connection between the steel pile shafts and the existing concrete caissons. One commonly used approach is to employ a reinforced concrete cap beam to provide an offset from the existing caissons, while connecting the new helical piles to the existing caissons. In this study, a depth of 1000 mm and a width of 800 mm was used to provide the required stability to the cap beam. A clear span of 700 mm was used between the caisson and the piles as per the geotechnical recommendations. This created a deep beam with a clear span-to-depth ratio of 0.7. Recall that deep beams do not satisfy the 'plane sections remain plane' hypothesis, and require a suitable formulation to capture the deep beam effects. The following sections present the verification studies using a nonlinear finite element method, while taking account of the deep beam effects.

2.2 Nonlinear Finite Element Modeling Guidelines

A two-dimensional nonlinear finite element analysis modeling was conducted using the computer program VecTor2, which incorporates constitutive models specifically developed for analyzing cracked reinforced concrete (Wong et al. 2013). VecTor2 employs a smeared rotating crack model based on the equilibrium, compatibility, and

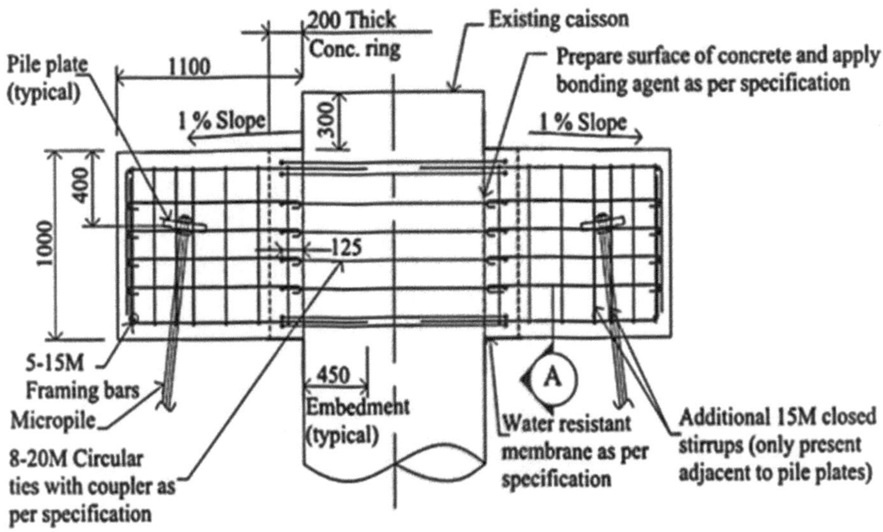

Fig. 3. Elevation of the proposed retrofit design

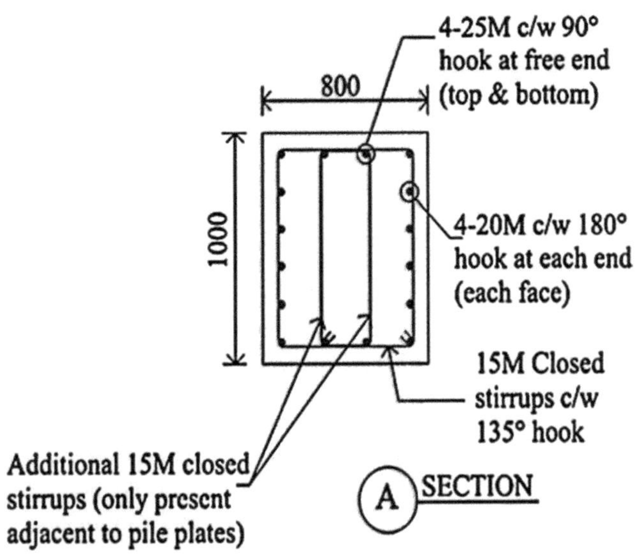

Fig. 4. Section of the proposed retrofit design

constitutive models of the Disturbed Field Model (Vecchio 2000), which is a refined version of the Modified Compression Field Theory (MCFT) (Vecchio and Collins 1986). Although other specialized programs such as ATENA (Cervenka 2016), WCOMD (Maekawa 2016), and DIANA (2016) could also be used for this purpose, the selection of VecTor2 was made because of two reasons: (1) it accounts for a large number of second-order material behaviors models relevant to this modeling study; and

Fig. 5. Construction of the proposed retrofit design (Guner and Carrière 2016)

(2) the MCFT has been recognized internationally and adopted by many design codes such as Canadian CSA A23.3 (2014) and American AASHTO LRFD (2016).

When modeling reinforced concrete structures, proper modeling of the constitutive response and important second-order material behaviors are crucial (Guner and Vecchio 2010a, b). The material models considered in this study are listed in Table 1. Among them, three models were found to be particularly important for the cap beam examined: *the concrete compression softening* (i.e., the reduction in the uniaxial compressive strength and stiffness due to transverse tensile cracking), *the concrete tension stiffening* (i.e., the ability of cracked reinforced concrete to transmit tensile stresses across cracks), and *the dowel action* (i.e., the additional shear strength provided by the main reinforcing bars). First of all, the low amounts of stirrup reinforcement present in the existing caisson makes it prone to transverse cracking under large axial forces, which requires the consideration of 'concrete compression softening'. Secondly, the cap beam is prone to cracking and its response is sensitive to the amount of tension transmitted across cracks, requiring the modeling of the 'concrete tension stiffening' effects. Finally, the shear force transfer at the beam-caisson interface may influence the response of the entire system, such that the additional shear resistance due to the 'dowel action' should be considered. More details on these material models can be found in Wong et al. (2013).

Table 1. Material behaviour models considered

Material behaviour	Default model
Compression base curve	Popovics (NSC)
Compression post-peak	Modified Park-Kent
Compression softening	Vecchio 1992-A
Tension stiffening	Modified Bentz 2003
Tension softening	Linear
Confinement strength	Kupfer/Richart
Concrete dilatation	Variable – Orthotropic
Cracking criterion	Mohr-Coulomb (Stress)
Crack width check	Agg/5 Max crack width
Concrete hysteresis	Nonlinear w/plastic offsets
Slip distortion	Walraven
Rebar hysteresis	Seckin w/Bauschinger
Rebar dowel action	Tassios (Crack slip)

2.3 Global Design Verification

The finite element mesh was created as a result of an iterative refinement process, starting with a coarse mesh and refining it gradually. The final mesh incorporated 50 × 50 mm, 8-degree-of-freedom quadrilateral elements, with a capability to account for the geometric nonlinearities. The uplift load was applied to the bearing plate at two nodes. The final mesh is presented in Fig. 6.

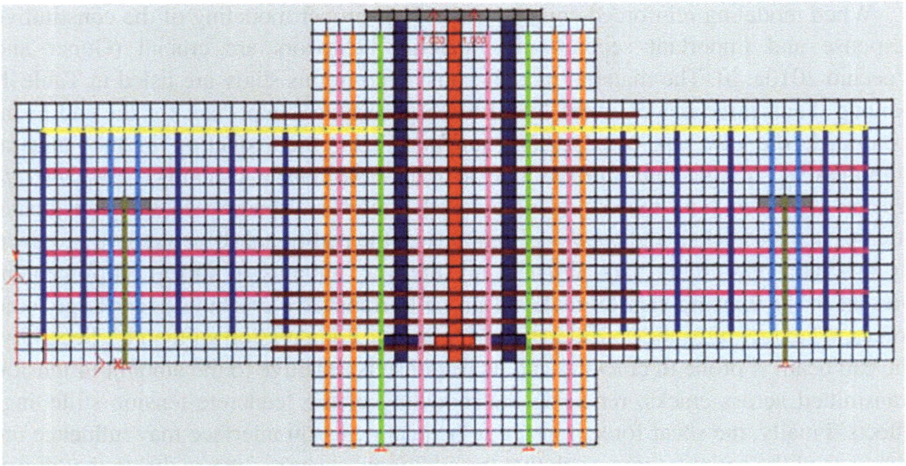

Fig. 6. Finite element model for the global design verification

Five different continuum regions were created based on the material properties as listed in Table 2, and shown in Fig. 6. To represent the 51 mm-dia anchor bolts, equivalent square areas were defined to match the finite element mesh. All reinforcing bars and the helical pile shafts were modelled using discrete truss bars to be able to observe their behaviour and to obtain their stress/strain conditions. Table 3 summarizes the material properties of the truss bars defined, which were obtained from the manufacturer specifications for the new bars and the original design drawings for the existing bars. The response of truss bars was modeled with a stress-strain curve including the Bauschinger effects, using the constitutive model of Seckin (1981) as shown in Fig. 7. The response of concrete was modelled using the plastic-offset-based nonlinear model of Palermo and Vecchio (2003) as shown in Fig. 8. This concrete model includes the nonlinear hysteresis rules for the unloading and reloading conditions. Note that some parts of the cap beam will unload, and some other parts will reload, as concrete cracking and reinforcement yielding take place.

Table 2. Continuum region properties

Region	Description	Color	*f'c* (MPa)	*fy* (MPa)	Thickness (mm)
1	Concrete		30	-	800
2	Anchor Bolt (Single)		-	414	40.5
3	Anchor Bolts (Double)		-	414	81
4	Pile Head Plate		-	400	200
5	Base Plate		-	414	554

Table 3. Truss bar properties

Truss	Description	Color	Area (mm^2)	*fy* (MPa)	Diameter (mm)
1	Stirrups (2-15M)		400	400	16
2	Double Stirrups (4-15M)		800	400	16
3	Main Bars (4-20M)		1200	400	19.5
4	Skin Bars (2-20M)		600	400	19.5
5	Hoop bars(2-20M)		600	400	19.5
6	Helical Pile Shaft (32 Dia)		806	830	32
7	Caisson Bars (2-#9)		1290	414	28.65
8	Caisson Bars (4-#9)		2580	414	28.65
9	Caisson Bars (3-#9)		1935	414	28.65

Definition of support conditions is a critical aspect of the modeling process. Four hinges were found to represent the actual support conditions reasonably well. Two hinges were defined to support the helical piles to create conservative loading conditions for the cap beam. The other two hinges were used to ensure that the existing caisson does not exceed its calculated ultimate capacity of 675 kN using the geotechnical resistance factor of 0.375. 6-#9 bars (shown with green color in Fig. 6) were restrained for this purpose. A displacement-controlled pushover analysis was performed using an increment equal to 0.25 mm.

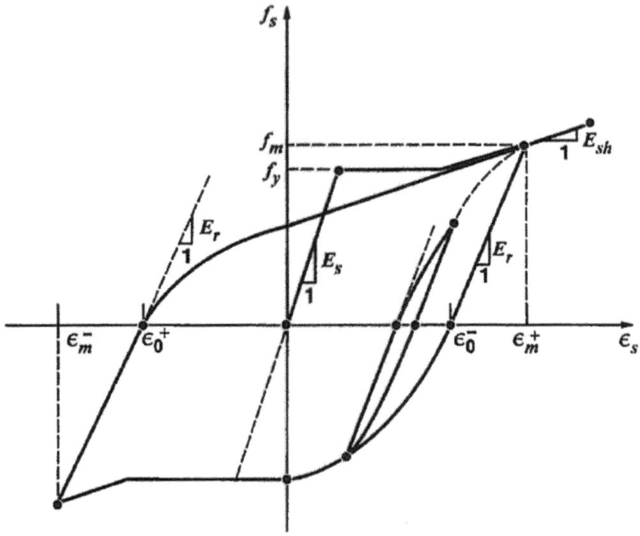

Fig. 7. Reinforcing bar response

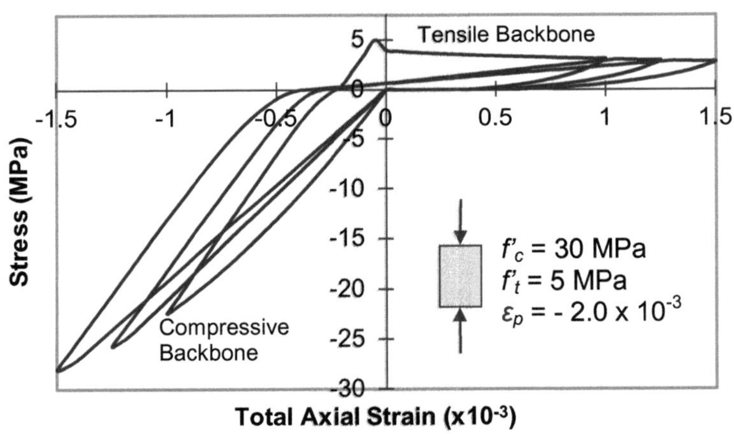

Fig. 8. Concrete response

The analysis indicated that the first concrete cracking occurs at a tensile leg load of 930 kN, as shown in Fig. 9, which is approximately equal to the service tension load. The retrofitted system exhibited a flexure-dominated response at the ultimate conditions as shown in Fig. 10. The failure mode involved yielding of the helical pile shafts at a leg load of approx. 3000 kN, as shown in Fig. 11. This is a desired failure mode, which indicates that the global response is acceptable. Figure 12 shows the load-deflection response of the global system. Since the required ultimate leg tension is 1530 kN, the global design capacity of 3000 kN is excessive. It will be seen in the following section

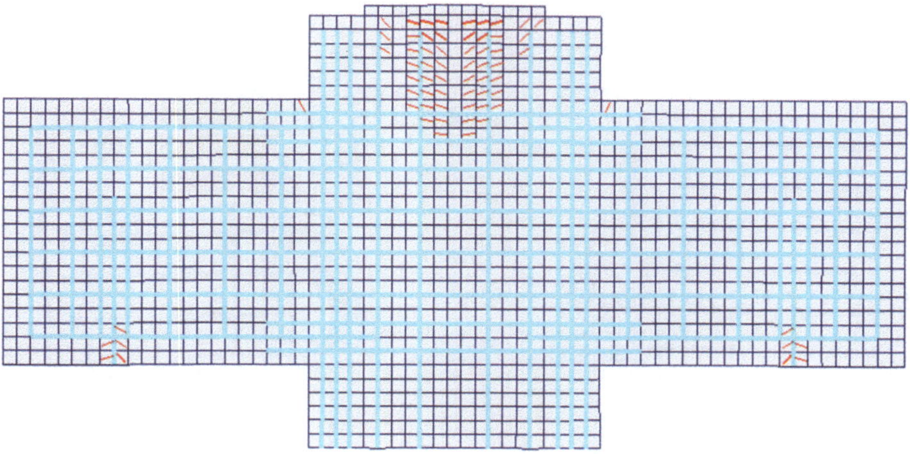

Fig. 9. Crack pattern at first cracking

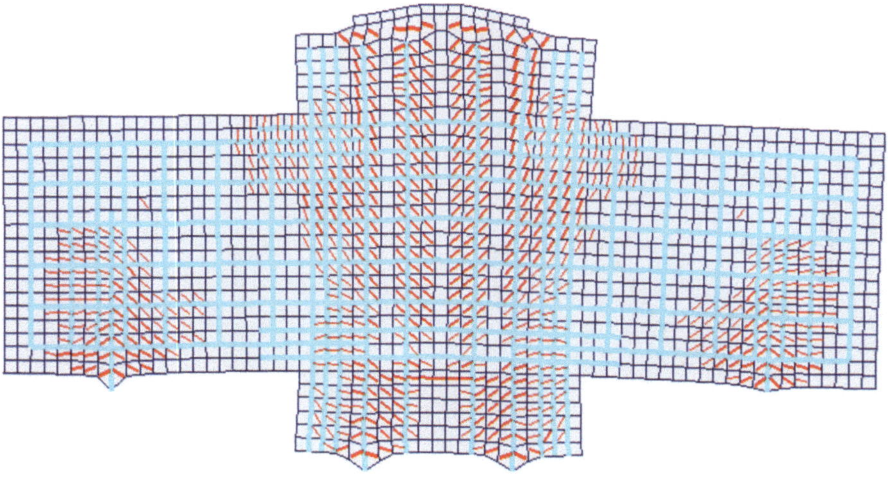

Fig. 10. Crack pattern at failure

that the design will be governed by the local response of the discontinuous dowel bars. Consequently, no change is necessary for the global design.

2.4 Local Design Verification

The site conditions and the presence of existing caisson's vertical reinforcing bars (i.e., 30-#9 – shown with orange, pink, and green bars in Fig. 6) makes it practically impossible to drill through the existing caisson to provide continuous main reinforcement to the new cap beam. As shown in Figs. 3 and 6, the main horizontal

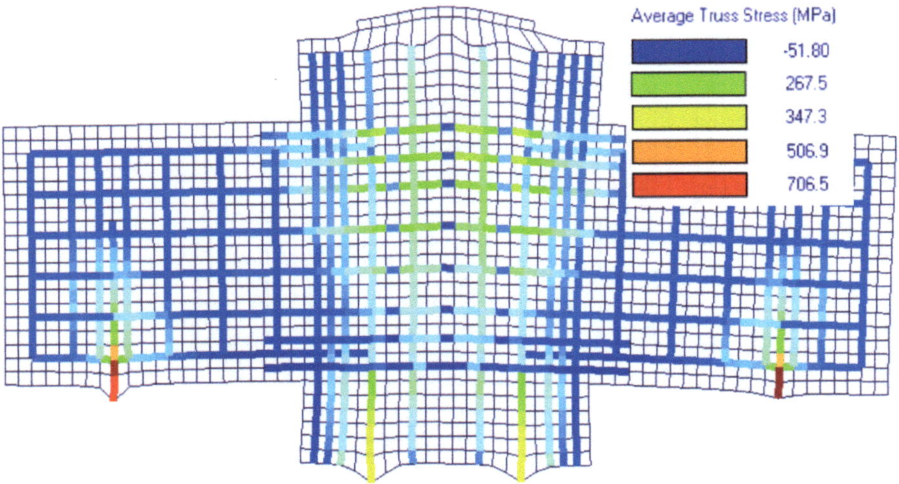

Fig. 11. Reinforcing bar stresses at failure

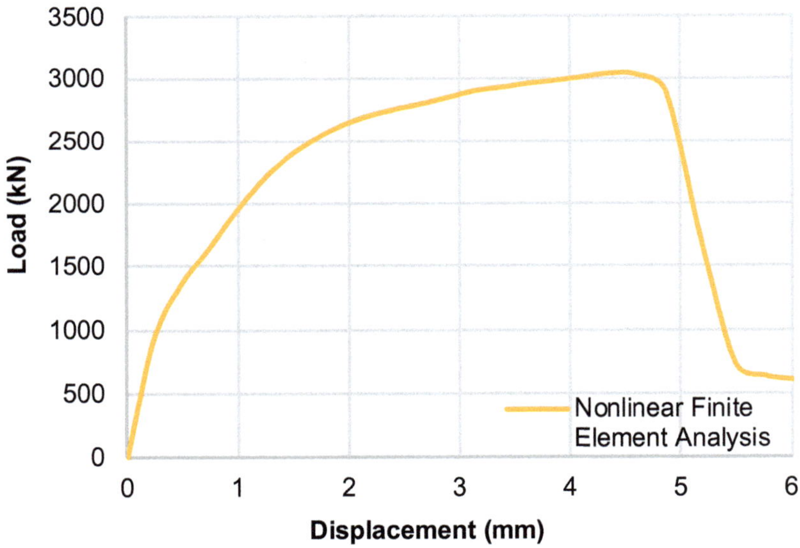

Fig. 12. Load-deflection response for the global response verification

reinforcing bars (shown with yellow bars in Fig. 6) is terminated inside the existing caisson through the use of an epoxy adhesive. The embedment length (shown in Fig. 3) required to develop the bond strength is typically provided by the adhesive manufacturer, which was in the range of 250 mm for the product that we selected. It should be noted that using the recommended bond development length will not ensure that the required bar tension can be successfully carried. A system of reinforcing bars or supports is required to carry this tension. To achieve this transfer, supplementary hoop reinforcement was used in the proposed design. Due to the obstruction of the existing tower legs, two half-circle hoops, connected with mechanical couplers, were employed. This is the most critical aspect of the proposed design; if not designed properly, it can render the entire retrofit design ineffective. To determine the required hoop quantity and to verify the resulting system response, a detailed local finite element analysis was undertaken using the program, VecTor2.

A finite element model was created using 3944 triangular elements (each with 6 degrees of freedom and 150 mm thickness) and 2054 nodes. The discontinuous reinforcing bars and the double-hoop reinforcement were modelled using perfectly-bonded discrete truss elements (each with two degrees of freedom at each node). The model was restrained with four hinges on one side, and the loading was applied uniformly on the other side with 0.1 mm displacement increments. A displacement-controlled analysis was employed to obtain the post-peak response, ductility, and failure mode. The finite element mesh is presented in Fig. 13.

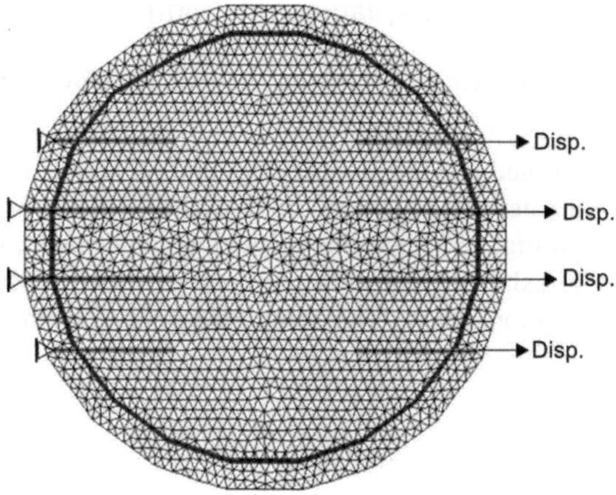

Fig. 13. Finite element model for the local design verification

In order to determine the required embedment length for the discontinuous dowel bars (shown in Fig. 3), six different models were created by varying the dowel bar embedment lengths: 650, 550, 450, 350, 250, and 170 mm for Models 1 to 6, respectively. The load-displacement responses for all six models are presented in Fig. 14. The responses of Models 1 to 5 exhibited similar behaviours: an initial peak load, followed by a sudden drop due to major cracking at the termination of the reinforcement, and a stiffening response due to the activation of the supplementary hoop reinforcement. Model 6, which had an embedment length less than the length recommended by the adhesive manufacturer, exhibited a brittle failure upon first cracking at an applied load of 200 kN. The hoops were ineffective in this model as evident from the suddenly dropping load capacity in Fig. 14.

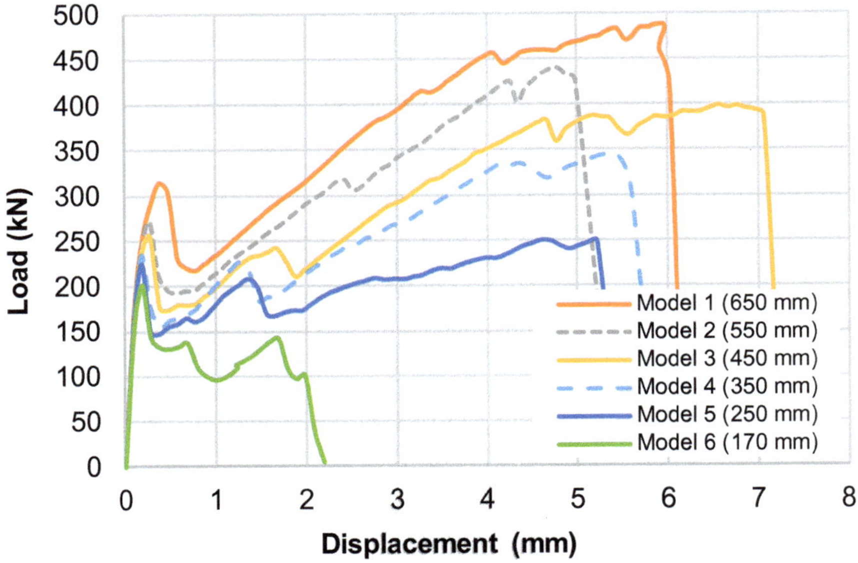

Fig. 14. Load-deflection responses for the local response verification

Analysis results indicated that the required minimum failure load of 400 kN, which corresponds to the ultimate capacity of the 4–20 M bars (shown with yellow bars in Fig. 6 and green bars in Fig. 13) was achieved with an embedment length of 450 mm (Model 3). This model exhibited a ductile response governed by the yielding of the supplementary hoop reinforcement. The three stages of cracking are presented in Fig. 15.

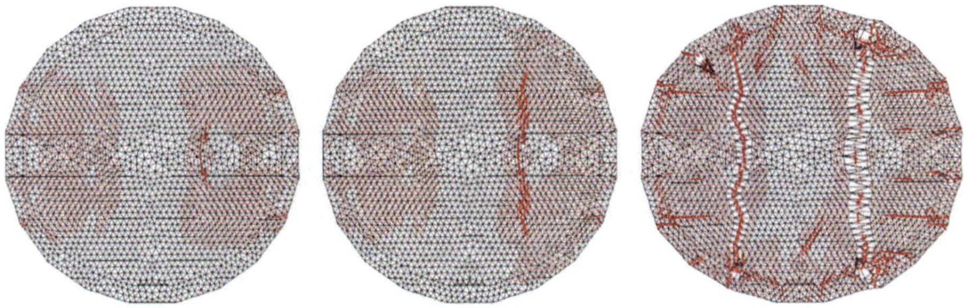

a. First peak b. Load drop c. Failure

Fig. 15. Crack pattern for Model 3

The change in the dowel bar embedment length (shown in Fig. 3) affected the load capacity and the failure mode of the caisson significantly as seen in Fig. 16. An embedment length of 250 mm, which is recommended by the adhesive manufacturer,

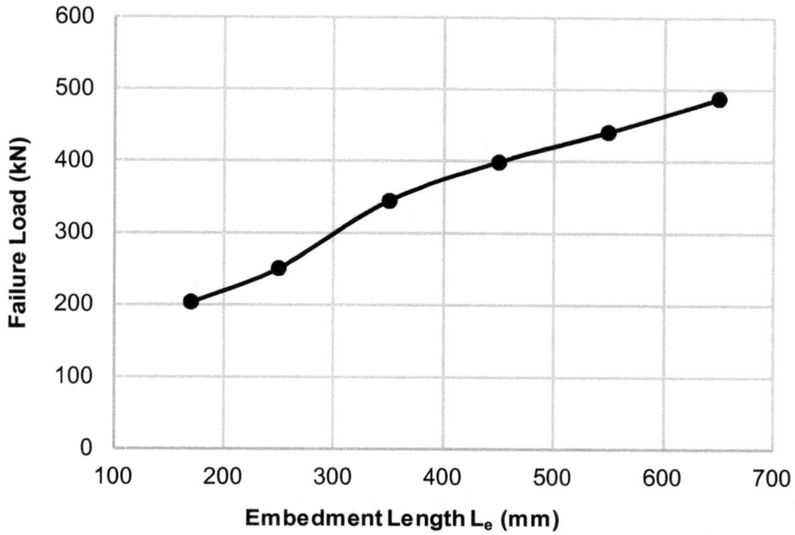

Fig. 16. Effect of the embedment length

resulted in an undesirable failure mode involving the local failure of concrete. The hoop steel was partially effective, and increased the load capacity by only 11% beyond the first peak load (as compared to 55% in Model 1). The failure load obtained was 250 kN, which is significantly lower than the required value of 400 kN.

3 Conclusions

(1) To significantly increase the tensile capacities of caisson foundations, addition of new structural elements is required.
(2) Helical piles are one viable element that can provide significant additional axial capacity in tension and compression. They are particularly useful for sites where this is limited space and limited head clearance.
(3) It is recommended that the actual load capacity of helical piles should be verified on site using at sacrificial pile tests. The complete load-deformation response should be obtained and provided to the structural design engineer for the design validation.
(4) This study demonstrated that helical piles can be connected to existing caisson foundations using reinforced concrete cap beams. It was found that the dimensions of the cap beams should be large enough to provide the required stiffness and stability.
(5) A proper analysis method must be employed to verify the global design. For deep beams, the analysis method must account for the nonlinear strain distribution. Simple sectional analysis methods with simply-supported slender beam approaches are not valid for deep beams.

(6) The critical aspects of the design (e.g., epoxy anchored bar embedment length in this study) must be verified by a proper local analysis method.
(7) Providing the recommended bond development length for epoxy anchored bars does not ensure that the bar tension can safely be carried. The designer must ensure that there are adjacent rebars available (or designed) to transfer the tension load of the terminated bars to a support point or other reinforcing bars.
(8) The analysis and design methodology proposed in this study was numerically shown to increase the uplift capacity of an existing caisson by a factor of 2.1. Overall behaviour, ductility, and the failure mode of the retrofitted system were found to be satisfactory.
(9) The proposed design has a general applicability and is suitable for applications where there is limited space around the existing caissons.

Acknowledgements. The author would like to thank Mr. Salim Khoso (a graduate student at the University of Toledo) and Mr. Sálvio Aragão Almeida Júnior (an undergraduate summer student funded by the Science Without Borders Scholarship Program) for assisting in the finite element analyses presented in this paper.

References

AASHTO: LRFD Bridge Design Specifications, Customary US Units, 7th Edn., with 2015 and 2016 Interim Revisions. American Association of State Highway and Transportation Officials, Washington, DC (2016)

Abdalla, H.A.: Assessment of damages and repair of antenna tower concrete foundations. Constr. Build. Mater. **16**(8), 527–534 (2002)

Cervenka, V.: ATENA: software for nonlinear analysis of reinforced concrete structures, Prague, Czech Republic (2016). http://www.cervenka.cz

CSA A23.3: Design of Concrete Structures. CSA Standard A23.3. Canadian Standards Association (CSA), Mississauga, ON, 290 pp. (2014)

CSA S37: Antennas, Towers, and Antenna-Supporting Structures. CSA Standard S37-01. Canadian Standards Association (CSA), Mississauga, ON, 118 pp. (2001)

DIANA: User's Manual. TNO DIANA BV. Delft, The Netherlands (2016). http://dianafea.com/manuals/d93/Diana.html

Guner, S., Carrière, J.: Analysis and strengthening of caisson foundations for uplift loads. Can. J. Civ. Eng. **43**(5), 411–419 (2016)

Guner, S., Vecchio, F.J.: Pushover analysis of shear-critical frames: formulation. Am. Concr. Inst. Struct. J. **107**(1), 63–71 (2010a). http://www.utoledo.edu/engineering/faculty/serhan-guner/docs/JP1-107-s07.pdf

Guner, S., Vecchio, F.J.: Pushover analysis of shear-critical frames: verification and application. Am. Concr. Inst. Struct. J. **107**(1), 72–81 (2010b). http://www.utoledo.edu/engineering/faculty/serhan-guner/docs/JP2-107-s08.pdf

Maekawa, K.: WCOMD: Nonlinear Analysis Program of Reinforced Concrete Structures. University of Tokyo, Tokyo (2016). http://www.forum8.co.jp/english/uc-win/wcomd-e.htm

Palermo, D., Vecchio, F.J.: Compression field modeling of reinforced concrete subjected to reversed loading: formulation. Am. Concr. Inst. Struct. J. **100**(5), 616–625 (2003)

Seckin, M.: Hysteretic behavior of cast-in-place exterior beam-column-slab assemblies. Ph.D. thesis, Department of Civil Engineering, University of Toronto, ON, Canada, 236 pp. (1981)
Vecchio, F.J.: Disturbed stress field model for reinforced concrete: formulation. ASCE J. Struct. Eng. **126**(8), 1070–1077 (2000)
Vecchio, F.J., Collins, M.P.: The modified compression field theory for reinforced concrete elements subject to shear. ACI Struct. J. **83**(2), 219–231 (1986)
Wong, P.S., Vecchio, F.J., Trommels, H.: VecTor2 and FormWorks user's manual. Technical Report, Department of Civil Engineering, University of Toronto, ON, Canada, 318 pp. (2013)

Theoretical Verification for Full-Scale Tests of Piled Raft Foundation

Hussein H. Hussein[3(✉)], Hussein H. Karim[1], and Kais T. Shlash[2]

[1] Al-Hussein University College, Karbala, Iraq
[2] Building and Construction Engineering Department, University of Technology, Baghdad, Iraq
[3] Civil Engineering Department, Al-Esraa University College, Baghdad, Iraq
enghhhpile@yahoo.com

Abstract. Piled-raft foundation is a three-dimensional problem, which requires three-dimensional modelling. Numerical modelling of experiments was carried out to get compatible agreement between the behavior of piled raft foundation and theoretical simulation. Two verification problems were simulated from the real field tests (1 × 1, 2 × 2) by using the program PLAXIS-3D. The soil behavior was modelled by using the HS-small model, while the piles and their cap were modelled by using linear elastic model. It can be noted that the soil stiffness and the strength model parameters have limited effects when remaining within acceptable range. It is believed that the stiffer behavior is due to installation effects that increase soil horizontal stresses and enable larger shear mobilization. This can be introduced in the model by artificially increasing K0ini. K0 should be increased to very high values (3–3.5) along with increasing dilatancy angle. It can be realized that there is a very good matching between the finite element and experimental results. Since this verification is based on real field tests, so it can be given a great reliability for the study of several variables associated with piled raft system, based on the material properties of the theoretical model that has been obtained from these simulations.

Keywords: Piled raft · Hardening · Finite element method · Plaxis-3D · Verification

1 Introduction

The complex piled raft behavior simulation in the analytical method is tedious because the problem is three dimensional in nature and the complexity is involved in the interaction process among its various components (O'Neill 2006).

The adoption of piled raft foundations concept in the design of pile groups is by no means new, and has been described by several authors, including Poulos and Davis (1972), Hooper (1973), Burland et al. (1977), Katzenbach et al. (1998), Prakoso and Kulhawy (2001), Reul and Randolph (2003), among many others. In the early years, because of the limited availability of computers memory and processing speed, the use of numerical methods was confined to simple problems. In the last two decades due to the rapid development in computer technologies, numerical methods such as full three-dimensional methods are often used to solve complex problems.

M. Abu-Farsakh et al. (eds.), *Advances in Analysis and Design of Deep Foundations*,
Sustainable Civil Infrastructures, DOI 10.1007/978-3-319-61642-1_16

The finite element method is considered the most powerful tool among the other methods of analysis. Therefore, the finite element method was selected in this study to develop a numerical model to predict the load-settlement relationship and the load sharing among the piles and the raft of the piled raft foundations. The developed model offers a considerable saving in computational effort and time while improving the accuracy of two-dimensional modelling of piled-raft systems.

2 The Concept of Piled Raft Foundations

The piled-raft foundation is an attractive choice for floating pile foundations where the underneath soil is very compressible and has a very low strength. The piled raft foundation is one of the concepts related to the foundation, which behaves as a construction of the composite and composes of three main elements of bearing, these are; subsoil, raft, and piles. Related to its stiffness S_{tot}, the structure loads are distributed and made over the piles as well as over contact pressure by the raft R_{raft} that is identified by the summation of pile resistance ($\sum R_{pile}$, i), in the ground as presented in Fig. 1. Therefore; the piled raft has a total resistance that is given by the Eq. (1):

$$R_{tot} = \left(R_{raft} + \sum\nolimits_{i=1}^{n} R_{pile}\right) \geq S_{tot} \quad (1)$$

Because of the need for basement below the structure, the positive effect of the raft is increasingly taken into consideration in the design of foundations particularly when the strength and stiffness of the pile foundation are not enough. As an example, the Emirate

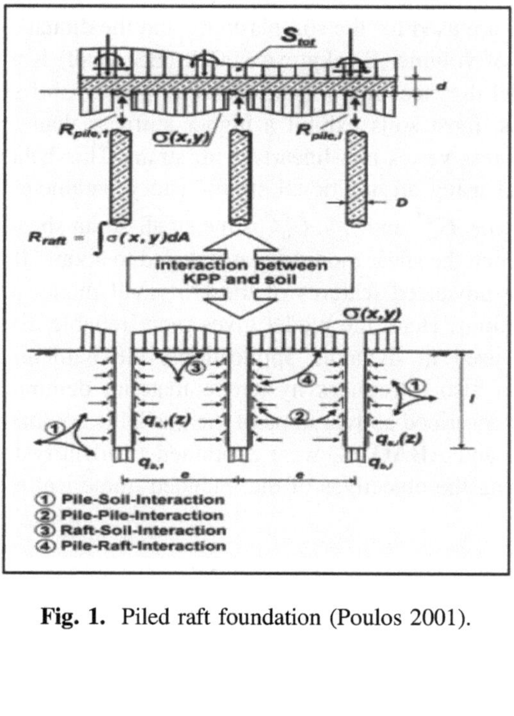

Fig. 1. Piled raft foundation (Poulos 2001).

Twin Towers in Dubai and the Twin Towers in Kuala Lumpur are designed with the concept of piled raft foundations (Russo 2011).

3 Finite Element Model

Foundation of piled-raft is a 3-D problem, which requires three-dimensional modelling. However, in three-dimensional models, the computational time and effort are excessive due to a large number of elements in the mesh. The time required for the computations depends on the number of elements used in the model. Katzenbach et al. (1998) reported that 3D finite element simulations of piled raft foundations with an average number of elements in the range of 10,000 to 25,000 elements need about 18 h of computational time on a Sun-Ultra 2 workstation. They also expected that increasing the number of elements and considering other issues in the simulations such as consolidation would lead to an enormous increase of computational time. Therefore, reducing the number of elements could save much time in the calculation process. However, reducing the number of elements in the mesh can affect the accuracy of the model. In the present paper, the soil was assumed to be homogenous sand soil. To predict the behavior of piled raft foundations at large settlements, a non-linear analysis is required. Therefore, the behavior of the soil was considered as non-linear. There are many constitutive models used to simulate the soil behavior such as the Cam Clay Model, Drucker-Prager, Mohr-Coulomb model, the Linear Elastic Model, and Hardening Soil Model. The Hardening soil model with small-strain stiffness (HS small) model was utilized to simulate the non-linear stress-strain behavior of the sand soil. (HS small) Model is a non-linear model that depends on the soil parameters that are known in the engineering practice. For this model, the modulus of elasticity of soil, Es, and the Poisson's ratio (μs) are used for the soil elasticity while the friction angle, φ, and the cohesion, c, are used for the soil plasticity and the dilatancy angle is needed to model the increase of volume (Brinkgreve 2002). (HS small) is a modification of the hardening soil model that accounts for the increased stiffness of soils at small strains. At low strain levels, most soils exhibit a higher stiffness than at engineering strain levels, and this stiffness varies non-linearly with strain. This behavior is described in the HS small model using an additional strain-history parameter and two additional material parameters, i.e. G_0^{ref} and $\gamma_{0.7}$. G_0^{ref} is the small-strain shear modulus and $\gamma_{0.7}$ is the strain level at which the shear modulus has reduced to about 70% of the small-strain shear modulus. The advanced features of the HS small model are most apparent in working load conditions. Here, the model gives more reliable displacements than the HS model. When used in dynamic applications, the hardening soil model with small-strain stiffness also introduces hysteretic material damping (Józsa 2011). To construct the model described above, some of the available commercial programs, such as PLAXIS, FLAC and ABAQUS, were examined to identify the most appropriate program for achieving the objectives of the intended numerical model.

3.1 Parameters of the (HS Small) Model

The HS small-model requires several parameters which are generally familiar to most geotechnical engineers. The parameters can be obtained from basic tests on soil samples, these parameters with their standard units are listed below.

E_{50}^{ref} Secant stiffness from triaxial test at reference pressure (kN/m^2)
E_{oed}^{ref} Tangent stiffness from oedometer test at p pressure (kN/m^2)
E_{ur}^{ref} Reference stiffness in unloading/reloading (kN/m^2)
G_O^{ref} Reference shear stiffness at small strains (HS small only) (kN/m^2)
$\gamma_{0.7}$ Shear strain at which G has reduced to 70% (HS small) (kN/m^2)
m Rate of stress dependency in stiffness behavior
p^{ref} Reference pressure (100 kPa) (kN/m^2)
v_{ur} Poisson's ratio in unloading/ reloading
c′ Cohesion (kN/m^2)
Φ' Friction angle (°)
ψ Dilatancy angle (°)
Rf Failure ratio qf/qa like in Duncan-Chang model (0.9)
K_0^{nc} Stress ratio σσ′xx/σ′yy in 1D primary compression.

4 Soil Properties of the Site

The soil profile was determined based on the results of a geotechnical investigation including two boreholes and a standard penetration tests (SPT) within the area of the study. The soil stratigraphy consists of two main layers. The upper layer consists of very dense yellowish to brown sandy soil from (0 to −5 m) depth and the second layer consisting of medium dense light brown to brownish yellow sandy soil from (−5 m to −15 m) depth.

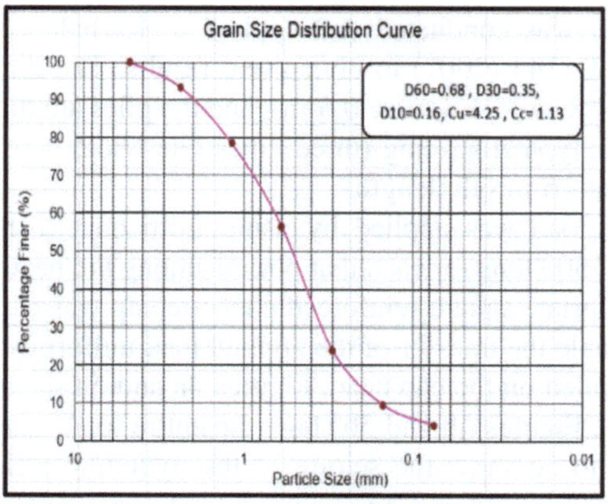

Fig. 2. Grain size distribution curve.

Table 1. Summary of soil properties

Property	Value
Upper layer	
Very dense dry sand from 0 to −5 m	
Unight weight (kN/m^3)	20
Angle of the friction φ	41
Angle of the dilatancy ψ	11
Cohesion (kPa)	0.1
Lower layer	
Medium dense drysand from −5 to −15 m	
Unight weight (kN/m^3)	20
Angle of the Friction φ	35
Angle of the Dilatancy ψ	5
Cohesion (kPa)	0.1

Laboratory tests included direct shear tests were carried out on disturbed samples. The particle size distribution was determined using the dry sieving method according ASTM D422 (2001) specifications and results are shown in Fig. 2. The uniformity coefficient (Cu) and coefficient of curvature (Cc) for the sand were 4.25 and 1.13, respectively. According to the ASTM standard soil classification, the soil is classified as poorly graded sand (SP).

The main soil properties of each soil layer, were derived by geotechnical investigation, and the evaluation of in situ and laboratory tests are presented in Table 1, and the evaluation of geotechnical soil profile at the project area is presented in Fig. 3.

5 Static Load Test

A full-scale field test was conducted and applied to Karbala's soil by load testing according to ASTM D1143 (2007) to obtain real results. The dimension of the piled raft foundation was (1.8 m × 1.8 m × 0.6 m), and the piles are distributed in the same manner maintaining a typical spacing (three times pile width = 0.9 m) between every two adjacent piles with 6 m pile length.

The compressive load was applied by using from one to three hydraulic jacks having a capability of 500 tons for each that placed among the head of cap and the main steel beam. An appropriate square reinforced concrete cap was cast on the head of the test foundation to enable the transfer of the applied load uniformly. Also, 25 mm thick steel plates were installed on the cap head. Figures 4a and 5a show the load-settlement behavior of piled raft. Figures 4b and 5b show the entire load frame and load concrete blocks. Figures 4c and 5c show the setting of the hydraulic jacks and displacement gages over the cap. Figures 4d and 5d show the load gage controlling.

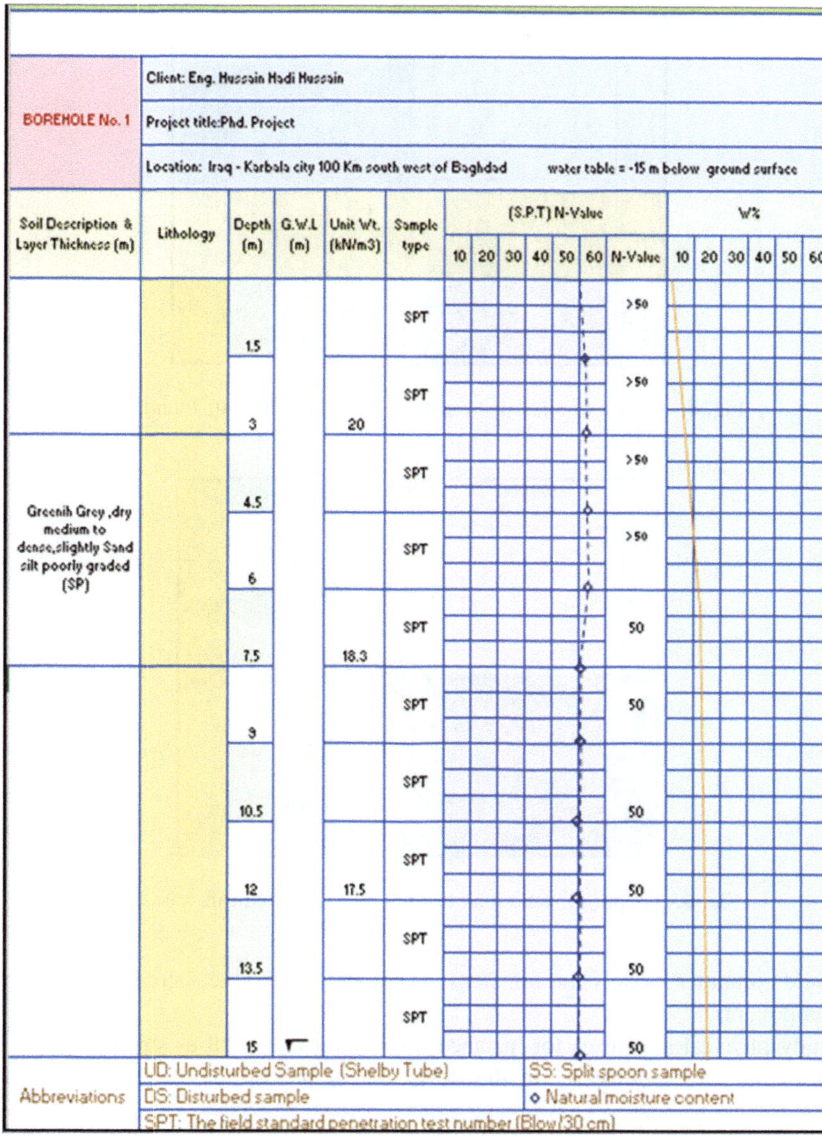

Fig. 3. Geotechnical soil profile at the project area.

6 Plaxis-3D Foundation Program

Plaxis-3D Foundation is one of the package software that is developed by Brinkgreve and Vermeer (1995) for numerical modelling of the structural response (Brinkgreve and Vermeer 2012). Also, Plaxis-3D Foundation is finite element analysis software that enables to do different tasks as follows:

Fig. 4. Output of static load test for (1 × 1), piled raft foundation.

Fig. 5. Output of static load test for (2 × 2), piled raft foundation

- Build computer models and transfer CAD models of products structures, systems or components.
- Study physical responses for instance distributions, as well as stress levels.
- Optimising a design early in the development process to decrease costs of the production.

PLAXIS software is based on the finite element method, and it is intended especially for analysing geotechnical problems. It can be considered as a special-purpose program. PLAXIS can be used as a tool for practical analysis for most areas of geotechnical engineering. Therefore, PLAXIS was selected to be used for developing the three-dimensional finite element model for this study.

7 Verification with Experimental Piled Raft Problem (Field Test)

As explained in static load test (Sect. 3), a real scale model for the foundation of piled raft has been executed and tested in the Karbala soil. The raft and piles are made of concrete, while the soil was dry sand. The results of (2 × 2) and (1 × 1) piled raft foundation (Fig. 6) depicts the layout of the piled-raft foundation considered in this analysis as a reference for checking the numerical solution implemented by the PLAXIS-3D program. The model sand ground was modelled during utilizing the (HS small) model having the parameters listed in the Table 2. Figure 7 shows the quarter of the problem analyzed by PLAXIS-3D program. While, Fig. 8 presents the mesh of the finite element of the vertical loading, taking into account the elastic behaviour of the piled raft and the elastoplastic behaviour of sandy soil by incorporating the (HS small) model.

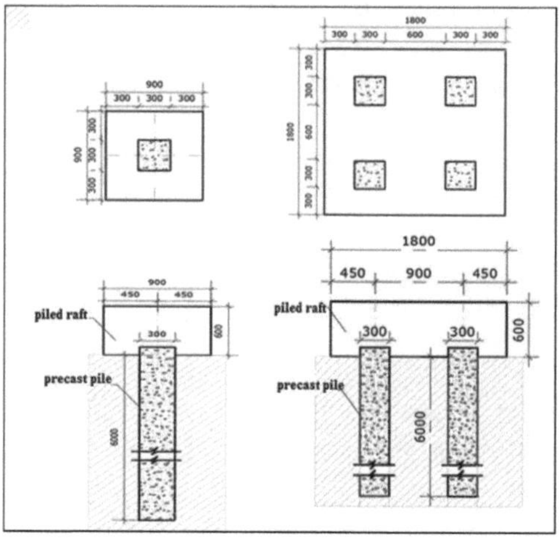

Fig. 6. The Problems of piled-raft (2 × 2) and (1 × 1) used for verification (all dimensions are in mm)

7.1 The Factors Affecting the Simulations Between Practical and Theoretical Results

- The soil stiffness and the strength model parameters have limited effects when remaining within acceptable range. As an example when changing the values of the angle of friction (∅) of 30° to 45° or change the modulus of elasticity (E) of 50000 to 81000 kPa (Swiss Standard 1999), note that the theoretical results had not close to the real results.
- It is believed that the stiffer behavior is due to installation effects that increase soil horizontal stresses and enable larger shear mobilization. This can be introduced in

Table 2. Material properties of the sand adopting soil model (HS small)

	Very dense sand	Medium dense sand
Unit weight (kN m^{-3})	20	20
Drainage type	Drained	Drained
$E_{50,ref}$ (kPa)	60000	35000
$E_{oed,ref}$ (kPa)	60000	35000
$E_{ur,ref}$ (kPa)	180000	105000
m	0.4	0.5
v_{ur}	0.15	0.15
p_{ref} (kPa)	100	100
$\gamma_{0.7}$	0.15E−4	0.15E−4
$G_{0,ref}$ (kPa)	130000	100000
Cohesion c(kPa)	0.1	0.1
Friction angle φ	41	35
Dilatancy angle ψ	11	5
Tension cut-off (kPa)	0	0
$K0_{NC} = 1 - \sin\varphi$	0.344	0.426
$K0_{ini} = K0_{NC}$	0.344	0.426

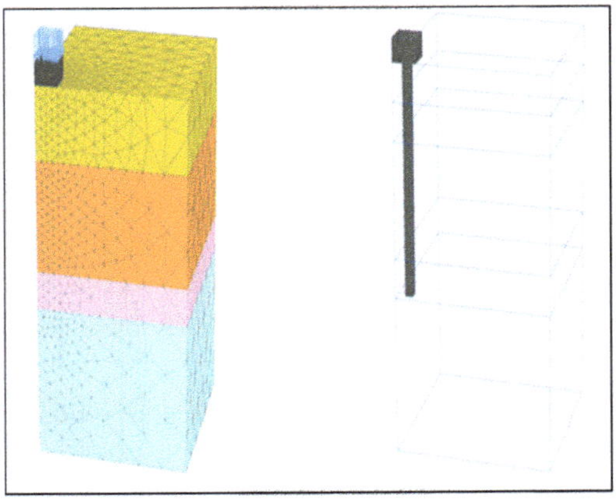

Fig. 7. Quarter of the problem of piled raft (2 × 2) as executed by PLAXIS-3D

the model by artificially increasing K_{oini}. Ko should be increased to very high values (3–3.5) along with increasing dilatancy angle to obtain a good match with the experimental outcomes. Pile driving causes densification of the sand around the pile. Figure 9 shows how the increased soil horizontal stresses make the theoretical results closer to experimental results.

The value of Ko is critical to the evaluation of the skin friction and is the most difficult to determine reliably because it is dependent on the stress history of the soil and the changes which take place during installation of the pile. In the case of driven piles

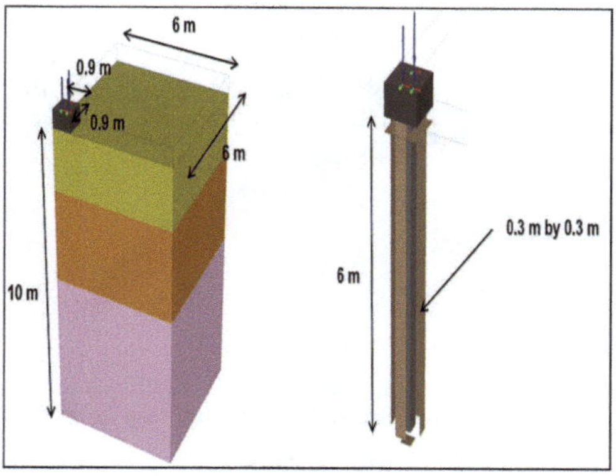

Fig. 8. Mesh of the finite element of piled raft model (2 × 2) for vertical loading.

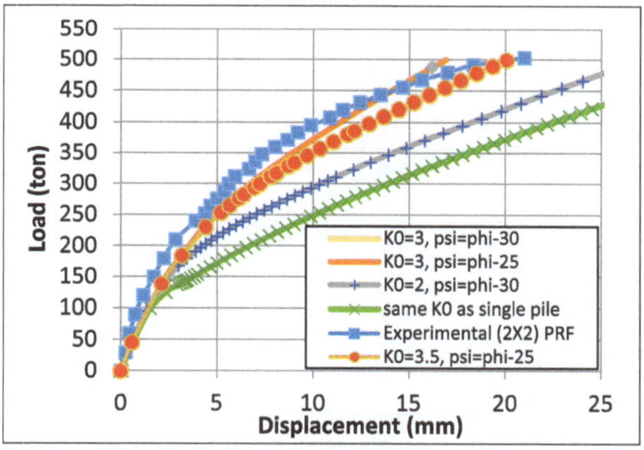

Fig. 9. The effect of increasing soil horizontal stresses on a simulation the theoretical results to the experimental results.

displacement of the soil increases the horizontal soil stress from the original Ko value. The range value of the coefficient of horizontal soil stress for driven piles equal to (1–2) or more (Tomlinson 2015).

The final result by plotting a load-settlement curve for the results, it can be realized that there is a excellent matching between the finite element and experimental results. This indicates that these results are in agreement. Figures 10 and 11 present load-settlement curve behavior for the piled raft foundation with a comparison between the experimental and PLAXIS-3D results for the cases (2 × 2) and (1 × 1) respectively.

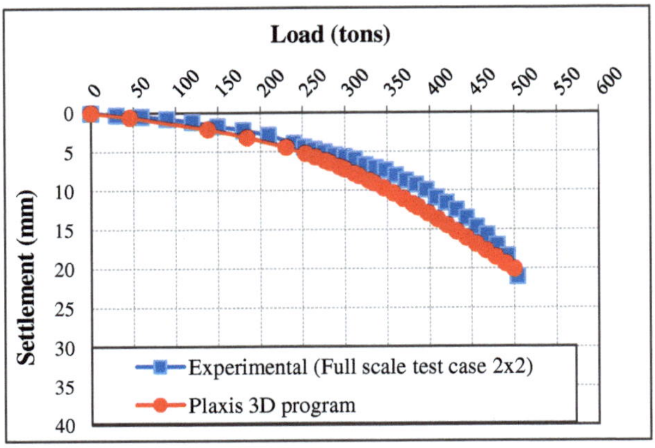

Fig. 10. Load-settlement curve showing a comparison between experimental and Plaxis results (k = 3.5) for case (2 × 2) PRF.

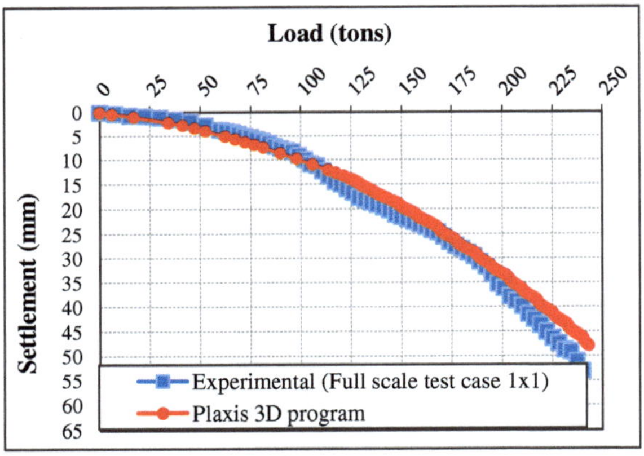

Fig. 11. Load-settlement curve showing a comparison between experimental and Plaxis results for case (1 × 1) PRF.

8 Conclusions

1. The soil stiffness and the strength model parameters have limited effects when remaining within acceptable range.
2. The stiffer behavior is due to installation effects that increase soil horizontal stresses and enable larger shear mobilization. This can be introduced in the model by artificially increasing K_{oini}. K_o should be increased to very high values (3–3.5) to obtain an excellent match with field tests.
3. Increasing dilatancy angle to get a good match with the experimental outcomes.

4. The PLAXIS-3D program is considered good because it can represent the system and execute the simulation depending on the type of foundation and the effect of implemented method.
5. Because of this verification is based on real field tests, it can give a high reliability for studying several variables associated with piled raft system, based on the material properties of the theoretical model that has been obtained from these simulations.

References

ASTM D422: Standard Test Method for Particle Size-Analysis of Soils. American Society of Testing and Materials, West Conshohocken (2001)

ASTM D1143: Standard Test Method for Piles Under Static Axial Compressive Load. American Society for Testing and Materials, West Conshohocken (2007)

Brinkgreve, R.B., Vermeer, P.A.: PLAXIS User's Manual Version 6.1. Balkema, Rotterdam (1995)

Brinkgreve, R.B.J.: PLAXIS User's Manual. 2D-Version 8. Balkema, Rotterdam (2002)

Brinkgreve, R.B., Vermeer, P.A.: PLAXIS 3D-Finite Element Code for Soil and Rocks Analysis. Balkema, Rotterdam (2012)

Burland, J.B., Brooms, B.B., De Mello, V.F.B.: Behavior of foundation and structures. In: Proceedings of the 9th International Conference on Soil Mechanics and Foundation Engineering, Tokyo, Japan, vol. 2, pp. 495–546 (1977)

Hooper, J.: Observations on the behavior of a piled-raft foundation on London clay. Proc. Inst. Civ. Eng. **55**(2), 855–877 (1973)

Jozsa, V.: Effects of rarely analyzed soil parameters for fem analysis of embedded retaining structures. In: Proceedings of the 21st European Young Geotechnical Engineers Conference, Rotterdam, vol. 4, no. 7, pp. 15–20 (2011)

Katzenbach, R., Arslan, U., Moormann, C., Reul, O.: Piled Raft Foundation Interaction Between Piles and Raft. Darmstadt Geotechnics, No. 4, pp. 279–296. Darmstadt University of Technology, Darmstadt (1998)

O'Neill, M.W.: Piled rafts in geotechnical practice – an international perspective. In: Proceedings of 10th International Conference on Piling and Deep Foundations, Amsterdam, pp. 197–208 (2006)

PLAXIS: Vermeer and Brinkgreve. Manual 2005

Poulos, H.G., Davis, E.H.: Rate of settlement under two-and three dimensional conditions. Geotechnique **22**(1), 95–114 (1972)

Prakoso, W.A., Kulhawy, F.H.: Contribution to piled raft foundation design. J. Geotech. Geoenviron. Eng. **127**(1), 17–24 (2001)

Poulos, H.G.: Methods of analysis of piled raft foundations. A report prepared on behalf of technical committee TC18 on piled foundations (2001)

Reul, O., Randolph, M.F.: Piled rafts in over consolidated clay: comparison of in situ measurements and numerical analyses. Géotechnique **53**(3), 301–315 (2003)

Russo, G.: Re-assessment of foundation settlement for the burj khalifa, Dubai. University of Naples, Italy (1986). Piled Raft Foundations on stiff clays. Contribution to design philosophy. Geotechnique **35**(2), 169–203 (2011)

Tomlinson, M.J.: Foundation Design and Construction, 7th edn. Pitman, London (2015)

Swiss Standard SN 670 010b: Characteristic Coefficients of Soils. Association of Swiss Road and Traffic Engineers, Zurich (1999)

Comparative Analysis of the Behavior of a Piled Raft and Corresponding Pile Groups

Hugo Pereira[1(✉)] and António Viana da Fonseca[2]

[1] FEUP - Faculdade de Engenharia da Universidade do Porto, Porto, Portugal
hugo.mendonca.pereira@gmail.com
[2] CONSTRUCT-GEO, Faculty of Engineering, University of Porto, Porto, Portugal
viana@fe.up.pt

Abstract. The aim of the present work is to compare the behavior between a piled raft and a group of piles. The analysis of the addition of a raft into a group of piles allows understanding the influence of the raft–soil interaction in the improvement of the piles capacity. Calculations and corresponding analysis are made regarding the piles and raft capacities, giving significant accuracy, while the association of piles with a raft gives a large benefit on the global foundation performance. This better understanding of the behavior, has been followed by analyzing how the interaction between raft-pile and pile-soil increases the stress state in the soil confined between the piles in the upper soil strata levels and how this improves the shaft resistance in these more superficial horizons. The results obtained shows that these interactions improves the performance of the foundation system, resulting from the increase of the soil confinement due to the loads applied. These multiple analyses were obtained by cross-comparing results, for two numerical codes: the RS3® from RockScience and FB-Multipier® from BSI which revealed a very important tool for this type of assessment.

Keywords: Foundation · Piled raft · Group of piles · Raft · Finite element method · RS3 · FB-Multipier

1 Introduction

In the past many solutions were pursued to arrive at the safe and economical way to design a foundation. One of the adopted solutions was the use of piles placed strategically underneath a raft, in an attempt to reduce the total and differential settlements of the structure. This new kind of foundation is called piled raft foundation and it has been the subject of many studies from different authors (Randolph 1994; Hemsley 2000; Mandolini 2003; Poulos 2001). The piled raft foundation is a combination of two elements: by combining these elements it is possible to increase the load capacity and reduce the settlements in an economical perspective. The design is made taking into account the behavior of this as a global foundation in order to understand the interactions between elements of the foundation. The understanding of the way as these elements interact in the system will allow obtaining a sustainable and economical design, and ensure that the safety parameters are met.

M. Abu-Farsakh et al. (eds.), *Advances in Analysis and Design of Deep Foundations*, Sustainable Civil Infrastructures, DOI 10.1007/978-3-319-61642-1_17

The modelling of the piled raft through computational methods allows to cross-check the results obtained with the traditional methods calculations. Whilst for a raft or a pile foundation it is possible to calculate the load capacity of each element using theoretical formulations, for the piled raft foundation there is still no general formulation where it can be possible to calculate the load capacity and the predicted settlements. Numerical and computational modulation assumes a big part of the design because it will permit the prediction of the performance of the combined foundation.

The main aim of the present work is to understand the behavior of the piled raft and assess the factors that can influence its behavior. To achieve this, two software were used for the modelling of these foundations – FB-Multipier® (Bridge Software Institute 2010) and RS3® (Rocscience 2013) with the goal of produce the foundation system on both the programs and compare the results between them. The results obtained through these software are also compared with the analytically available formulations.

1.1 Piled Raft

The piled raft is a system combining a raft acting together with a group of piles. When it comes to the first approach of designing in geotechnical engineering, firstly it is considered the use of shallow foundations (raft), which it is not always the best solution. If not, the second option should be the use of deep foundations (piles). Of course these preliminary solutions will depend on many factors as type of structure, geology of the area and also economical/sustainable. However, in the last years the use of a piled raft system has become popular. This combined foundation system has two main functions: the use of piles in order to reduce significantly the settlements, differential and total, and the utilization of the raft as structural element.

When this system is loaded with vertical load (Q_{PR}), the raft distributes the effort, depending on the stiffness of the soil between the soil and the piles. Part of the load acts directly to the soil that is present between the base of the raft and the piles, resulting from this interaction a force (Q_R), and throughout the piles the remaining force of the system (Q_P):

$$\alpha_{PR} = \frac{\sum_{i=1}^{n} Q_{pile,i}}{Q_{PR}} \tag{1}$$

Where,

$Q_{pile,i}$ is the applied load to the pile;

Q_{PR} is the total load applied on the piled raft system.

This coefficient presents the load sharing between piles and the raft. This coefficient assumes the value equal to 0 when the system is only a shallow foundation and equal to 1 when the system comprises only deep foundations. Figure 1 presents the situations described before whenever it is a shallow or a deep foundation and the last situation as the piled raft foundation with the coefficient ranging between 0 and 1 (Mandolini 2003). Randolph (1994) refers to the fact that in most cases, the main reason to use this combined system foundation is to reduce the settlements. However, if this option is chosen, piles are designed for the full structural load – ultimate resistance pile design.

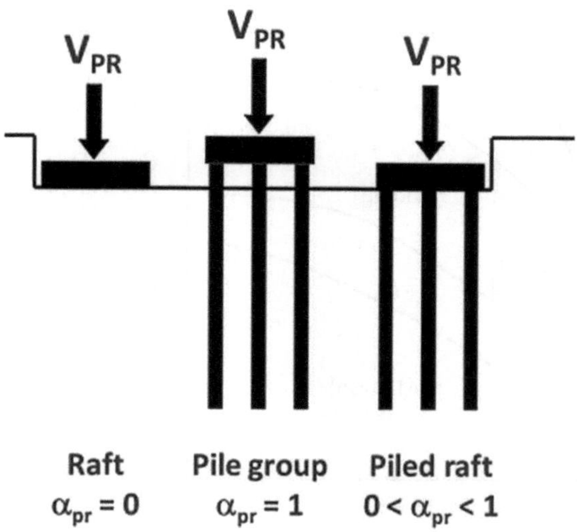

Fig. 1. Foundation systems – Mandolini (2003)

Soil-pile interaction takes an important part of the pile total resistance design, as it will conduct to lower settlements on the foundation.

Randolph (1994) also defines three possible approaches to this piled raft system design:

A. The conventional approach: the piles are designed in group to support great part of the load, where the raft will also contribute.
B. Creep-Piling: the piles are designed to support loads where significant deformations will occur in the pile-soil interface, normally ranges between 70% and 80% of the maximum pile load. Some piles are included in the design to reduce the pressure contact between raft and the soil.
C. Differential settlement control: the piles are strategically positioned in the center with the principal objective of reducing the differential settlements.

Poulos (2001) concluded that there is a greater deformation on piles using this system of piled raft, where the maximum capacity of the piles is used and all of the piles reach that maximum value. This means that the piles on this combined foundation systems works like settlement reducer but also will contribute to the increase of the total capacity load of the system.

Figure 2 represents the behavior of the settlement-load curves with different approaches. Curve 0 shows the behavior of a shallow foundation – raft, with excessive settlement. Curve 1 represents the conventional approach, where the settlements are lower and the behavior of the curve is commanded by the piles. The total load in the system on this approach is supported by the piles. Curve 2 defines the creep-piling approach where the piles are designed with a lower factor of safety, but with the decrease of the number of piles comparing with the curve 1, the raft will carry more loads coming from the structural system. The last curve, number 3, it shows the

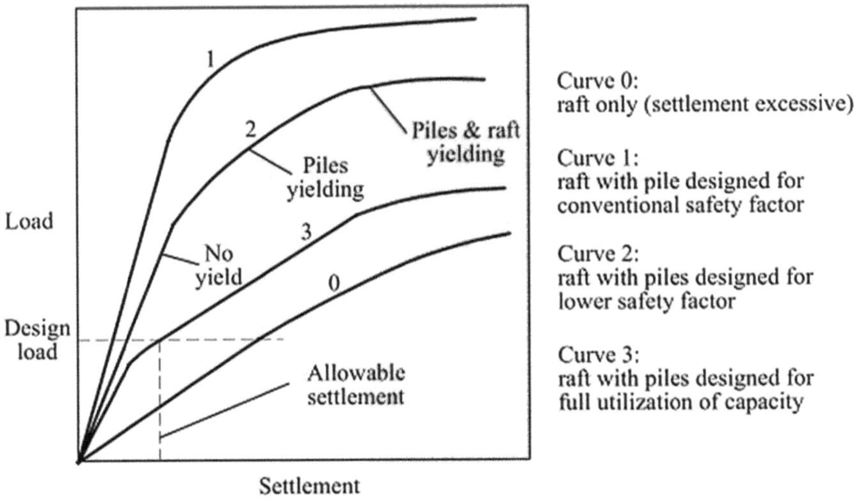

Fig. 2. Load-settlement curve for different approaches (Poulos 2001)

behavior of the system when settlement reduction is the priority. The system is optimized in order to reach the plastic stage faster than curve 1 and 2 but settlement criteria it is met.

The piled raft system is more effective when the raft has the capacity to support the total load from the structure but the verified total and differential settlements are higher than the admissible.

Poulos (2001) studied a different group of soils which could fit for this type of foundation and his conclusions on what were the most favorable are:

i. Stiff clays;
ii. Dense sands.

On these two different soil geologies, the raft can support a significant part of the load transmitted by the structure and where the piles will increase the total capacity and will improve the behavior of the whole foundation system.

On other side, Poulos (1991) also refers that there are some situations where the use of a piled raft will be unfavorable:

i. Soft clays;
ii. Loose sands;
iii. Shallow loose and compressible soil layers;
iv. Soils that can suffer settlements by consolidation due to external reasons;
v. Soils that are susceptible of collapsing or expanding due to external causes.

The first two cases (i and ii) the raft could not support the load. In the third case (iii) the settlement on the long-term of the compressible layers can decrease the stiffness of the foundation (structure-soil interaction) which will reduce the contribution of the raft. The last two cases, case (iv) where settlements by consolidation can occur (drainage or shrinkage), the loss of contact between raft and soil can increase the load carried by the piles which takes into an increase of the settlements of the foundation, while in the situation (v) when the soil expands, it will occur an increasing of the soil stress state which can affect piles performance.

1.2 Design Process

Design of this type of foundations can be described into two distinct phases:

(i) Preliminary phase to assess the viability of use of a piled raft solution and to decide the number of piles to satisfy the project requirements;
(ii) Design process phase with details through different aspects like number, type and location of the piles, prediction of the settlements and structural forces to be applied in the foundation system.

The preliminary phase begins with a simple design calculation regarding vertical, horizontal and bending moment forces. In the same line of calculations, settlements are also predicted through theoretical and conventional methods. This way allows the designer to have a prediction of the load types to be applied in the foundation system and what methodology to adopt in the design regarding the type of foundation.

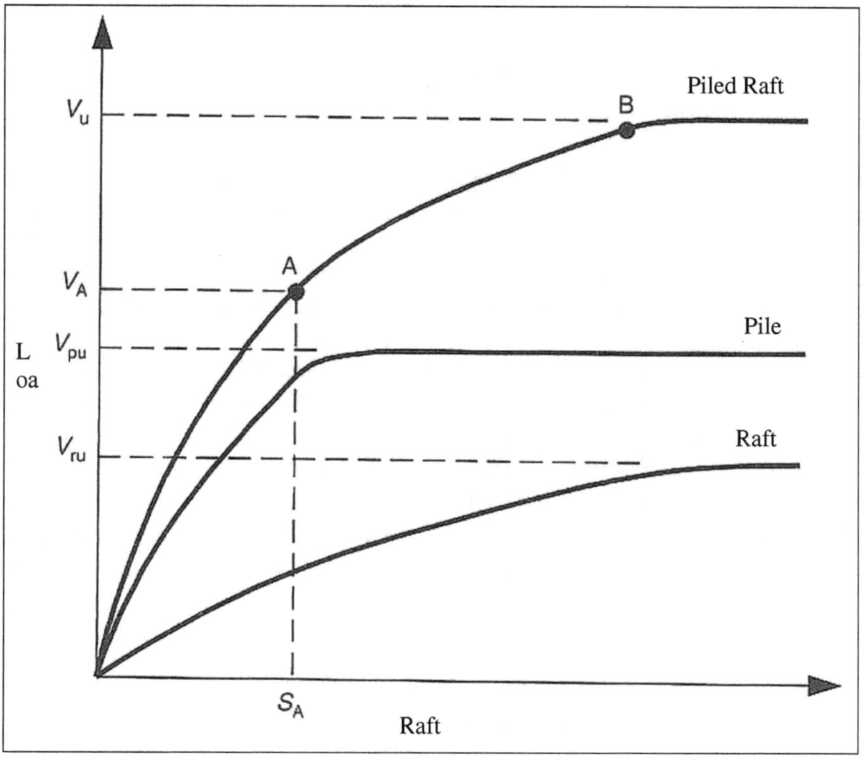

Fig. 3. Pile raft load-settlement curve (Hemsley 2000)

Therefore, if the loads can only be supported partially by the raft, a conventional approach should be adopted (Fig. 2, curve 1). In the case of the raft can support the loads but the settlements of the foundation are not within the limits, i.e. the allowable settlement defined for the project is reached, the methodology to be adopt should be creep-piling (Fig. 2, curve 2). If the main concern in the design is to control total and differential settlements, this should be the methodology to be adopted (Fig. 2, curve 3).

Hemsley (2000) describes the piled raft system combining the approaches from other authors (Randolph 1994; Poulos and Davis 1980). The method combines the prediction of load sharing between raft and the piles. Such combination can be interpreted through the load-settlement graphic for the piles and also for the raft, which will result a trilinear analysis for the piled raft plot (Fig. 3).

Figure 3 presents the behavior of the load-settlement curve for the three elements – raft, pile and piled raft. The curves for the pile and the raft were obtained through a hyperbolic modelling analysis. The piled raft curve is trilinear which can be easily understood as the combination of the previous two curves. It is also perceptible that the response to the load-settlement behavior is improved.

The contribution of the raft in this system takes a good advantage for the global response of the system, once it is clear not only the initial stiffness on the piles is improved but also it is seen that the pile ultimate resistance reached an additional load capacity (segment AB on Fig. 3), which is the combination of the piles and the raft systems.

2 Comparative Analysis Between Numerical and Analytical Formulations

Initially, it was studied the different types of foundation described in this work – raft, pile and pile raft – individually, calculating for each of those elements the load capacity and the prediction of the settlements either by the conventional methods or using the software referred in this paper. This analysis allowed to understand the behavior of each foundation elements and also to validate the results obtained with the software, comparing the results.

The soil strata considered in this study it is a homogeneous granular soil and characteristics can be found in Table 1. The characteristics of this soil allows to reach a safe bearing capacity for the foundation. Even that this soil should be considered "theoretical" and can be discussed the fact of its characteristics allows the use of the conventional foundations, the aims for this work is to interpret the interaction between soil-foundation in piled-raft solutions and consequently take advantage from the good properties of more superficial soil, in the system response. An economical-sustainable designing can therefore be undertaken.

Table 1. Soil geotechnical characteristics.

Υ (kN/m^3)	E (kPa)	ν	ϕ (°)
20	50000	0, 3	40

2.1 Analytical Solutions

The raft load capacity was calculated through the classical formulation (Terzaghi 1943, Eq. 2). It was defined as a raft with 1.05 by 1.05 meters where this value appears as the relation where the width of the raft corresponds to 3 times the value of the pile diameter (L = 3D).

$$q_{ult} = c \cdot N_c \cdot s_c + q \cdot N_q \cdot s_q + \frac{1}{2} \cdot \gamma \cdot B \cdot N_\gamma \cdot s_\gamma \quad (2)$$

Where,

c is the soil cohesive intercept (cohesion);
N_c is a adimensional factor for cohesion;
s_c is a correction factor;
q is the vertical stress in the base;
N_q is an adimensional factor for taking into account the stress in the base of the embedded foundation;
s_q is a correction factor;
γ is the unit weight of the soil;
B is the width of the raft;
N_γ is a adimensional factor;
s_γ is the correction factor (function of the soil characteristics).

The calculations of settlements predictions were obtained following the Schmertmann (1978) expression presented in Eq. 3, including the non-linearity of the stiffness modulus of the soil (Fig. 4).

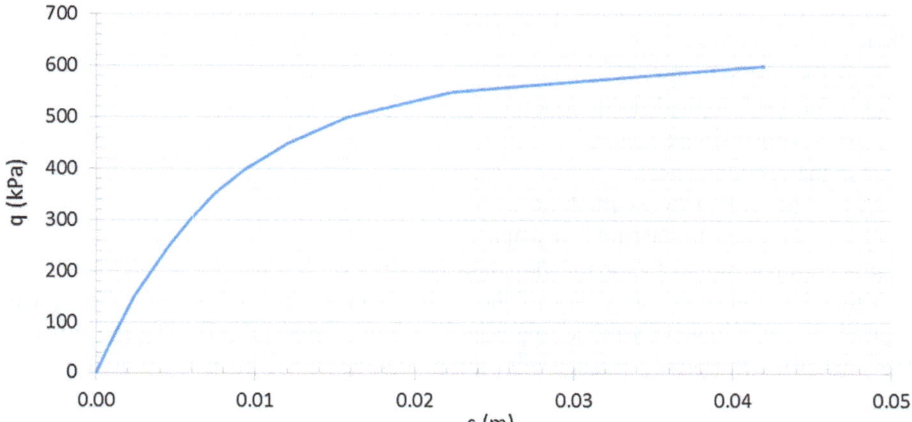

Fig. 4. Raft bearing resistance-settlement curve.

$$s_i = \Delta q_s \cdot B \frac{(1 - \upsilon^2)}{E} . I_s \tag{3}$$

Where,

Δq_s is the load applied (kPa);
I_s is a adimensional factor;
B is the raft width (m);
v is the Poisson coefficient;
E is the soil stiffness modulus.

The ultimate resistance of the piles where calculated through the classic method Beta (effective stress). The value of the unitary shaft resistance it is an average along the depth of the pile (Eq. 4).

$$Q = Q_b + Q_s = A_b \cdot q_b + \sum A_{si} \cdot q_{si} \tag{4}$$

Where,

A_b is the pile base area (m^2);
q_b is the pile base resistance (kPa);
A_{si} is the pile shaft area for the layer i;
q_{si} is the pile shaft resistance (kPa).

The pile was designed with 350 mm diameter and 20 m deep. The settlements (Fig. 5) were calculated following Randolph (1994) methodology (Eq. 5).

$$s_p = \frac{Q \cdot I_\rho}{D \cdot E_{max} \cdot \left[1 - \left(\frac{Q}{Q_{ult}}\right)\right]^n} \tag{5}$$

Where,

s_p is the pile settlement;
Q is the load applied on the pile;
I_ρ is a adimensional factor;
D is the pile diameter;
E_{max} is the soil stiffness modulus;
Q_{ult} is the pile ultimate total resistance;
n is a adimensional factor for the soilss (sands – 0, 5).

Table 2 presents the summary of the main characteristics of the pile. These calculations were followed by modelling both elements – raft and pile – on the software RS3® and FB-Multipier®, finished with the modelling of the piled raft solution.

2.2 Numerical Software – RS3® and FB-Multipier®

The RS3® it is tridimensional finite element software that can easily generate a tridimensional geometry. For this study, a mesh was generated automatically, formed by tetrahedrons with 4 nodes per element. The mesh was refined in the essential points to the analysis.

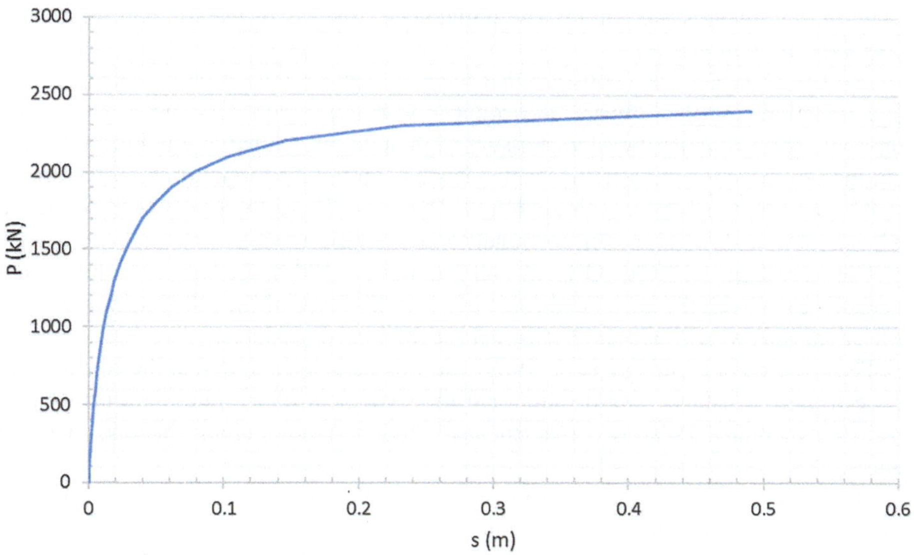

Fig. 5. Pile load-settlement curve.

Table 2. Pile characteristics.

Υ (kN/m^3)	E (MPa)	ν	D (mm)	L (m)	Ab (m^2)
25	30000	0, 25	350	20	0, 09621

The pile was modelled as a beam with 6 degrees of freedom (three of displacements and three of rotation). The nodal points of the pile can or cannot be in conformity with the mesh. If this option is active, the pile will be embedded within the mesh. The pile is also defined by resistance properties and pile-soil interactions. This interaction is defined by the stiffness modulus of the soil and corresponding distortion modulus. The pile-soil interaction is defined by an interface through the parameters of the friction angle and cohesion. The pile tip is the only element that it is not embedded in the mesh, and so, it is defined as a stiffness spring. To calculate this value, it was performed an adjustment between the base resistance of the pile and the corresponding settlement value for the force in order to obtain an equivalent value for the tip stiffness. This was undertaken using analytical methods where settlement and pile base load resistance were calculated.

The Mohr-Coulomb criterion for resistance was adopted, using the isotropic and hyperbolic model to simulate the behavior stress-strain of the soil. The isotropic model adopts an analysis considering that the mechanical parameters of the soil like stiffness modulus, Poisson coefficient, friction angle and cohesion does not change through the loading process of the foundation. On the other side, the hyperbolic model (Duncan and Chang 1970) aims to a non-linear relationship between stress and strain of the soil, this means that the results from this analysis can be more precise. The soil was considered elastic-plastic towards the failure. The initial state of stress of the soil and weight are considered in both analyses.

The FB-Multipier® is a software that focuses more on the structural calculation of the foundation. In this software, the connection between soil-structure is made with non-linear springs. After the characteristics of the pile are defined, it will proceed to the definition of the soil characteristics which is made through the input of the soil stiffness modulus and friction angle, or throughout SPT values. There are 4 models to define the soil: lateral model, axial model, torsional model and tip model.

The lateral model used (O'Neill and Dunnavant 1984) correlates the horizontal displacement y with the horizontal force per unit of length of the pile P with the angle of friction, unit weight and the horizontal reaction sand modulus.

The axial model (McVay et al. 1989) uses, as parameter to define the relationship between axial resistance mobilization and the settlement of the pile, the unit weight, distortion modulus, Poisson coefficient and shaft resistance along the pile (average value).

The torsional model uses as parameters to define the relation between torsional moment and the friction angle, the unit weight, distortion modulus and the friction angle of the soil. The hyperbolic curve of this model is defined by a non-linear analysis regarding the soil behavior during the load stage.

The last model, tip model (McVay et al. 1989) uses, as parameters to achieve a relation between the base resistance of the pile and the settlement, the distortion modulus, Poisson coefficient of the soil and the value of the base resistance of the pile. This is modelled as a spring (non-linear analysis).

2.3 Validation of the Software

Figure 6 presents the comparison of the results from both software and also the analytical curve obtained through classical methodologies for the pile.

As it is perceptible, the FB-M shows a better curve for the pile with a better performance comparing to the RS3 and the classic methodology both in the elastic phase and plastic phase. The initial stiffness of the pile is higher on the FB-M model which gives this better response when compared with the other solutions. This analysis will be the starting point to understand the benefits to the system when a raft is placed on the top of the pile and in direct contact with the soil.

Figure 7 presents the pile curve modelled in the FB-M software (once the FB-M pile curve presented the best performance as seen in Fig. 6, it is used now as comparison) and both piled raft cures modelled in FB-M and RS3. It is noted that the curves that define the piled raft present a better performance on load capacity when the raft is placed on the pile. On the other hand, it is also noted that a better performance of the piled raft on RS3 when compared to the piled raft on FB-M.

Both curves modelled in the different software as a piled raft presents different results with better performances when compared to the curve for the single pile. As it is also shown, the RS3 software presented a big incremental on the load capacity when the pile is associated with the raft which leads to the conclusion that the soil of the foundation has a higher contribution in this situation. This means that the increasing of

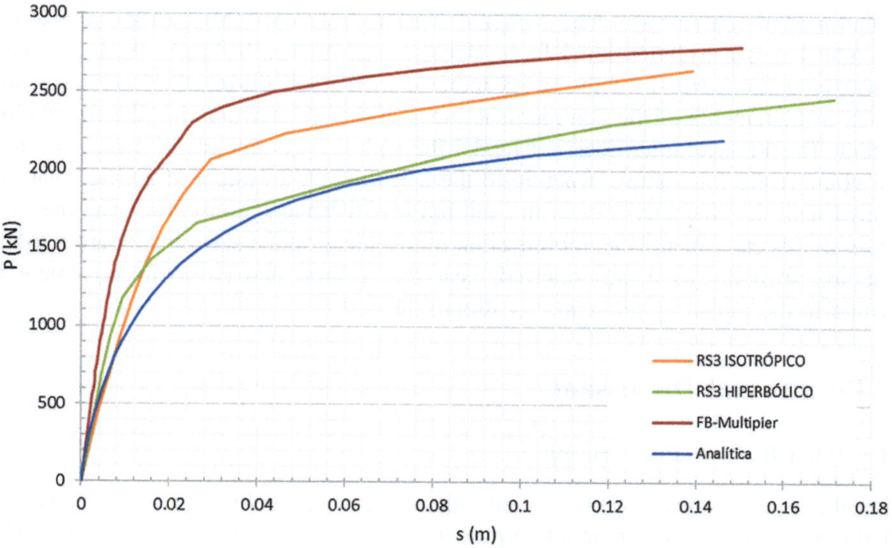

Fig. 6. Pile load-settlement curve compared between software and the analytical curve.

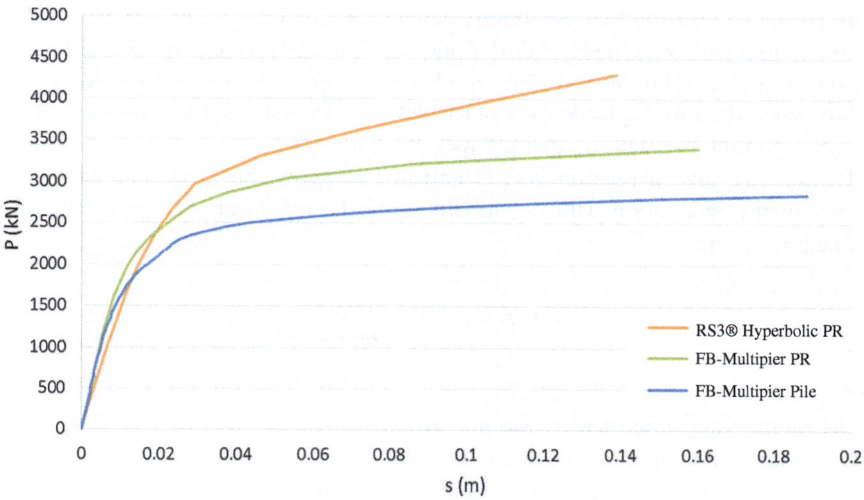

Fig. 7. Piled raft load-settlement curve compared between RS3® and FB-Multipier®.

the stress of the soil during the load stage phase on the foundation, especially in the surface, results in a big improvement of the load capacity of the system this happens due to the positive a confinement of the soil in the surrounding of the foundation having favorable effects to the system performance.

This difference is clear in the FEM numerical code (RS3) used in this study and can be explained by the fact that in the RS3 the soil properties affected during the load phase result in an increasing of the soil resistance and contributes to the improvement of the foundation in terms of load capacity. The reason for this observation is due to the fact that this is a tridimensional finite element code with a tridimensional geometry reproducing the increase of the stress induced in the soil during the load stage and results in increments of the confinement stress in the soil surrounding the foundation system. The FB-Multipier does not model the soil as a continuous system, instead it will connect the foundation structure to the soil through a system of springs that are defined with the characteristics of the soil but it will remains constant during the load phase. In this case, there is no mobilization of the stress on the surrounding areas to the structure, specially as it was referred, near the surface, and so no improvement is verified in the global system capacity due to the soil.

3 Results and Discussion

3.1 Piled Raft vs. Pile Groups

After the initial studies of each element individually and combined, two more realistic piled raft geometrically comprising a pile group for 4 and 9 piles were modelled. The aim of this study in this phase was to clearly understand the contribution of the raft when placed on the top of the piles and in direct contact with the underlying soil, in terms of load distribution (raft-piles) and the benefits of the soil increasing confinement. To distinguish the two conditions, where the raft was in total contact with the soil, one simulation was made with the pile cap 1 m above the ground and the other with direct contact with the soil. This is what will allow a better understanding of the benefits when the raft is placed in contact with soil, or only placed as a pile cap 1 m above the ground and with no contact with the soil.

To quantify this, a relation was established to show the increasing of the load capacity of the foundation with or without using the raft in contact with the soil, in percentage (Eq. 6):

$$EN(\%) = 1 - \frac{Q_{PG.}}{Q_{PR}} \quad (6)$$

Where,

Q_{PG} is the load supported by the pile group;

Q_{PR} is the load supported by the piled raft.

For both cases analyzed, with 4 and 9 piles, the results are compared in the Tables 3, 4, 5 and 6.

The results obtained with FB-Multipier® code present an increase in the capacity of the piled raft foundation in relation to the corresponding piles group by 22.2% for 4 piles and 30.4% for 9 piles. These values obtained for these two cases, are a good indicator that there is a contribution of the ground on load carrying capacity of the system, as the ground between piles has an important contribution to the structure strength. Despite of this, there is no clear stress dependency in raft-soil relationship and pile-soil, specially the increase in confinement stress, in the most superficial layers which is not as well reproduced in FB-Multipier as in FEM RS3® program. On FB-Multipier® this relationship does not depend on the increase in the stress of the

Table 3. Comparison of the results (RS3®) between the piled raft (PR) and the group of piles (GP) for 4 piles.

RS3	P (kN)	s (m)
PR (n = 4)	12150	0.053
GP (n = 4)	6804	0.051
Desvio	44%	

Table 4. Comparison of the results (FB-Multipier®) between the piled raft (PR) and the group of piles (GP) for 4 piles.

FB-Multipier	P (kN)	s (m)
PR (n = 4)	12912	0.092
GP (n = 4)	10043	0.096
Desvio	22%	

Table 5. Comparison of the results (RS3®) between the piled raft (PR) and the group of piles (GP) for 9 piles.

RS3	P (kN)	s (m)
PR (n = 9)	32490	0.119
GP (n = 9)	18275	0.111
Desvio	43%	

Table 6. Comparison of the results (FB-Multipier®) between the piled raft (PR) and the group of piles (GP) for 9 piles.

FB-Multipier	P (kN)	s (m)
PR (n = 9)	33500	0.098
GP (n = 9)	23314	0.091
Desvio	30%	

ground and so there is not representative interaction with the structural elements. The increase in stress of the soil occurs in the underlying soil to the raft that will benefit increasingly confining pressure around the piles.

The analyses obtained in RS3® have slightly different results. The analysis of the piled raft and pile group with 4 piles, it appears that the deviation is 44% higher than the same case analyzed in FB-Multipier®. These 22% more in load capacity for the same case when the modeling is done in RS3® code will be largely due to the underlying soil contribution between the raft and the piles in shallower horizons. This means that the soil in this 3D finite element analysis contributes (particularly in case of soil with medium to high density under the piled raft) by increasing this confinement and, consequently, increasing stiffness and strength during the load stage. For the other studied case, the piled raft foundation with 9 piles the deviation value remains very similar to that observed in the previous case, approximately 43.7%.

Assuming that the ground modelled on FB-Multipier® with the raft foundation of 4 piles, in contact with the soil, there was a lower strength increase than in the result, obtained with RS3, it can be concluded that the increase in number of piles in a piled raft foundation or a pile group, it does not give a increase in load capacity. That is the increase of the foundation load capacity after certain point. The stiffness of the foundation soil experiences a gradual increasing during the load stage, which will result after a certain point, in a global movement of the foundation similar to a rigid block.

3.2 Individual Analysis of the Piles Within the Piled Raft Foundation

Individual analyses were conducted for the piles that contained in the different system modelled – piled raft of 1, 4 and 9 piles – and compared between them, aiming to understand the benefit of the soil confinement to the piles and to the structure. Theses analysis consisted in the observation of the load-settlements curves of the piles contained in the piled-raft systems. The results were compared with the two different codes. There is symmetry in the different pile designs, which results confirmed for the two first systems, where the predicted results are the same in all of the elements. However, within the piled-raft system with 9 elements, the results are different depending on pile position. It was observed 3 different behaviors according to their locations: central pile, the piles that are positioned in the corners and the piles positioned in the middle of the side.

Figure 8 it represents the case of 1 pile with the raft it can be observed that the increase of the load capacity is not significant.

Figure 9 presents the results for the piled raft with 4 piles, where some differences are now noted and there is an improvement on the pile load-settlement curve behavior. As already referred, due to the symmetry it was only represented one curve from the RS3® software which reflects the curve for the 4 piles. It is also relevant that the piles

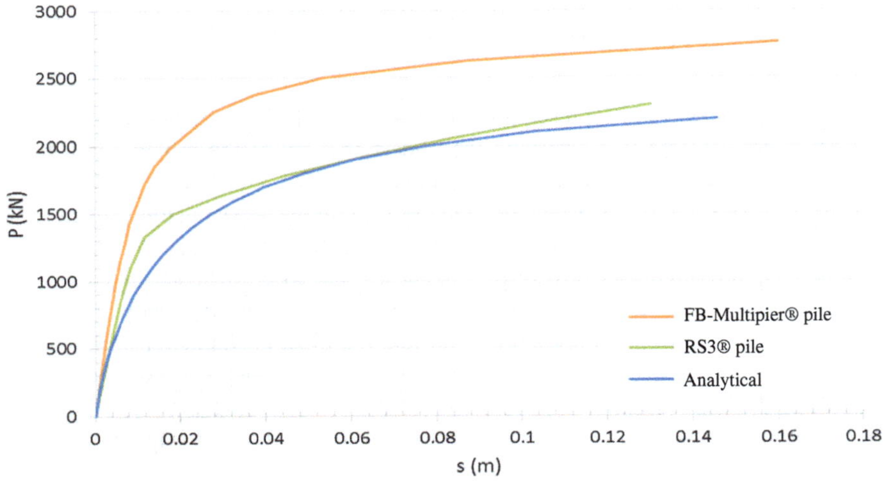

Fig. 8. Comparison between the behavior of the load-settlement curve of the pile contained within the pile raft of one pile (RS3® and FB-Multipier®).

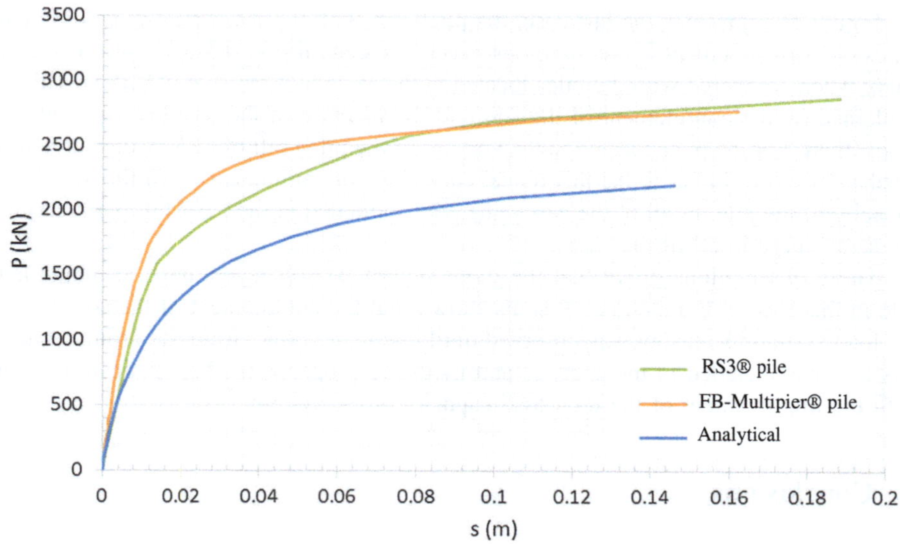

Fig. 9. Comparison between the behavior of the load-settlement curve of the pile contained within the pile raft of four piles (RS3® and FB-Multipier®).

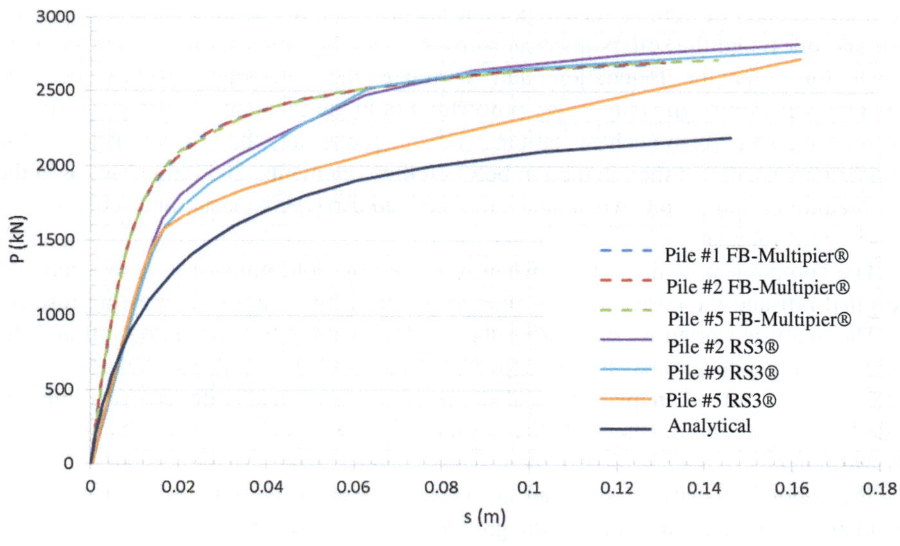

Fig. 10. Comparison between the behavior of the load-settlement curve of the pile contained within the pile raft of nine piles (RS3® and FB-Multipier®).

modelled in the FB-Multipier maintain the same curve as the Fig. 8, where no changes are observed in the load capacity improvement. In the RS3®, the improvement of the total load capacity of the piles is now significant, once the curve approaches the FB-Multipier results.

Figure 10 represents the last case, the piled raft with 9 piles, where three different types of load-settlement curve were observed. The central pile (#5 RS3) has a similar curve when compared with the piles located in the middle of the side (#2 RS3). This is explained by the confinement of the soil and the increase of the in-piles soil confinement stress. For the piles located in the vertices of the piled raft (#9 RS3), the curve has a behavior more similar to the theoretical curve but with an increasing of the total load capacity of the pile. In all of the cases, the piles modelled in the FB-Multipier® shown the same curve in all of the cases.

These observations prove that the more superficial soil layers have an important role in this type of foundations. Results shown that the confinement of the soil due to the load system increasing was observed until one meter down from the ground. Once this factor is included in the analysis and modelled properly, the benefits can be seen with the improvement of the total load capacity of the system.

4 Conclusions

It was shown in this study that the piled raft system develop a strong interaction between the foundation elements and the soil. Not having an analytical calculation methodology and design being an approximation of several results to the reality of the proposes of the project, is important that the designer take in account many aspects to the final decision design. A very important factor is the interactions between structure (raft and piles) and the soil. Numerical software has a big impact, once it is possible to model this type of foundation and correlate the interaction factors between structure-soil. Although being a very powerful tool in the calculations, designers should always have some care such as defining the soil properties, the characteristics of the foundation system and the interaction between these elements; and critical sense in the interpretation of the results when analyzing the load carrying capacity and settlement of the different foundations.

The comparison of the results, when modelled on both numerical codes, between the piled raft and the corresponding pile groups lead this study to interesting results.

The analysis started by comparing the results obtained between piled raft and the corresponding pile groups. It was expected, from previous analysis, that the contribution of the raft when placed on the top of the piles and with contact with the underlying soil should have better results in the RS3® software than in the FB-Multipier®.

The following analysis consisted in the assessment of the individual piles of each foundation and the contribution of the ground during the load phase. It was proved that as the number of piles increases in the system, the load capacity of the foundation as well as the soil, improves becoming insignificant after a certain number of piles.

There is an increase of stiffness, a lot due to the incremental stress in the soil mobilized in the pile shaft caused by stress increasing on the ground by the raft, which it will be negligible from the moment that the foundation behaves like a rigid block.

Acknowledgments. I would like to thank to Professor Paulo Lopes Pinto from FCTUC (Coimbra, Portugal) for the invaluable help and guidance of this work.

References

Bridge Software Institute: FB-Multipier. Programa e Manual. University of Florida, Gainesville Florida (2010)

Duncan, J.M., Chang, C.Y.: Nonlinear analysis of stress and strain in soils. J. Soil. Mech. Found. Div. **96**(5), 1629–1653 (1970)

Hemsley, J.A.: Design Applications of Raft Foundations. Thomas Telford, London (2000)

Mandolini, A.: Design of piled raft foundations: practice and development. In: Proceedings of Deep Foundations on Bored and Auger Piles–BAP IV, Ghent, Belgium, pp. 2–4 (2003)

McVay, M.C., Townsend, F.C., Bloomquist, D.G., O'Brien, M.O., Caliendo, J.A.: Numerical analysis of vertically loaded pile groups. Comunicação apresentada em Foundation Engineering@ sCurrent Principles and Practices (1989)

O'Neill, M.W., Dunnavant, T.W.: A study of the effects of scale, velocity, and cyclic degradability on laterally loaded single piles in overconsolidated clay. University of Houston, Department of Civil Engineering (1984)

Poulos, H.G.: Analysis of piled strip foundations. In: Proceedings of the 7th International Conference on Computer Methods and Advances in Geomechanics, Cairns, Australia, vol. 1, pp. 183–191 (1991)

Poulos, H.G.: Methods of analysis of piled raft foundations. A Report Prepared on Behalf of Technical Committee TC18 on Piled Foundations, 46 pages (2001)

Poulos, H.G., Davis, E.H.: Pile foundation analysis and design. Wiley, 397 pages, ISBN-10: 0471020842

Randolph, M.F.: Design methods for pile groups and piled rafts. In: Proceedings of 13th ICSMFE, vol. 5, pp. 61–82 (1994)

Rocscience: RS3 - Rock and Soil 3-dimensional analysis program (2013)

Schmertmann, J.H.: Guidelines for Cone Penetration Test Performance and J Design, Report No. FHWA-TS-78-209, U.S. Department of Transportation, Federal Highway Administration, Washington, D.C. (1978)

Terzaghi, K.: Theoretical Soil Mechanics. Wiley, New York (1943)

Field Monitoring of Concrete Piles of an Integral Abutment Bridge

Khalid Alshibli[1(✉)], Andrew Druckrey[1], and George Z. Voyiadjis[2]

[1] University of Tennessee, Knoxville, TN, USA
[2] Civil and Environmental Engineering, Louisiana State University, Baton Rouge, LA, USA

Abstract. Integral abutment bridges (IABs) are a cost-effective design method for bridges. Recently, Louisiana Department of Transportation and Development (LA-DOTD) constructed their first two IABs; one was supported by HP steel piles driven in clayey soil while the other was supported by pre-stressed precast concrete (PPC) piles driven into fine sand deposit. The use of PPC piles has long been recognized as a good option for Louisiana bridges. However, there are concerns that the rigidity of the piles driven in sandy soils might cause excessive stresses in the bridge superstructure. This paper presents the instrumentation plan of two 36 in (914.4 mm) square PPC piles and the bent-soil interaction. Sisterbar strain gauges were attached to the pre-stressing strands in the piles along with nine pressure cells that were attached to the face of the bent supporting the piles. The bridge deformations were mainly controlled by the piles' rigidity, soil resistances surrounding the piles, and connection behaviors between the pile-bent. Based on the observed temperature effects, the design of the piles of the Caminada Bay Bridge is very conservative. The piles experienced very low bending moments and very small amount of pressure on the backfill soil.

1 Introduction

Integral abutment bridge (IAB) systems have become a cost-effective alternative in the last two decades. They are constructed without deck joints, particularly at the abutments. IAB have also been referred to as integral, jointless, rigid-frame, and U-frame bridges. First built in the United States in the 1930s, by the 1990s, IABs were extensively used worldwide. IABs can be single or multiple spans, offer several advantages over conventional structures, and are currently used in more than 30 US states and Canadian provinces (e.g., Wolde-Tinsae and Greimann 1988; Russell and Gerken 1994; Kunin and Alampalli 2000; Arockiasamy et al. 2004) and in other countries. Benefits offered by integral bridges include reduced initial costs, lower long-term maintenance expenses, elimination of problematic expansion joints and bearings, less deterioration, lower impact loads, improved riding quality, simple construction procedures, and structural continuity to resist overloads. The disadvantages of such construction include subjecting the superstructures to large secondary stresses caused by the response of continuous superstructures to thermal and moisture changes. These cyclic movements and stresses must be addressed at the bridge abutment. US Federal Highway Administration (FHWA) is promoting the usage of integral

M. Abu-Farsakh et al. (eds.), *Advances in Analysis and Design of Deep Foundations*, Sustainable Civil Infrastructures, DOI 10.1007/978-3-319-61642-1_18

abutment and jointless bridges (IAB) and a large number of IABs have been constructed in many states. However, due to Louisiana's unique soil conditions and the complexity of the pile and soil interaction in the integral abutment bridges, no full integral bridge was ever considered in Louisiana until 2011.

Few studies investigated the behavior of IAB support piles (e.g., Jorgenson 1983; Card and Carder 1993; Abendroth et al. 2005; Arsoy 2000; Frosch et al. 2005, Girton et al. 1991; Lawver et al. 2007; Yannotti, et al. 2005). Selecting and installing an appropriate pile is important in the design of integral abutment bridge since the pile must withstand annual thermal displacements. When the piles are embedded into the abutment, the monolithic nature of the structure would cause them to translate and rotate with the superstructure when the temperature changes. The rigid pile-to-abutment connection and fixity at the pile base makes the pile perform as fixed-fixed columns. Square PPC piles are common foundation piles for Louisiana bridges, but the state has concerns about the rigidity of the piles, especially if they are driven in sandy soils. This paper presents the field performance of pile foundation for the first IAB in Louisiana with emphasis on the behavior of the piles and the bent-soil interaction. The behavior of the super-structure will be presented in another paper.

2 Instrumentation Plan for Caminada Bay Bridge Site

The bridge is located at Grand Isle, Louisiana, USA (29°15'48" N 89°57'24" W), about 100 miles (160 km) to the south of New Orleans. The total length of the bridge is 3945 ft (1202 m). The monitoring program was conducted on the first 11 spans. The width of the bridge is 50 ft (15.2 m) consisting of two 21 ft (6.4 m) lanes and a 7 ft (2.1 m sidewalk on the northern side. The slabs are fully integrated with the first bent (Bent 1) at the left end, simply supported on the eleventh bent (Bent 11) at the right end, and rigidly connected with all the interior bents. Each bent is supported by a single row of four PPC piles with a square cross section of 36 in (914 mm, Fig. 1). In addition, the material properties designed for this bridge are summarized as follows: (a) Class AA (M) concrete, with a compressive strength of 4060 psi (28 MPa), was used for the slabs and bents; (b) Class P (M) high performance concrete, with a minimum compressive strength of 6000 psi (41 MPa) at 28 days, and an average compressive strength of 10,000 psi (69 MPa) at 56 days, was used for the PPC piles; (c) Type 316LN stainless steel, with an elastic modulus of 29000 ksi, a tensile strength of 75 ksi, and a yield stress of 30 ksi, was used for the deformed reinforcing steels in the bents and slabs; (d) Grade 60 black steel, with a 60 ksi yield stress, was used for all the other deformed reinforcing steels; (e) Grade 270 steel, with a 270 ksi yield strength, was used for the pre-stressing strands.

Table 1 lists the properties of soils in Borehole 1, which was drilled at Bent 1. The piles are 60 ft (18.3 m) long that were mainly driven into the fine sand deposit. Figure 2 shows the instrumentation plan for the IAB. It included two piles, one abutment, and a backfill on one side of the bridge. Bridge Diagnostics, Inc. (BDI) There are other sensor measurements not reported in this paper due to page limitations and the scope of the paper. BDI installed all the sensors. Before the piles were cast, 32 sisterbar strain gauges (GeoKon model 4911) were attached to pre-stressing steel of two piles (16 gauges in

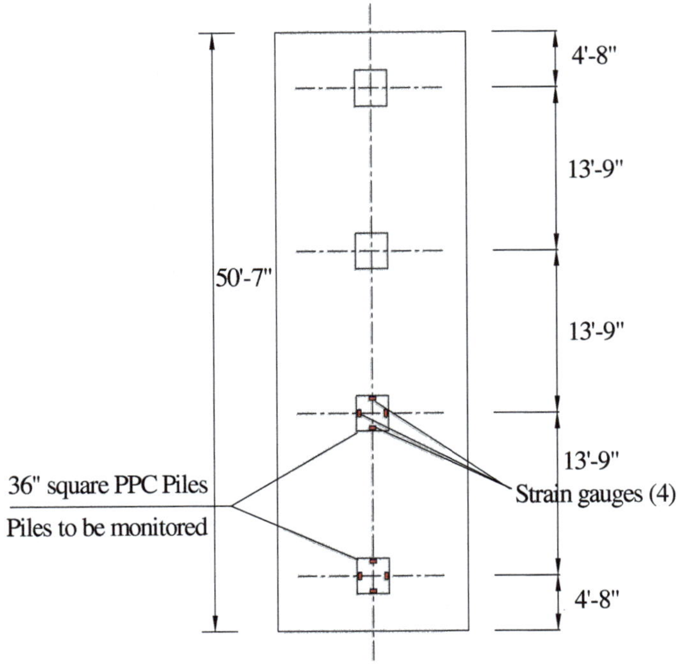

Fig. 1. Plan view of piles layout and Bent 1 top view dimensions.

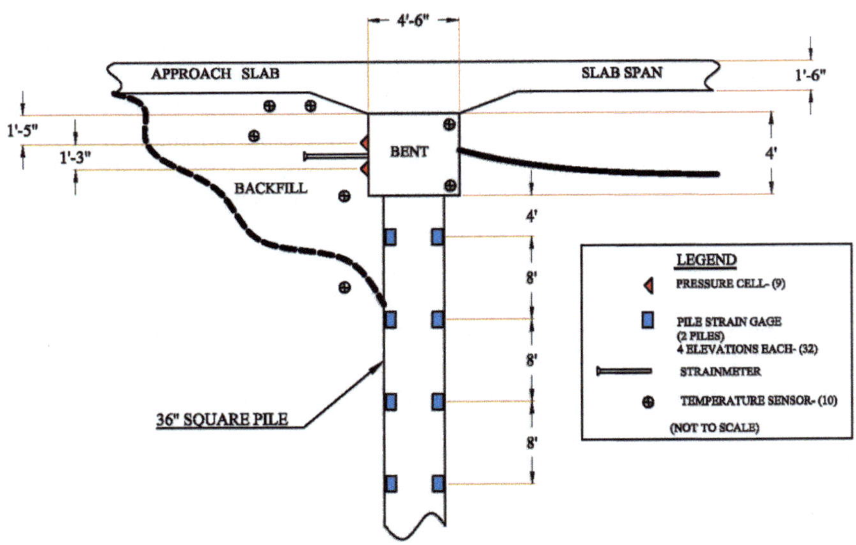

Fig. 2. Instrumentation plan for Bent 1 of Caminada Bay bridge-substructure.

Table 1. Soil properties in Borehole 1 drilled at Bent 1.

Soil Description	Wet Density (lb/ft^3)	Moisture Content (%)	Liquid Limit	Plasticity Index	q_u (tsf)	SPT or UU (tsf)	Elevation (ft)
Tan fine sand with traces of silt						N-11 N-13 N-15	0 −5 −10
Grey fine sand with traces of silt						N-18 N-11	−15
Grey fine sand						N-13 N-6 N-9	−20 −25
Grey fine sand with traces of silt						N-21	
Grey silty fine sand						N-22 N-17 N-13	−30 −35
Tan and grey fine sand	101	31				0.67 at 17.8 psi	−40
Grey fine sand with traces of silt						N-10 N-16	−45
Grey fine sand	112	25				3.44 at 21.3 psi	−50
Grey fine sand with traces of silt						N-44 N-42 N-33 N-35	−55 −60
Grey clay	106	51	86	55	0.92	N-8	−65 −70
Grey clay with silt lenses	112	28			0.68		−75
Grey clay	113 112	48	51	29	1.35		−80 −85
Grey slightly silty clay	118	33	42	24	0.7		−90
Grey clay with silt lenses	109	44			1.44		−95
Grey very sandy clay	123	24	31	13	1.10		−100
Grey sandy clay	120	27			1.28		−105 −110

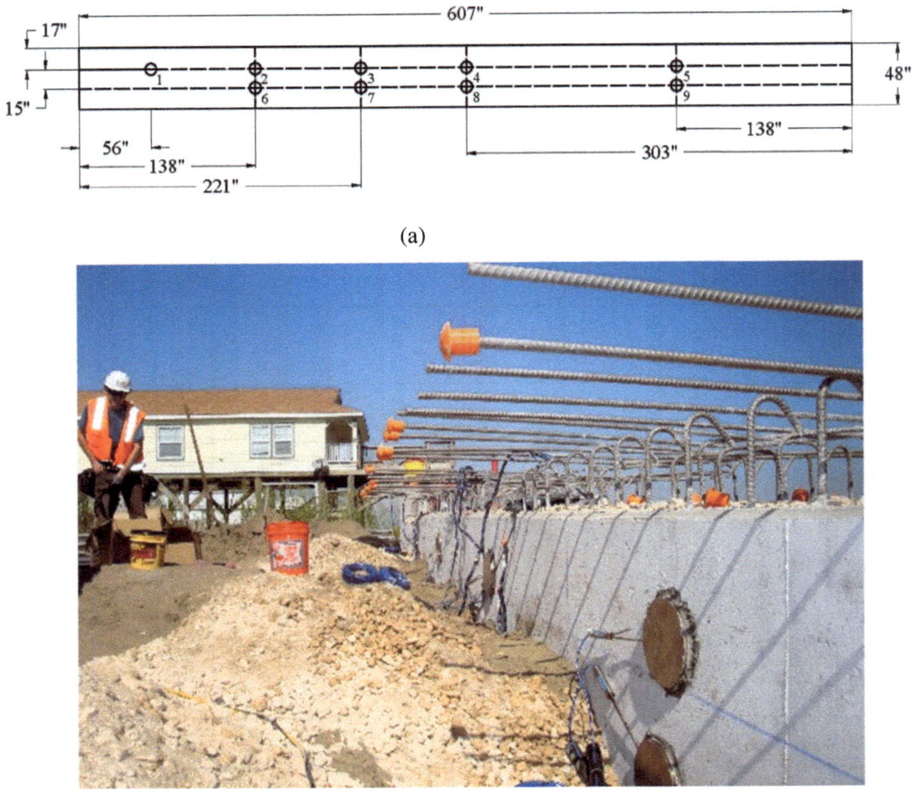

(a)

(b)

Fig. 3. (a) Schematic of location of pressure cells on the face of bent 1. (b) Photo of pressure cells mounted on Bent 1 face.

each pile) in the concrete cast yard. After applying the pre-stressing tension to the tendons, the concrete was cast, and the piles were shipped to the bridge site. Driving the piles did not damage the sisterbar strain gauges cast inside the piles. Strain gauge integrity was checked before shipping, after arrival, and after pile driving. The pressure cells (GeoKon model 4810) were mounted at the locations shown in Fig. 2 using stainless steel mounting hardware. In addition, BDI provided a mason to place a small pad of mortar behind each cell during installation to ensure that it would make uniform contact with the concrete surface (Fig. 3).

3 Behavior of Piles and Bent-Soil Interaction

The soil pressure measurements were recorded using the nine pressure cells that were mounted on the face of the bent (Fig. 3). Figure 4 shows a summary of the soil pressure measurements. A zoom in of the measurements for few days in January (cold days) and July of 2012 (hot days) are depicted in Figs. 5 and 6, respectively. One can notice that

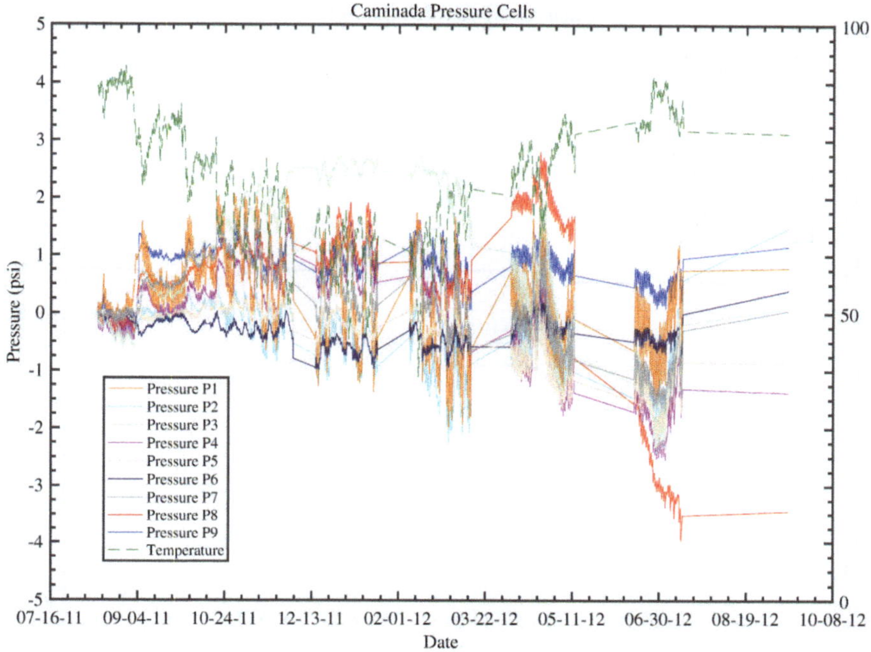

Fig. 4. Pressures from all Caminada pressure cells for entire dataset.

the bridge shrunk in cold weather which caused the soil pressure to drop to zero or a small negative value and expanded in hot weather which results in a maximum of 3 psi (20.7 kPa) passive pressure on soil.

To calculate the bending moments in the piles, the micro-strains for each of the sisterbar strain gauges were normalized based on the reading collected at the time of installation and were corrected for temperature effects. The following are the strains that were used to calculate the moments in the respective direction:

$$\varepsilon_t = (\varepsilon_1 - \varepsilon_2 + \varepsilon_3 - \varepsilon_4)/4 \tag{1}$$

$$\varepsilon_x = (\varepsilon_1 + \varepsilon_2 - \varepsilon_3 - \varepsilon_4)/4 \tag{2}$$

$$\varepsilon_y = (-\varepsilon_1 + \varepsilon_2 + \varepsilon_3 - \varepsilon_4)/4 \tag{3}$$

where $\varepsilon_1 - \varepsilon_4$ are the measured strains in the four corners of the pile and ε_t, ε_x, and ε_y are the strains in the transverse, x, and y directions (Fig. 7), respectively. The moments were calculated as:

$$M = \frac{E \times I \times \varepsilon}{h} \tag{4}$$

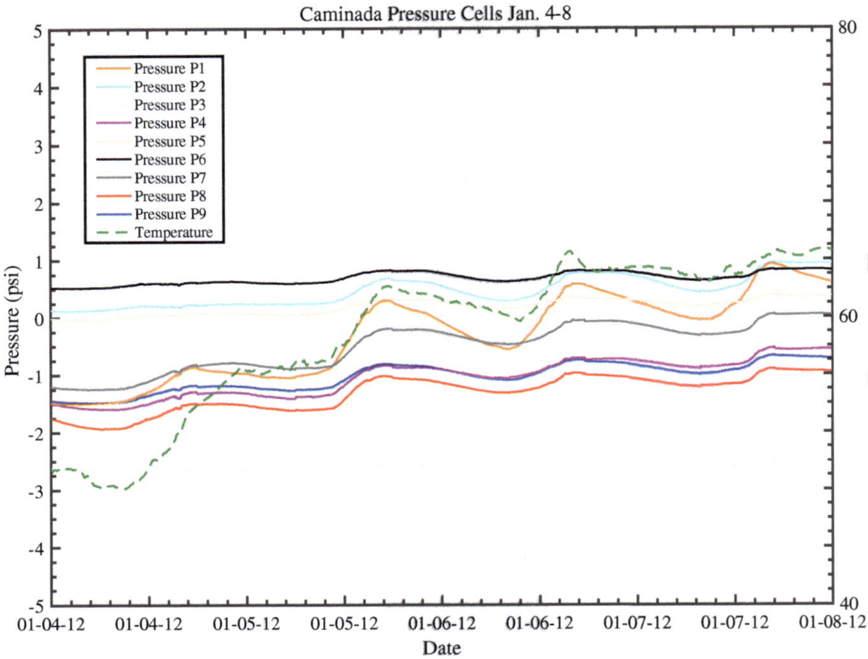

Fig. 5. Pressures from all Caminada pressure cells for Jan. 4, 2012 Jan. 8, 2012.

where M is the directional bending moment, E is the modulus of elasticity, I is the mass moment of inertia, ε is the directional strain, and h is the distance between the sensors. The modulus of elasticity for the piles was computed as:

$$E = 57,000\sqrt{f_c'} \tag{5}$$

where f_c' is the compressive strength of the concrete after 28 days (6000 psi = 41,370 kPa). The mass moment of inertia was calculated based on the gross area of the pile. The moment in each direction was then calculated for each section of the pile based on the strain in the representative direction. Moments were calculated for the interior and exterior piles on October 1, 2011; January 1, 2012; April 6, 2012; and July 7, 2012 using the average strain during a particular day. The cracking moment for the pile was calculated assuming no pre-tension. Since concrete cracks in tension, the modulus of rupture was used to determine when the pile will crack. The modulus of rupture (f_r) is equal to 7.5 times the square root of the concrete compressive strength per ACI code. The cracking moment was calculated using the following equation:

$$M_{cr} = \frac{f_r \times I}{c} \tag{6}$$

where M_{cr} is the cracking moment, c is the distance of the tensile fibers to the neutral axis (half the pile width = 18 in. (457.2 mm)), and I is the mass moment of inertia of

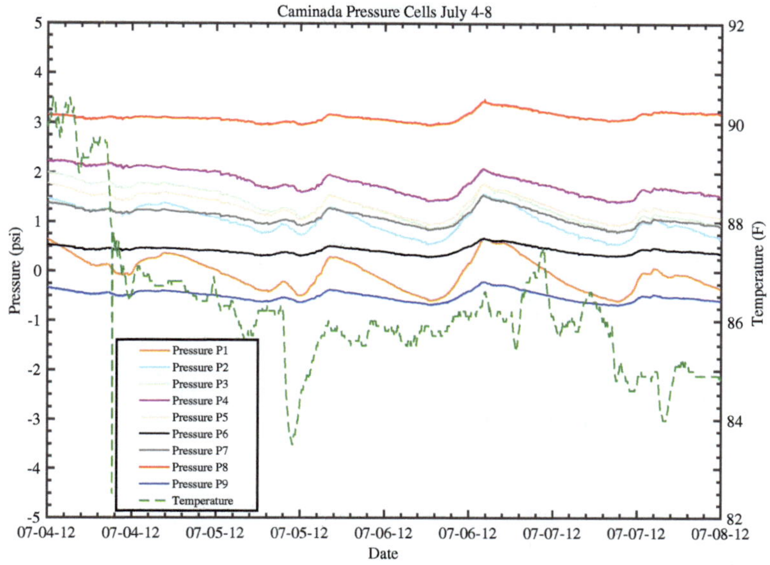

Fig. 6. Pressures from all Caminada pressure cells for July 4, 2012-July 8, 2012.

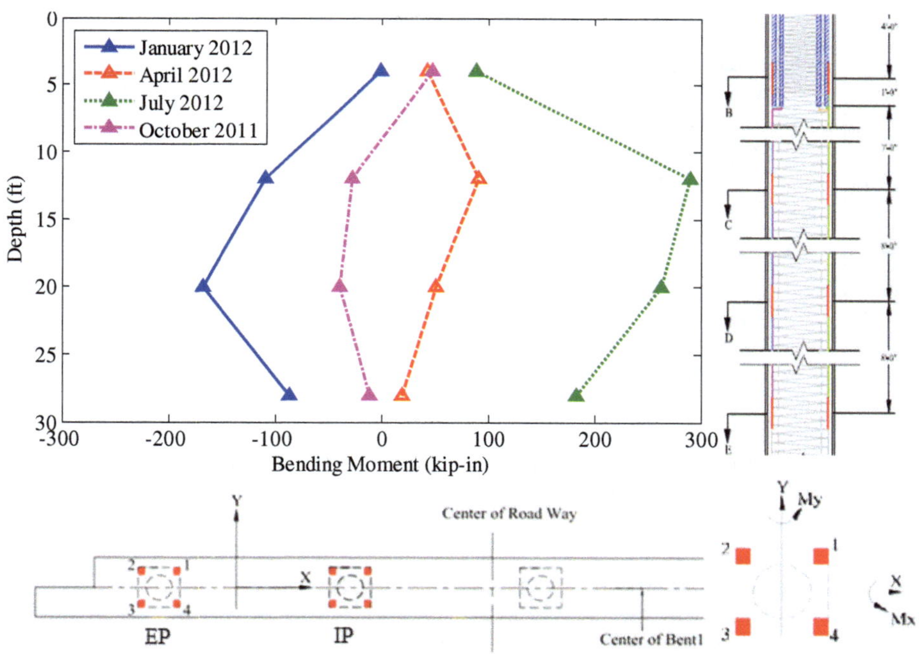

Fig. 7. Moment about x-axis, M_x in the interior pile obtained from strain gauge data. Theoretical cracking moment is 4518 kip-in.

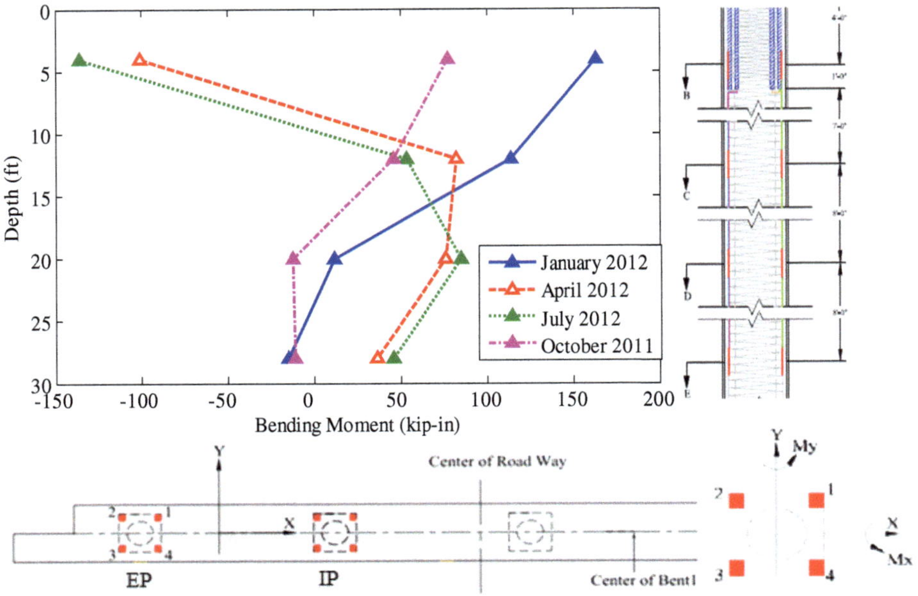

Fig. 8. Moment about y-axis, M_y in interior pile obtained from strain gauge data. Theoretical cracking moment is 4518 kip-in.

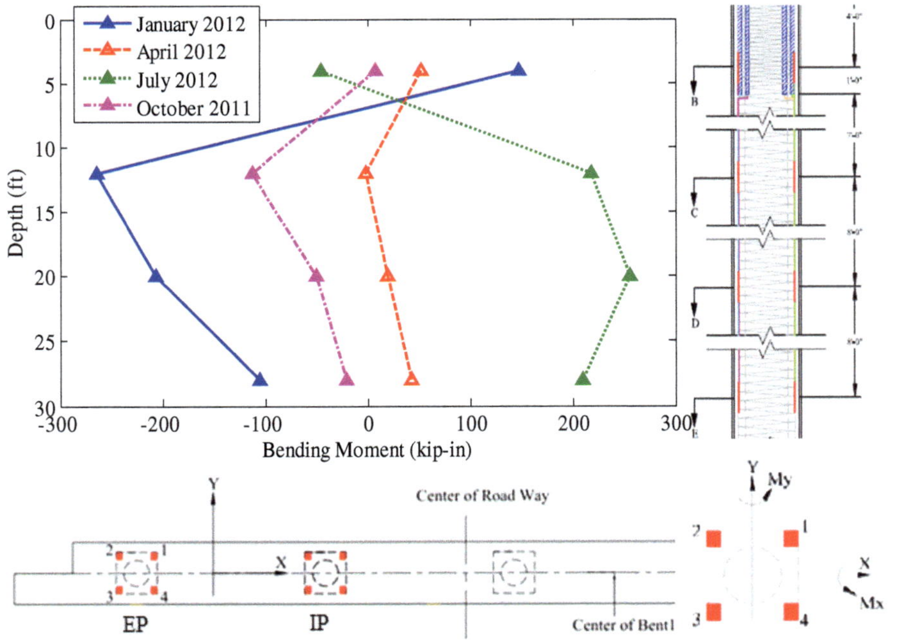

Fig. 9. Moment about x-axis, M_x in the exterior pile obtained from strain gauge data. Theoretical cracking moment is 4518 kip-in.

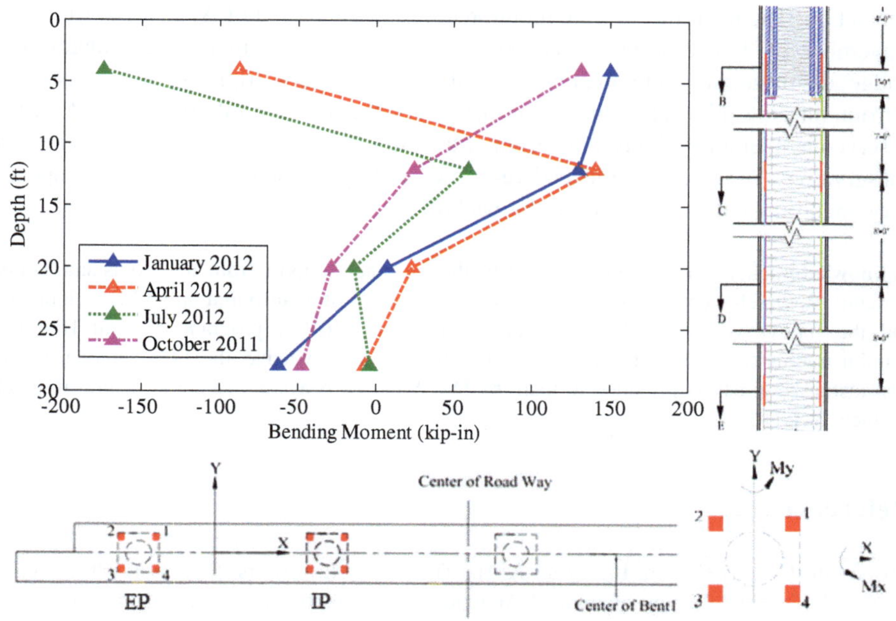

Fig. 10. Moment about y-axis, M_y in exterior pile obtained from strain gauge data. Theoretical cracking moment is 4518 kip-in.

pile cross-section (139,968 in^4 = 5.8 × 10^{10} mm^2). The bending moments in the interior and exterior piles are shown in Figs. 7 through 10. The moments in the direction of traffic (Figs. 7 and 9) show typical trends with values that range from −200 kip-in to 200 kip-in. Systematic change is evident with temperature fluctuations; for example, January and July represent cold and hot sessions, respectively. Moment trends in piles due to bridge deck thermal expansion and contraction resulted in negative moments in January (contraction) and positive moments in July. Moderate temperatures in April and October resulted in smaller moments. Moments in the transverse direction (90° to the direction of traffic, Fig. 10) do not show a clear trend with season but rather similar trends for April through July and January through October with overall moment ranges less than the moments in the direction of traffic. There are fewer thermal expansions in the transverse direction. and south Louisiana is warm with no major temperature changes from April to July and October to January. The concrete cracking of the pile is 4518 kip-in. Clearly, the piles experienced small moments, and the design is very conservative (Fig. 8).

4 Conclusions

The paper presented an instrumentation plan for two piles and the supporting bent of Caminada Bay bridge, the first IAB bridge constructed in Louisiana. The PPC square piles are 60 ft (18.3 m) long and were driven into mainly fine sand deposit. Two piles were instrumented with sisterbar strain gauges along with pressure cells at bent-soil interaction. The instruments were monitored during the period of August 2011 through September 2012. The bridge expanded and contracted with temperature changes. The bridge deformations were mainly controlled by the piles' rigidity, soil resistances surrounding the piles, and connection behaviors between the pile-bent. Based on the observed temperature effects, the design of the piles of Caminada Bay Bridge were very conservative. The piles experienced very low bending moments, and the bent imposed very small amount of pressure on the backfill soil.

Acknowledgments. The authors appreciate the financial support from the Louisiana Transportation Research Center (LTRC), Louisiana Department of Transportation and Development, and the Innovative Bridge Research and Deployment (IBRD) program, Federal Highway Administration (FHWA). The authors would like to acknowledge the help, guidance, and administrative direction provided to them by Dr. Walid R. Alaywan, senior structural research engineer at LTRC.

References

Abendroth, R.E., Greimann, L.F., Lim K.-H., Thomas, M.E., Sayers, B.H., Kirkpatrick, C.L., Ng, W.C.: Field Testing of Integral Abutments. Final report, Iowa State, University, Iowa Department of Transportation, 819 pages (2005)

Arockiasamy, M., Butrieng, N., Sivakumar, M.: State-of-the-art of integral abutment bridges: design and practice. J. Bridge Eng. **9**(5), 497–506 (2004)

Arsoy, S.: Experimental and analytical investigations of piles and abutments of integral bridges. Ph.D. dissertation, Virginia Polytechnic and State University (2000)

Card, G.B., Carder, D.R.: A literature review of the geotechnical aspects of the design of integral bridge abutments. Proj. Rpt. 52, Trans. Res. Lab., U.K. (1993)

Frosch, R.J., Wenning, M., Chovichien, V.: The in-service behavior of integral abutment bridges: abutment-pile response. In: Proceedings of the Integral Abutment and Jointless Bridges Conference, FHWA, Baltimore, MD, pp. 30–40 (2005)

Girton, D.D., Hawkinson, T.R., Greimann, L.F.: Validation of Design Recommendations for Integral-Abutment Piles. J. Struct. Eng. ASCE **117**(7), 2117–2134 (1991)

Jorgenson, J.L.: Behavior of abutment piles in an integral abutment in response to bridge movements. In: Transportation Research Record: Journal of the Transportation Research Board, No. 903, Transportation Research Board of the National Academies, Washington, DC, pp. 72–79 (1983)

Kunin, J., Alampalli, S.: Integral abutment bridges: current practice in United States and Canada. J. Perform. Constr. Facil. **14**(3), 104–111 (2000)

Lawver, A., French, C., Shield, C.K.: Field performance of integral abutment bridge. In: Transportation Research Record: Journal of the Transportation Research Board, No. 1740, Transportation Research Board of the National Academies, Washington, DC, pp. 108–117 (2007)

Russell, H.G., Gerken, L.J.: Jointless bridges-the knowns and the unknowns. Concret Int. **16**(4), 44–48 (1994)

Wolde-Tinsae, A.M., Greimann, L.F.: General design details for integral abutment bridges. Civ. Eng. Pract. **3**(2, 7–20), 199–210 (1988)

Yannotti, P.A., Alampalli, S., White, H.L.: New York State Department of transportation's experience with integral abutment bridge. In: Proceedings of the Integral Abutment and Jointless Bridges Conference, FHWA, Baltimore, Maryland, pp. 41–49 (2005)

Numerical Modeling of Pile Groups Composed of Two Open-Ended Steel Piles

Khalid Abdel-Rahman(✉) and Martin Achmus

Institute for Geotechnical Engineering, Leibniz University of Hannover, Hannover, Germany
idkhalid@igth.uni-hannover.de

Abstract. Piles are often used as foundation elements to support different offshore structures founded by jacket foundation system, which consist mainly of space truss, supported in the four corners by driven steel piles. In each corner of the jacket construction, a single pile has to be driven through the pile sleeve. In the case of platforms with high vertical loads is could be favourable to use two piles which are closely placed to each other in each corner of the jacket. In this case, the expected pile group effect should be examined. This paper deals with the numerical modeling of pile groups composed of two open tubular steel piles embedded in mainly non-cohesive soil. A three-dimensional numerical model using the finite element system ABAQUS will be developed. In this model the material behaviour of the subsoil is described using an elasto-plastic constitutive model with Mohr-Coulomb failure criterion. Parametric studies will be done to investigate the structural behaviour under different conditions regarding the pile spacing. The results will be presented to examine the overall behaviour of pile groups.

1 Introduction

Pile foundations have been used as load carrying and load transferring systems for different structures. Steel pipe pile foundations are the part of a structure used to carry and transfer the load of the structure to the bearing ground located below ground surface. If the pile spacing exceeds a certain distance, then the capacity of pile group is the sum of the individual capacities of piles. However, if the spacing between piles is too close, in this case the distribution of skin friction and base resistance around the pile may overlap and the ultimate load of the group can be less than the sum of the individual pile capacities especially in the case of friction piles. The efficiency of pile group depends mainly on the following factors: type of the soil, the method of pile installation, spacing of piles, total number of piles in a row and number of rows in a group and the dimensions of pile (diameter and length). In this paper, the results of the numerical modeling for single and pile group (2 × 1) used as foundation system for jacket structure for a platform founded on a subsoil typical for German North Sea conditions will be presented. Firstly single pile model will be modeled then a parametric study for pile group (2 × 1) will be executed. The pile spacing will be varied in order to find out the dependency of pile group efficiency on the pile spacing and consequently the pile group capacity.

M. Abu-Farsakh et al. (eds.), *Advances in Analysis and Design of Deep Foundations*,
Sustainable Civil Infrastructures, DOI 10.1007/978-3-319-61642-1_19

2 Overview of Previous Work

The response of an individual pile in a pile group, where the piles are situated closely enough to one another, may be influenced by the response and geometry of neighboring piles. Piles in such groups may interact with one another through the surrounding soil, resulting in what is called the pile group action. The efficiency of a pile group is defined as the ratio of the actual capacity of the group to the summation of the capacities of the individual piles in the group when tested as single piles.

Cambefort (1953) examined steel single piles as well as pile groups. His experiments were carried out mainly in non-cohesive soil. The steel closed-ended piles had a diameter D = 50 mm and a length L = 2.50 m. The pile spacing (a) was chosen to 2 D. From his experiments, the pile group efficiency in the ultimate stage was about 1.7 and in the working stage was less than one (almost 0.96).

Kézdi (1957) demonstrated that piles groups in sand can possess a higher ultimate load than the sum of the individual piles (efficiency > 1). Meyerhof (1959) related the group pile capacity to the sand compaction during driving giving pile group efficiency higher than one. Laboratory tests by Walker (1964) indicated similar trends, confirming that stress interactions during the installation of pile groups can result in profound changes in pile behavior.

Vesic (1969, 1975) performed high-quality load tests on piles jacked mainly in medium-dense soil. The model piles were closed-ended aluminum pipes 0.1 m in outer diameter with a length of embedment of 1.53 m. The results show that the geometric efficiency increases from 0.96 to 2.23 when spacing-to-diameter ratio (a/D) increases from 2 up to 6 for the 2 × 2 pile group, and from 0.71 to 0.99 when spacing-to-diameter ratio (a/D) increases from 2 up to 3 for the 3 × 3 pile group, where a = the center-to-center pile spacing, and D = the diameter of the pile. Broms (1981) showed that the efficiency of a pile group in sand will be, in general, larger than unity and stated that when the initial relative density of the sand is low, then the sand surrounding the piles will be compacted during driving process if "a/D" is less than 5 to 6.

O'Neill (1983) reported that tests on model piles in loose sand indicate that efficiency in compression always exceeds unity, with the highest values occurring at a spacing-to-diameter ratio (a/D) of 2. According to O'Neill (1983), pile group efficiency in dense sand may be either greater or less than unity. Most pile load tests indicate values of pile efficiency greater than unity in pile groups of different sizes with (a/D) ratio between 2 and 4.

Briaud and Tucker (1989) performed field load tests on a single pile and a five pile group in medium dense sand. These tests showed an overall group efficiency of 0.99, but a significantly higher shaft efficiency (1.83) and lower base efficiency (0.67) when compared with the single pile.

The aim of these investigations is to determine numerically the pile group efficiency for two pile groups placed close to each other in order to quantify the effect of pile spacing on overall pile group capacity.

3 Numerical Modeling

3.1 Finite Element Model

A three-dimensional finite element model using ABAQUS to investigate the behavior of piles were established. Due to symmetry, only one half of the model was modelled. The elements used in the model for both the soil and the piles are 8-noded (C3D8) elements. Close to the pile, the soil mesh is very fine so that the thickness of soil elements have nearly the same dimensions like the wall thickness of the pile in order to get more accurate results. The numerical modeling was performed for a steel tube pile with length (L) of 57.0 m, diameter (D) of 2.60 m and wall thickness (t_p) of 8.0 cm. To investigate the effect of pile spacing on the behavior of the pile group and consequently on the pile group efficiency another finite element model was developed. The spacing between the two piles was varied from 2.0 D up to 3.0 D. The boundaries of the mesh were at a radius of 18.0 m from the pile vertical axis and 30.0 m below the base of the pile. With these dimensions, it was verified that the calculated behavior of the pile is not affected by the boundary conditions (see Fig. 1).

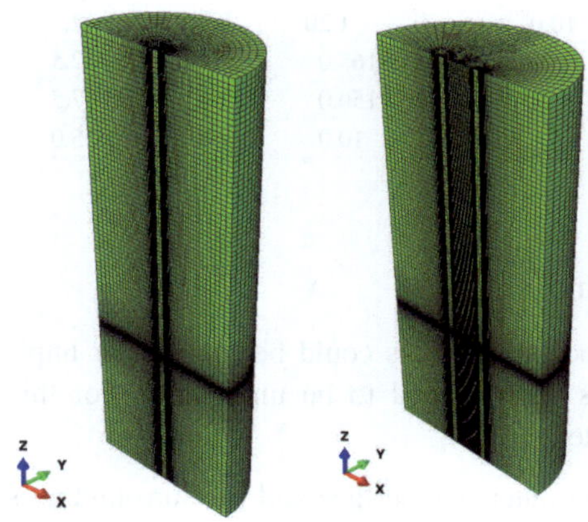

Fig. 1. Finite element model for single pile & pile group

A linear elastic material behavior of the piles was assumed with the parameters E = 210 GPa (Young's modulus) and ν = 0.20 (Poisson's ratio) for steel. To account for the non-linear soil behavior, elasto-plastic material behavior was assumed for the soil elements. The elasto-plastic material law with Mohr-Coulomb failure criterion was used to describe the behavior of the different soil layers. For the elastic region, E-modulus and Poisson's ratio are required. For the plastic region, the angle of internal friction (φ'), angle of dilatation (ψ) and cohesion (c) will be implemented (see Table 1). For the planned investigation, Mohr-Coulomb constitutive model can be considered sufficient as one of the important aspects of the numerical modeling is the comparison between single pile and pile group behavior in order to find out the pile group efficiency also the modeling procedure is mainly monotonic without any unloading or reloading stages. The material properties are based on different borings and cone penetration tests (CPTs). From the skin friction and base resistance of the cone, the properties for each soil layer (unit weight, E-Modulus, friction angle and cohesion) can be determined. Most of the soil layers consist of medium to very dense sand except one layer from 69.0 till 71.30 which is semi-solid clay layer. The main soil properties for 15 soil layers are summarized in Table 1.

Table 1. Material parameters used for different soil layers

Depth below sea bed level [m]	Unit weight γ [kN/m^3]	Submerged unit weight γ' [kN/m^3]	E-modulus E [MN/m^2]	Poisson's ratio ν [−]	Friction angle ϕ' [°]	Dilatation angle ψ [°]	Cohesion c' [kN/m^2]
0.0–1.2	19.0	9.0	6.0	0.3	27.5	0.1	1.0
1.2–5.5	19.0	9.0	38.0	0.3	35.0	5.0	1.0
5.5–9.5	19.0	9.0	45.0	0.3	35.0	5.0	1.0
9.5–12.5	19.5	9.5	90.0	0.3	37.5	7.5	1.0
12.5–31.5	20.5	10.5	140.0	0.3	42.5	12.5	1.0
31.5–34.5	19.5	9.5	110.0	0.3	40.0	10.0	1.0
34.5–46.5	20.5	10.5	140.0	0.3	42.5	12.5	1.0
46.5–69.0	21.0	11.0	165.0	0.3	42.5	12.5	1.0
69.0–71.5	20.0	10.0	1.20	0.47	22.5	0.1	40.0
71.5–78.0	21.0	11.0	165.0	0.3	42.5	12.5	1.0
78.0–83.0	20.5	10.5	150.0	0.3	37.5	7.5	1.0
83.0–86.0	20.0	10.0	10.0	0.47	25.0	0.1	80.0

3.2 Contact Modeling

During the driving process, the piles could be plugged or unplugged. In our investigations, the steel piles are assumed to be unplugged. For this case, three different contact pairs are adopted:

1. Contact between the outer pile surface and the surrounding soil outside the pile;
2. Contact between the inner pile surface and the surrounding soil inside the pile;
3. Contact between the pile base (annulus) and the soil beneath the pile.

For the contact behavior between pile (master) and soil (slave) an elasto-plastic model was used. The maximum frictional shear stress is dependent on the normal stress σ_n and a coefficient of friction μ. In the numerical simulations presented $\mu = \tan(2/3\ \varphi')$ was implemented, where φ' is the angle of internal friction of the corresponding soil layer (see Table 1).

For full mobilization of the limit frictional stress the relative displacement (elastic slip) between the pile and the surrounding soil was set to $\Delta u_{el,slip} = 1$ mm.

3.3 Modeling Procedure

The numerical modeling was performed in three different steps. In the first step the initialization of geostatic stresses (primary stress conditions) in the soil mass was applied by activating only the soil elements. The effect of pile installation (driving process) on the stress conditions in the soil will be not considered in the modeling. Based on this assumption, in the second step, the soil elements located at the pile position were removed and replaced by pile elements (wished-in-place concept) and the pile own weight was activated. Also the contact conditions between the pile and the surrounding soil were activated.

Finally, the pile loading was applied displacement controlled up to 26.0 cm downwards to find out the pile capacity under working stage and ultimate stage.

4 Numerical Modeling Results

4.1 Single Pile

By applying downward vertical displacement (26.0 cm) on the pile the deformations shown in Fig. 2 were calculated. Figure 2 shows that the vertical deformation inside the pile varies from 26.0 cm at the top down to 18.0 cm at the pile tip and the main deformation of the soil is localized at the pile tip.

In this case, the pile capacity is carried by three different components: outer skin friction, inner skin friction and base resistance. The pile deformation curves for these three different components are shown in Fig. 3. These diagrams show that the outer skin friction between the outer pile surface and the surrounding soil reaches a high value of 26.0 MN, after relatively small vertical displacement of around 4.50 cm.

The other two components (inner skin friction between the inner pile surface & soil inside the pile and the base resistance between the annulus and the soil) increase very slowly with the downward displacement and by vertical displacement of 4.50 cm, a kink in the two curves is observed and no maximum value for both components can be reached.

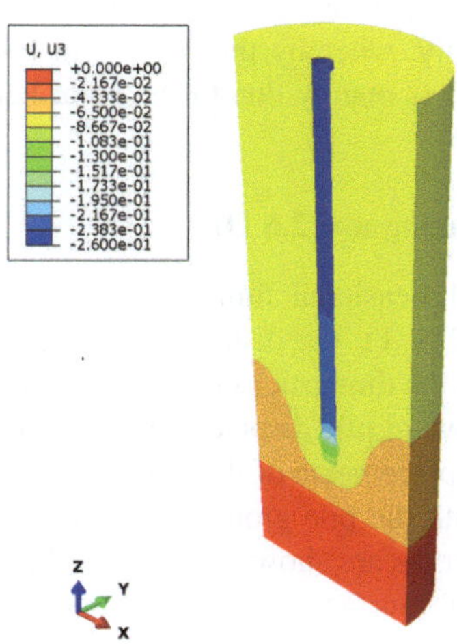

Fig. 2. Vertical displacement (u_3) for single pile after 25.0 cm downward

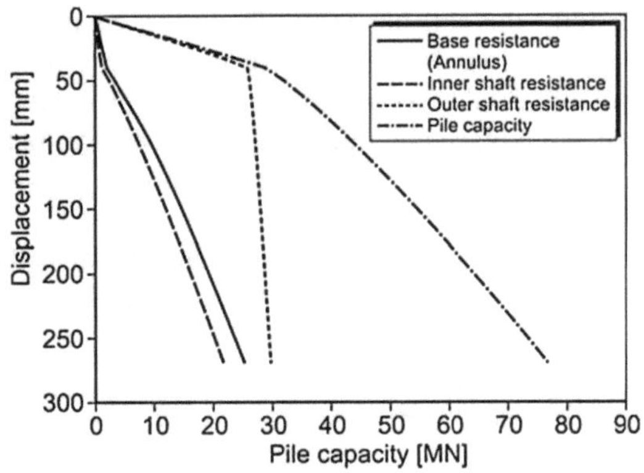

Fig. 3. Pile displacement curves for single pile

Table 2. Pile capacity for different pile displacement (single pile)

Pile displacement [cm]	Inner shaft resistance [MN]	Outer shaft resistance [MN]	Base resistance (Annulus) [MN]	Pile capacity [MN]
0.02 D = 5.20	2.70	26.12	3.48	32.39
0.04 D = 10.40	7.83	27.12	9.91	44.87
0.05 D = 13.0	10.16	27.56	12.62	50.34
0.1 D = 26.0	20.80	29.50	24.20	74.45

Table 2 summarizes the main results for different pile displacement. It is obvious that the pile capacity is governed mainly by the outer shaft resistance, which is about 40% of the total pile capacity, whereby the inner skin friction is about 28% and the base resistance is 32%. This is mainly due to the great length of the pile (57.0 m).

4.2 Pile Group (Pile Spacing a = 2.6 D)

Figure 1 shows the three-dimensional finite element model used to investigate the behavior of the pile group (2 × 1). The distance between the two piles was taken firstly 6.74 m (2.6 D) according to the dimensions of the jacket structure for the platform. The soil deformation after downward pile displacement of 26.0 cm is shown in Fig. 4. It is obvious that the vertical displacements of the two piles are interconnected and the soil block in-between moves with the pile group downward together.

The pile deformation curves are shown in Fig. 5. The diagram shows that the skin friction between the outer pile surface and the surrounding soil reaches a value of 53.0 MN, after vertical pile displacement of almost 5.0 cm. Table 3 shows that the other two components (the skin friction between the inner pile surface & soil inside the pile and

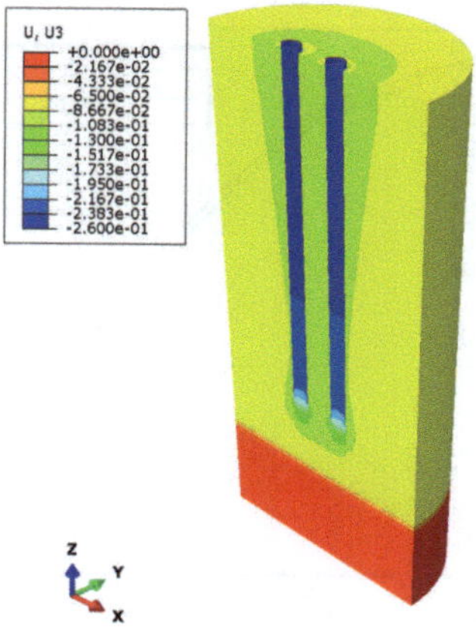

Fig. 4. Vertical displacement (u_3) for pile group after 25.0 cm downward displacement

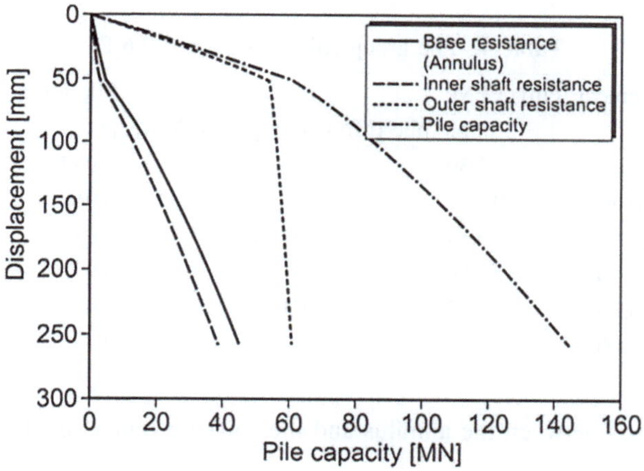

Fig. 5. Pile displacement curves for pile group (a = 2.6 D)

Table 3. Pile group capacity for different pile displacement (a = 2.6 D)

Pile displacement [cm]	Inner shaft resistance [MN]	Outer shaft resistance [MN]	Base resistance (Annulus) [MN]	Pile capacity [MN]
0.02 D = 5.20	2.94	53.61	4.82	61.37
0.04 D = 10.40	8.51	55.14	11.55	65.20
0.05 D = 13.0	17.96	57.05	22.82	97.83
0.10 D = 26.0	38.60	60.76	44.93	144.30

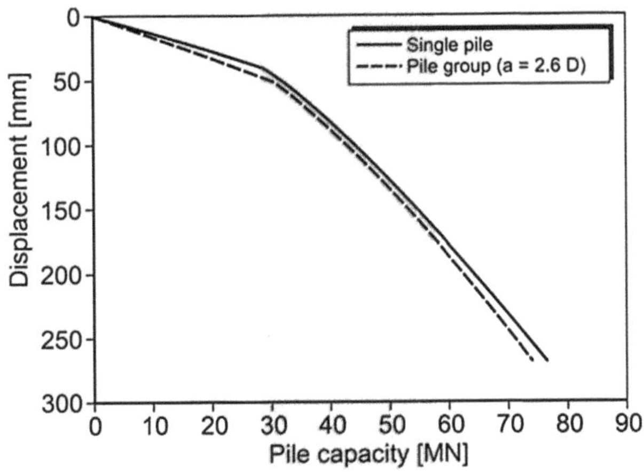

Fig. 6. Comparison between single & pile group behavior (a = 2.6 D)

Table 4. Pile group efficiency for a = 2.6 D

Pile displacement [cm]	FEM		Pile group efficiency
	Single pile [MN]	Pile group [MN]	[1]
0.02 D = 5.20	32.39	30.69	0.947
0.04 D = 10.40	44.87	43.50	0.969
0.05 D = 13.00	50.34	48.91	0.972
0.06 D = 15.60	55.50	54.04	0.972
0.08 D = 20.80	65.57	63.75	0.973
0.10 D = 26.00	74.45	72.43	0.973

the base resistance between the annulus and the soil) are much smaller than the outer shaft resistance.

Figure 6 compares between the capacity of single pile and pile group under vertical loading. From this figure the single pile shows slightly higher pile capacity compared to the pile group capacity divided by the number of the piles (n = 2).

The pile group efficiency is shown in Table 4. This table shows that the pile group efficiency increases gradually from 0.94 (in the working stage) up to 0.97 in the limit state.

4.3 Pile Group (Pile Spacing a = 2.0 D)

The numerical results are summarized in Figs. 7 and 8, which show the development of pile resistance (outer skin friction, inner skin friction and base resistance) with the vertical pile displacement.

From Table 5, it is obvious that the pile capacity is governed mainly by skin friction, which means that the pile behaves like a "friction pile", which is very similar to the previous case (a = 2.6 D).

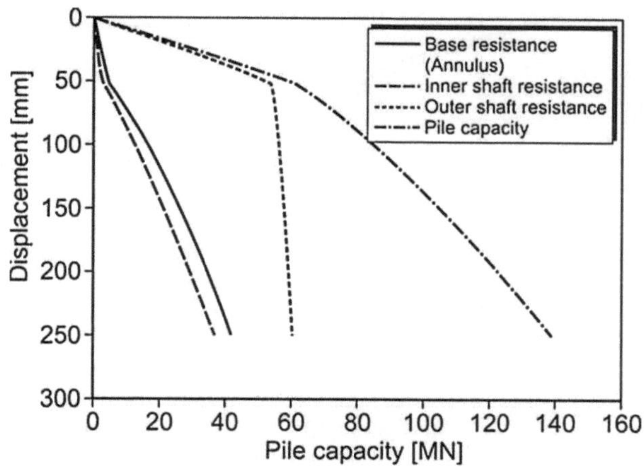

Fig. 7. Pile displacement curves for pile group (a = 2.0 D)

Table 5. Pile group capacity for different pile displacement (a = 2.0 D)

Pile displacement [cm]	Inner shaft resistance [MN]	Outer shaft resistance [MN]	Base resistance (Annulus) [MN]	Pile capacity [MN]
0.02 D = 5.20	2.91	53.38	4.75	61.04
0.04 D = 7.80	13.19	56.36	17.12	86.67
0.05 D = 13.0	17.72	57.24	22.33	97.29
0.10 D = 26.0	38.54	60.79	43.39	142.72

The pile group efficiency is tabulated in Table 6. This table shows that the pile group efficiency increases gradually from 0.94 (in the working stage) up to 0.965 in the limit state which is very similar to the previous case (a = 2.6 D).

Table 6. Pile group efficiency for a = 2.0 D

Pile displacement [cm]	FEM		Pile group efficiency
	Single pile [MN]	Pile group [MN]	[1]
0.02 D = 5.20	32.39	30.47	0.941
0.04 D = 10.40	44.87	43.33	0.966
0.05 D = 13.00	50.34	48.65	0.966
0.06 D = 15.60	55.60	53.65	0.965
0.08 D = 20.80	65.57	62.90	0.959
0.10 D = 26.00	75.05	71.40	0.952

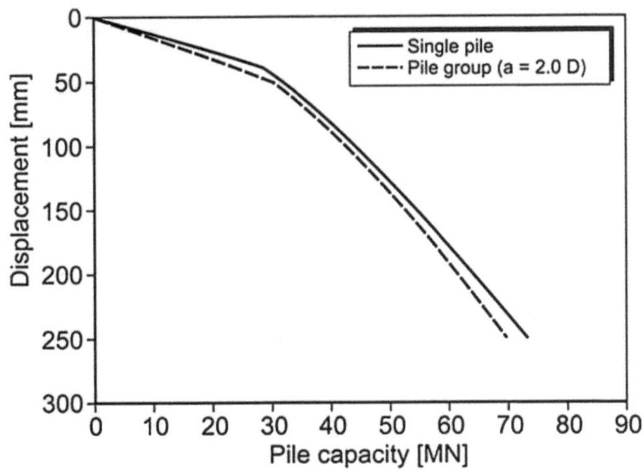

Fig. 8. Comparison between single & pile group behavior (a = 2.0 D)

4.4 Pile Spacing = 3.0 D

The numerical results for the pile spacing a = 3.0 D, are shown in Figs. 9 and 10. The diagrams show the pile resistance developed for this case.

Also for this pile spacing, the outer skin friction is the main factor governing the pile capacity, which is about 40% of the total pile capacity (see Table 7).

The pile group efficiency is tabulated in Table 8. This table shows that the pile group efficiency increases gradually from 0.95 (in the working stage) up to 0.98 in the limit state which shows a slight increase compared to the previous cases. This means that for the investigated cases the increase of pile spacing has a minor positive effect on the pile group efficiency.

Table 7. Pile group capacity for different pile displacement (a = 3.0 D)

Pile displacement [cm]	Inner shaft resistance [MN]	Outer shaft resistance [MN]	Base resistance (Annulus) [MN]	Pile capacity [MN]
0.02 D = 5.20	2.98	53.56	4.91	61.44
0.03 D = 7.80	8.56	55.04	11.68	75.28
0.05 D = 13.0	18.08	56.96	23.04	98.08
0.10 D = 26.0	39.43	60.84	45.99	146.26

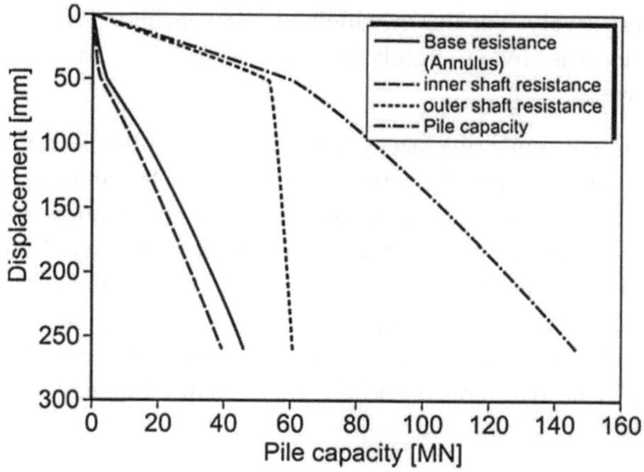

Fig. 9. Pile displacement curves for pile group (a = 3.0 D)

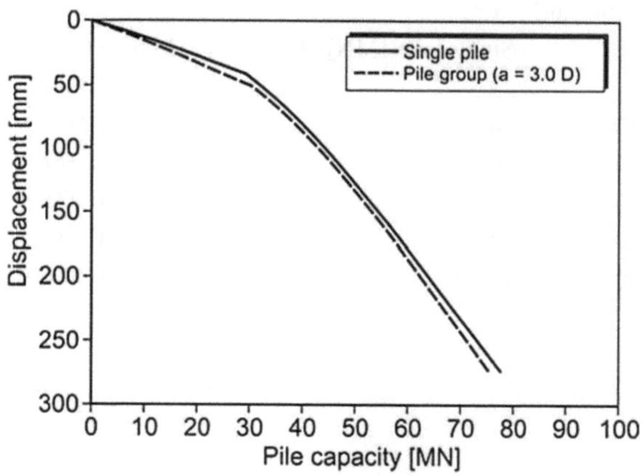

Fig. 10. Comparison between single & pile group behavior (a = 3.0 D)

4.5 Pile Group Efficiency

Based on the previous results, the pile group efficiency (G_r) will be evaluated using the following equation:

$$G_r = R_g/n \cdot R_e \qquad (2)$$

Where,

R_g: Pile group capacity;
R_e: Single pile capacity;
n: number of piles (n = 2)

Figure 11 shows the pile group efficiency (G_r) for different pile spacings. The three curves are very similar to each other and the increase of pile spacing (a) affects the pile group efficiency positively. For the investigated pile spacings (2.0 D up to 3.0 D), the pile group efficiency was always slightly less than one ($G_r = 0.96$). These results can be explained as follows:

(a) In the first part up to pile displacement 0.05D, the soil block captured between the piles moves downwards with the piles, which means that the skin friction between the outer pile surface and the surrounding soil will be somehow smaller than in the case of single pile, consequently the pile group efficiency will be smaller than 1.0.
(b) In the second part (pile displacement from 0.05 D till 0.08 D), the skin friction is fully mobilized and the pile group efficiency increases up to 0.97.
(c) Finally in the last part (0.08 D till 0.1 D), the base resistance of both piles in the pile group will be overlapped and the pile group efficiency will remain almost constant (0.96).

Table 8. Pile group efficiency for a = 3.0 D

Pile displacement [cm]	FEM		Pile group efficiency
	Single pile [MN]	Pile group [MN]	[1]
0.02 D = 5,20	32.39	30.72	0.949
0.04 D = 10,40	44.87	43.56	0.971
0.05 D = 13,00	50.34	49.04	0.974
0.06 D = 15,60	55.60	54.23	0.975
0.08 D = 20,80	65.57	63.99	0.980
0.10 D = 26,00	75.07	73.27	0.980

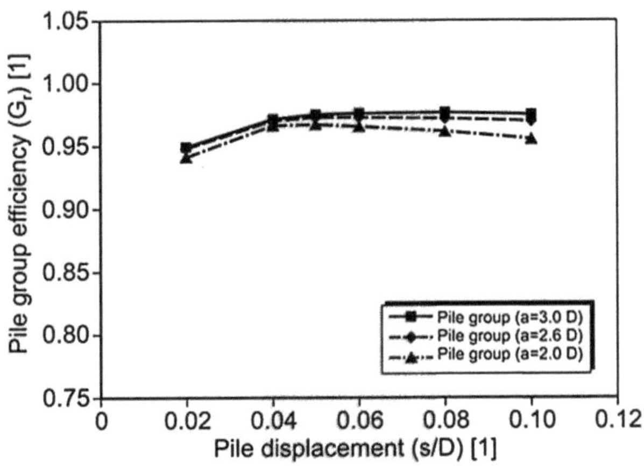

Fig. 11. Pile group efficiency for different pile spacing (a = 2.0 D, 2.6 D, 3.0 D)

5 Conclusions

In this paper a numerical modelling for single pile and pile group (2 × 1) has been developed. Firstly a single pile embedded in mainly non-cohesive soil was modelled. To investigate the effect of pile spacing on the total pile group capacity, different numerical models by varying the pile spacing from 2.0 D up 3.0 D were developed.

The numerical simulation presented here demonstrates the effect of pile spacing in the working stage and also in the ultimate stage. This effect is more pronounced in the working stage ($G_r = 0.95$) and less pronounced in the ultimate stage ($G_r = 0.98$). The calculated pile group efficiency was slightly less than one as the effect of driving process on the stress conditions and on the soil compaction was not considered in our investigations. In subsequent investigations, further parametric studies will be carried out for different pile dimensions and pile spacing.

References

Briaud, J.-L., Tucker, L.M.: Axially loaded 5 pile group and a single pile in sand. In: 12th International Conference of Soil Mechanics and Foundation Engineering, Rio de Janeiro, vol. 2, pp. 1121–1124 (1989)

Cambefort, H.: La force portante des groupes de pieux. In: Proceedings of the third International Conference on Soil Mechanics and Foundation Engineering, Switzerland 16th–27th August 1953, vol. 2, pp. 22–28 (1953)

Kézdi, A.: The bearing capacity of piles and pile groups. In: Proceedings Fourth International Conference on Soil Mechanics and Foundation Engineering, London, vol. II, pp. 46–51 (1957)

Meyerhof, G.G.: Compaction of sands and bearing capacity of piles. J. Soil Mech. Found. Div. ASCE, Part I **85**(SM6), 1–29 (1959)

O'Neill, M.W.: Group action in offshore piles. In: Proceedings of Conference on Geotechnical Practice in Offshore Engineering, pp. 25–64 (1983)

Walker: Experiments on model pile foundations in sand', Ph.D. thesis, Univ. of London imperial College. London. UK (1964)

Vesić, A.S. (1969): Experiments with instrumented pile groups in sand. In: Performance of Deep Foundations. ASTM Special Technical Publication vol. 444, pp. 177–222

Vesić, A.S.: Principles of Pile Foundation Design. Duke University Durham N.C., School of Engineering, Soil Mechanics series Nr. 38 (1975)

Adfreeze Strength and Creep Behavior of Pile Foundations in Warming Permafrost

Abdulghader A. Aldaeef(✉) and Mohammad T. Rayhani

Geoengineering Research Group, Carleton University, Ottawa, Canada
{abdulghader.abdulrahman, mohammad.rayhani}@carleton.ca

Abstract. An experimental investigation was carried out to evaluate the thermal exposure effect on load carrying capacity and creep behavior of steel piles embedded in ice-poor frozen soils using steel-soil interface tests. The interface testing was conducted in a walk-in cold room to enable testing at various temperatures below the freezing point. A series of stress-displacement curves were established at different temperatures and under various normal stresses. The results showed a significant reduction in adfreeze strength of the pile-soil interface as the exposure surface temperature increased. The interface strength decreased approximately 300% when the exposure temperature increased from −1.5 °C to 0 °C. Such condition may be witnessed in warm permafrost that experience temperature ranging from −3 °C to 0 °C. The shear stress–strain curves showed a brittle behavior followed by significant loss of bearing capacity. Pile creep rate in ice-poor soils increased by about 60% when the interface was exposed to warming from −10 °C to −5 °C and showed tertiary creep when reaching −4 °C.

1 Introduction

Permafrost, or perennially frozen ground, is the ground that remains frozen for at least two consecutive years. These frozen grounds can be classified into ice-reach and ice-poor soils, where ice-reach exhibits a frozen bulk density less than 1.7 gm/cm^3, while ice-poor soil shows higher frozen bulk density (Nixon and McRoberts 1976). Permafrost is considered a critical component of the Arctic system and extends over 24% of the terrestrial surface of the Northern Hemisphere. Ground materials in their frozen state are claimed to be stiffer than the ground materials in unfrozen state so they provide higher bearing capacities. In cold regions, however, frost heave and excessive settlement following ice melting in frozen soils are critical problems that can degrade the stability of infrastructure. Ice melting and permafrost degradation could happen in micro-scale due to improper design of structures or in macro-scale due to the effect of global warming.

The impact of global warming on permafrost has been investigated and numerically modeled by many researchers. Lawrence and Slater (2005) discussed the evolution of permafrost area in the Northern Hemisphere and predicted the present-day permafrost as well as future permafrost extent in the 21st century. Their model showed 25% reduction in the global permafrost area occurred between 1900 and 2000 when

M. Abu-Farsakh et al. (eds.), *Advances in Analysis and Design of Deep Foundations*,
Sustainable Civil Infrastructures, DOI 10.1007/978-3-319-61642-1_20

compared with the International Permafrost Association (IPA) map. A total permafrost disappearance was also predicted to occur by the year 2100.

Recently, cold region has become economically important after the exploration of enormous natural resources especially in the northern circumpolar. Therefore, the region has witnessed significant increase in population and expansion of infrastructure such as civil facilities, hydrocarbon extraction facilities, transportation networks, communication lines, and industrial projects. However, construction in cold region might be more expensive as it complies with a special design code for construction in harsh environments. Smith and McCarter (1997) stated that these expenses might be aggravated severely in both environmental and human terms by the effects of global warming on permafrost. Frozen ground could fail to maintain frozen condition in confrontation of global warming. In warm permafrost, a small temperature increase may be sufficient to cause extensive thawing settlement. Therefore, the assessment of the structures' performance situated on permafrost and the stability of their foundations are crucial under the simultaneous thermal and mechanical loadings.

Weaver and Morgenstern (1981) suggested that foundation design in frozen ground must satisfy both thermal and serviceability considerations. The thermal aspect associated with frozen ground is investigated using analytical solutions proposed by many researchers (e.g., Nixon 1978; Linell and Lobacz 1980). However, the serviceability and settlement of pile foundations in permafrost seems to lack clear guidelines. Therefore, efforts have been dedicated mostly to evaluate the bearing capacity and settlement behavior for piles in frozen grounds with scant attention to the thermal role.

Pile foundations have been widely used to support superstructures situated on permafrost in cold regions. Pile foundations are preferable over the shallow foundations since they can be installed to a greater depth and respectively provide larger resistance against structural load and any down-drag or frost heave loads subjected to the pile following frost thawing or soil freezing, respectively. Early studies have adopted a design method for piles in permafrost based on the rupture of the adfreeze bond corresponding to the ultimate load capacity of the piles. More recent investigations (e.g., Nixon and McRoberts 1976; Morgenstern et al. 1980) suggested that the adoption of bearing capacity criteria for pile design in permafrost without taking creep behavior in consideration may be improper, claiming that excessive settlement may occur over the design life time of the structures. Therefore, it was emphasised that ensuring tolerable pile displacements throughout the life of the structure is essential.

Bearing capacity of piles enhanced by adfreeze strength was previously studied by several researchers including Weaver and Morgenstern (1981) and Ladanyi and Theriault (1990). However, most of these studies were conducted at limited range of temperatures and do not show the behavior of load carrying capacity of piles at temperatures slightly below the freezing point. Weaver and Morgenstern, (1981) correlated the adfreeze shear strength to the long-term shear strength of the frozen soil using an "m" factor which characterizes the roughness of the pile surface. In this correlation, they neglected the contribution of the friction at the interface of the piles attributing that to the small confining pressure (P_n) acting on the pile shaft in frozen grounds. Therefore, the long-term adfreeze strength of the piles in frozen ground was expressed in the following form:

$$\tau_{a,lt} = m\ C_{lt} \tag{1}$$

where $\tau_{a,lt}$ is the long-term adfreeze strength, C_{lt} is the long-term cohesion of the frozen soil, and "m" is a factor that describes the surface roughness of the pile and has a value of 0.6 for steel, 0.7 for concrete and 1 for corrugated steel pipe piles.

Ladanyi and Theriault (1990) reported that the long-term capacity of the pile shaft in frozen ground does not solely depend on the long-term adfreeze, but also on the residual friction angle at the interface, and respectively on the total lateral ground stress. Therefore, Ladanyi and Theriault (1990) improved the Weaver and Morgenstern (1981) equation by adding the contribution of the residual friction angle at the pile–soil interface and proposed the following formula:

$$\tau_{a,lt} = mC_{lt} + \sigma_{n_{total}} \tan \emptyset_{lt} \tag{2}$$

In more recent study, Aldaeef and Rayhani (2016) showed the evolution of adfreeze strength of piles in ice-poor soils over a wider range of temperature including temperatures slightly below 0 °C. This study could lead to proposing a bearing capacity equation for piles in cold and warm permafrost considering the effect of temperature.

In terms of pile creep in permafrost, Nixon and McRoberts (1976) and Morgenstern et al. (1980) developed different analytical solutions to predict axial strain of ice-rich frozen soils using published creep data obtained from uniaxial creep test conducted on ice. They later used their ice creep laws to predict the steady pile displacement rate "u_a" in frozen ground based on the applied shaft shear stress (τ_a). Later, Weaver and Morgenstern (1981) conducted a comprehensive research on the pile performance in permafrost and stated that pile design in ice-rich soils should be governed by settlement while pile design in ice-poor soils should satisfy both strength criteria and settlement. Therefore, they proposed a creep law for ice-poor soils considering the effect of confining pressure on the primary creep as follow:

$$\epsilon_1 = D(\sigma_1 - j\sigma_3)^c t^b \tag{3}$$

Where

$$j = \frac{(1 + \sin \emptyset)}{(1 - \sin \emptyset)}$$

$$D = \left[\frac{1}{w(\theta + 1)^k}\right]^c$$

$\emptyset$ is the internal friction angle, t is the time elapsed after the application of the load (h), θ is the temperature below the freezing point of water (°C) and w, b, c, k are material dependent parameters. The authors later used the creep law model of the ice-poor soils to derive flow law model for piles in ice-poor condition and proposed the following equation:

$$\frac{u_a}{at^b} = \frac{3^{\frac{c+1}{2}} D\, \tau^c}{c - 1} \tag{4}$$

Based on the information available in the literature, most of the approaches adopted for pile foundation design in frozen ground are derived mainly based upon creep observations rather than load carrying capacity, although, some studies (e.g., Weaver and Morgenstern 1981) have proven that the design of piles in ice-poor soils should satisfy both strength and settlement criteria. There is real need to understand the behavior of adfreeze strength and pile creep for piles installed in permafrost that undergoes warming. The existing design approaches for piles in ice-rich and ice-poor soils were always derived using flow law models that incorporate creep parameters of ice material solely due to the lack of reliable creep data on ice-rich and ice-poor soils. The model proposed by Nixon and McRoberts (1976), for example, was mainly built upon published creep data on ice and then was idealized to describe creep behavior in ice-rich soils claiming that this will be a conservative measure in the absence of reliable creep data on ice-rich soils. Nevertheless, ice-rich soils might behave differently compared to the ice due to the presence of unfrozen water content in ice-rich soils even at relatively low temperatures. In addition, Weaver and Morgenstern (1981) revealed that the use of published creep data on ice and ice-poor soils were short of reliable information in the temperature range between 0 and −5 °C.

The current paper presents the results of a preliminary experimental investigation on the behavior of the adfreeze strength of steel piles in warming ice-poor frozen soils. The results were then used to assess the creep behavior of steel piles in warming ice-poor permafrost. The study was carried out in a fully controlled walk-in cold room. The adfreeze strength and creep behaviors of pile-soil interfaces were determined at different levels of temperature ranged from −10 °C to 0 °C, and at room temperature.

2 Physical Properties of the Test Soil

Poorly graded sand (SP in USCS classification system, ASTM D2487, 2005) was utilized as the interface soil for this study. The maximum dry density ($\rho_d$max) of the soil was found to be 1.85 g/cm^3 at an optimum gravimetric moisture contents (wopt) of 10% in accordance to the Standard Proctor test (ASTM D698 2012). The particle size distribution for the sand was determined in accordance to ASTM D422-63 (2007). The sand was non-plastic, and contained approximately 5% fines passing through the 0.075 mm sieve.

3 Pile Interface

A common pile material used in permafrost is structural steel. It is manufactured in different pile shapes such as pipe piles, H-section piles, and helical piers. Steel piles have some disadvantages including the relatively high steel costs and the vulnerability to corrosion in harsh environments. However, Linell and Johnston (1973) reported that steel piles embedded in permafrost would be well preserved from corrosion, but small amount of corrosion could form on the shaft along the active layer depth. In this experiment, steel plates were used to simulate the shaft surface of a typical steel pile. The interface test specimens were 90 mm by 90 mm square steel plates with a thickness 25.4 mm, machined to couple with the upper half of the direct shear box apparatus and provide a soil-steel interface area of 65 mm by 65 mm (Fig. 1). The total and average surface roughness values for this particular type of steel were reported by Giraldo and Rayhani (2013) to be 9.7 μm and 11.3 μm respectively. The steel plates were equipped with thermocouples inserted in tiny holes underneath the upper surface of the plates in order to track the temperature change at the soil-steel interface.

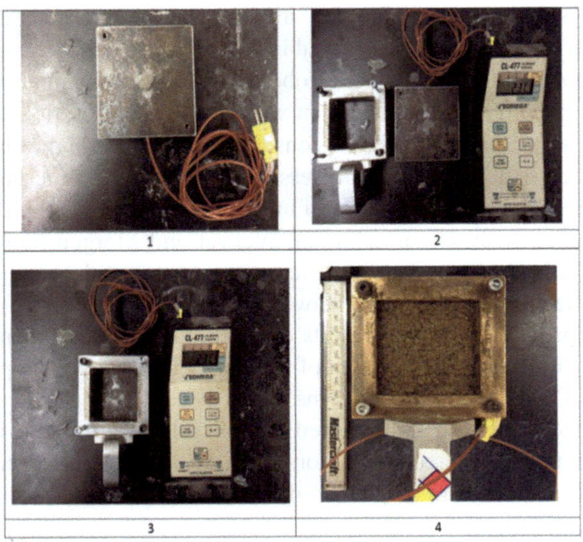

Fig. 1. Modified shear box for pile-soil interface test.

4 Experimental Procedure

The adfreeze bond was characterized using a direct shear test apparatus in accordance to ASTM D3080/D3080M-11 and ASTM D5321-12. The shear box was modified by replacing its lower part by a steel plate to simulate the pile-soil interface. The apparatus was placed inside a walk-in cold room to enable shear testing at different temperatures below the freezing point of water. The direct shear test apparatus consists of an electrical motor that enables applying a constant displacement to the lower part of the shear box (steel plate), while the upper part, that contains the soil sample, is restrained by a digital load cell connected horizontally to it. Horizontal and vertical displacements are measured through linear variable differential transducers connected to a digital logging station using LabView software. In order to mimic the confining pressure, a vertical load was applied to the top of the soil sample by a steel bearing arm. The shear stresses are measured by dividing shear force, collected by the load cell, by the interface area. A picture of the shear box modification is illustrated in Fig. 1.

The soil was prepared at a bulk density corresponding to its field capacity moisture content. The bulk density at field capacity was determined by pouring the sand-water slurry into a known volume container and permitting water drainage while the soil settles under gravity. When water stops coming out of the sample, the soil sample was then weighed and its density was determined to be 2060 kg/m^3 at field capacity moisture content of 13.5%. To prepare the interface test sample, dry sand was hydrated to the predefined field capacity and compacted to the corresponding density in the upper part of the shear box which was already assembled to the steel plate. The shear box was then mounted on the direct shear test apparatus and the desired normal stress was respectively applied simultaneously with setting the temperature of the chamber at the target level to permit soil consolidation and freezing.

4.1 Ultimate Adfreeze Capacity Test

After keeping the interface setup for 24 h in the cold room, the screws that attach the upper part of the shear box to the steel plate were removed and the shear load was applied at a constant displacement rate of 0.00208 mm/min (≅3 mm/day). This loading rate was the lowest rate that could have been achieved by the used apparatus. However, for the purpose of investigating the ultimate and residual adfreeze strengths, this rate is believed to be suitable to capture the strength characteristics of the interface. The shear tests were conducted at different temperature levels including −10 °C, −5 °C, −4 °C, −3 °C, −1.5 °C and +22 °C with an accuracy of ± 0.5 °C. At each temperature, the shear strength was determined under normal stresses of 25 kPa, 50 kPa, and 100 kPa.

4.2 Creep Test

Similar procedure was followed to prepare the samples for creep testing; however, the creep test was conducted differently. Typical creep test is performed by applying a constant load and recording the corresponding displacement over the time. However, a constant load application cannot be attained using the existing shear test apparatuses, thus, most creep tests are conducted using full scale or small scale pile creep set up in field or in laboratory respectively. In this study, the direct shear test apparatus was upgraded to serve creep testing. A pulley system was incorporated to the apparatus which allowed applying constant loads to the steel plate (representing the pile) by attaching galvanized wire to the shear box to be parallel to the shear plane. The wire then was passed to the pulleys and connected to a dead load hanger (see Fig. 2). A dead load is applied to the hanger resulting in vertical pulling force applied to the wire. The pulleys are responsible to transform the vertical pulling load to a horizontal pulling motion and applying it to the shear box to generate a constant shear loading at the interface. The test began by preparing the sample and leaving it in the cold room at −10 °C for 24 h under a confining pressure of 100 kPa. A constant load, equivalent to half of the maximum adfreeze strength calculated at −10 °C, was applied in one step. When a steady state creep rate observed, the temperature was increased to the next level. This procedure was

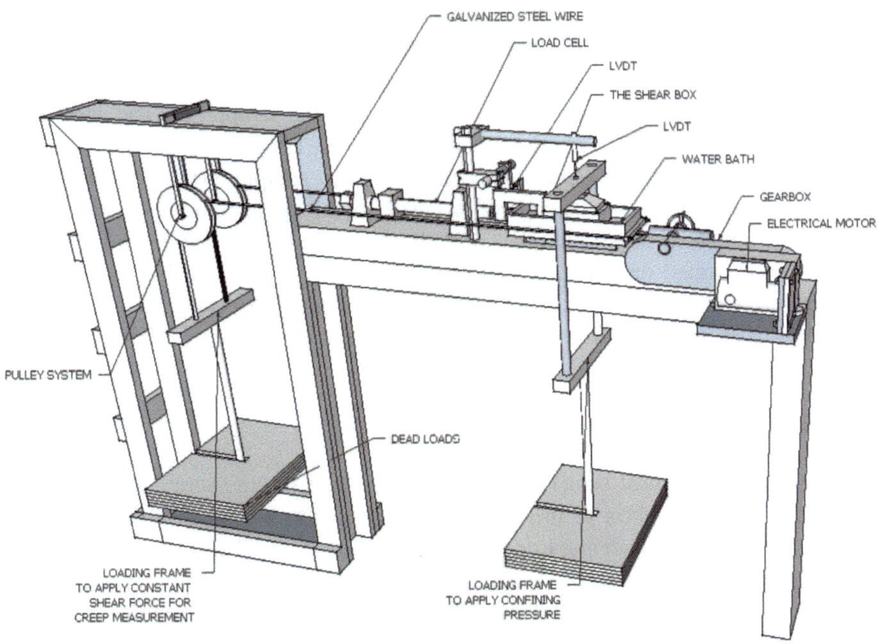

Fig. 2. Modified direct shear apparatus for creep testing.

repeated until pile-soil rupture occurred giving that a steady state creep rate was always attained before increasing the temperature to the warmer level.

5 Test Results and Discussion

5.1 Effects of Temperature Change on Adfreeze Strength

The stress-displacement curves for all tests are illustrated in Fig. 3a, b, and c. Adfreeze failure (peak strength) in all tests occurred at a displacement ranged from 0.3 to 1.3 mm depending on the test temperature and applied normal stress. The sample tested at room temperature, however, failed at lower shear strain showing elastoplastic failure mode. The frozen samples, on the other hand, showed a brittle failure mode where a significant strength loss recorded right beyond the peak. This failure mode might be attributed to the viscoelastic behavior of frozen soils. The recorded residual adfreeze strength was comparable to the residual strength of the unfrozen sample. Most of the tests were continued after the peak to a displacement of 2.5 mm in order to characterize the residual strength. It is found that, at −10 °C, the peak adfreeze strength was around 1 MPa, however, it drops significantly corresponding to the temperature increase to be around 0.2 MPa at −1.5 °C. The interface strength becomes the lowest at the room temperature (+22 °C) with a reduction of 92% in shear strength compared to what has been measured at −10 °C. This could be attributed to the ice melting and respective loss of ice bonding. If a warm ice-poor permafrost at a temperature of −1.5 was

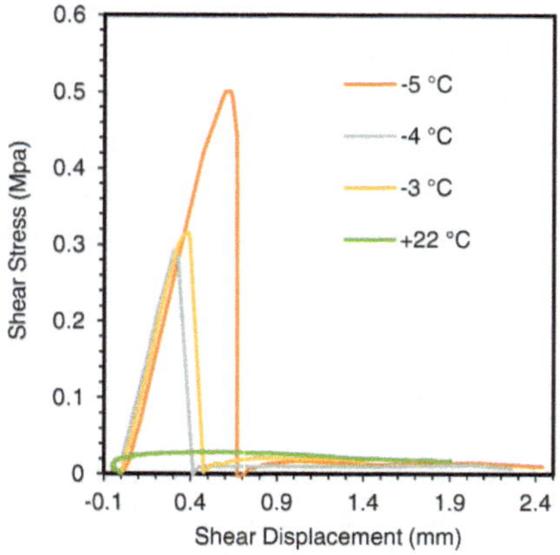

a. **Stress-displacement curves under normal stress of 25kPa.**

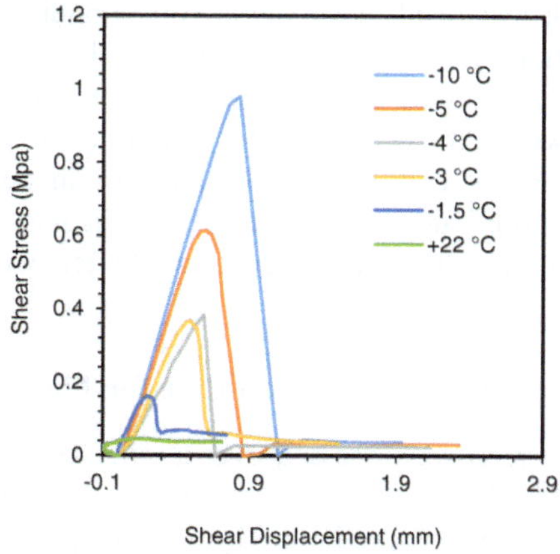

b. **Stress-displacement curves under normal stress of 50kPa.**

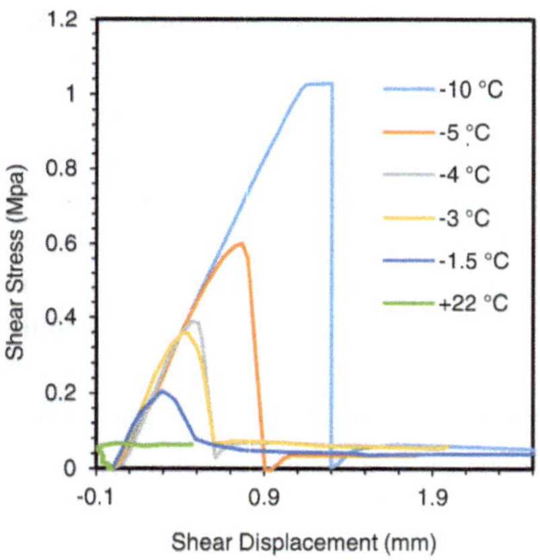

c. **Stress-displacement curves under normal stress of 100kPa.**

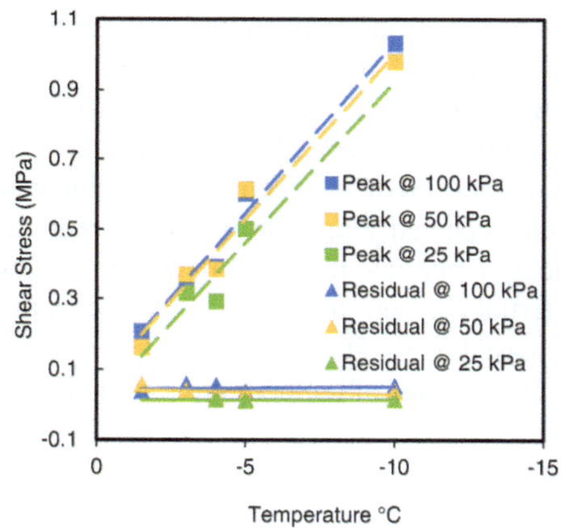

d. **Variation of peak and residual adfreeze bonds with temperature.**

Fig. 3. Effect of permafrost warming on the pile-soils adfreeze strength.

exposed to warming to a temperature higher than the freezing point, its strength (based on the current experiment) will drop from 0.2 MPa to around 0.065 MPa showing a reduction of 300% which represents a factor of safety of 3. This means a failure can occur to a structure supported by piles designed with a factor of safety of 3, if a total degradation of the permafrost encountered. This highlights the importance of understanding the quantitative change of the ultimate capacity of the piles in warming permafrost due to the thermal effects.

To quantify the change in adfreeze capacity of the piles in ice-poor soils as a function of temperature, Mohr Coulomb failure criterion was employed (Fig. 3d). The plot includes three dashed and three solid trend lines representing the variation of the peak and residual strengths corresponding to the change in temperature at different normal stresses (i.e., 100 kPa, 50 kPa, and 25 kPa) (Fig. 3d). The average peak strength decreased significantly as the temperature increased toward the freezing point. However, the residual strength at different temperatures remains constant and shows no significant correlation. The increase in peak strength corresponding to temperature reduction might be attributed to the increase of ice content at the interface level and the increase in ice shear strength as the temperature decreases (Haynes 1978)

5.2 Variation of Creep Rate for Piles in Warming Permafrost

Pile creep was assessed over a range of temperatures under a constant shear stress of 0.535 MPa. The result of pile creep assessment is presented in Fig. 4. The creep behaviors observed were typical for piles in ice-poor soils where the primary creep dominated pile movement with a slow steady state creep rate. A primary creep of 0.44 mm was recorded at temperature of −10 °C right after the load application. The specimen then showed a secondary creep with a constant creep rate equal to $8E^{-6}$ mm/h. When the temperature was increased to −7 °C, the specimen exhibited a primary creep of 0.0047 mm representing about 1% of the primary creep recorded at −10 °C. However, as the temperature gets warmer, the primary creep increases recording 2.7% and 3.1% of the primary creep recorded at −10 °C for temperatures of −5 °C and −4 °C respectively. This indicates that piles installed in ice-poor permafrost could repeatedly undergo primary creep incidences after experiencing a constant secondary creep rate if the permafrost undergoes warming. The result, in addition, showed a tertiary creep and total adfreeze rupture when the temperature was gradually increased to −4 °C for piles designed to work safely at −10 °C. This clearly indicates the importance of temperature role in designing pile foundation in warming permafrost.

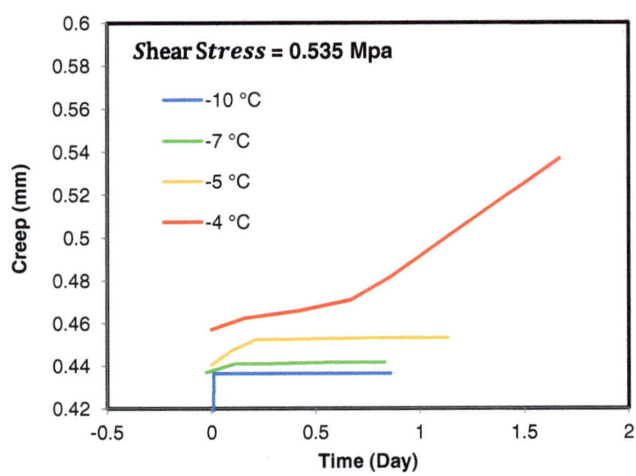

Fig. 4. Variation of pile creep rate with temperature.

Secondary creep rate recorded at different temperatures was always slow showing a typical condition for piles in ice-poor soils. However, the secondary creep rate showed a positive correlation with permafrost warming, where a faster secondary creep rate recorded at higher temperature (Fig. 5). The steady state creep rate showed 60% increase from $8E^{-6}$ mm/h to $3E^{-5}$ mm/h when the temperature increased from −10 °C to −5 °C under same loading condition. This significant increase could cause an outstanding subsidence for the supported structure over its life time and respectively degrade its overall stability.

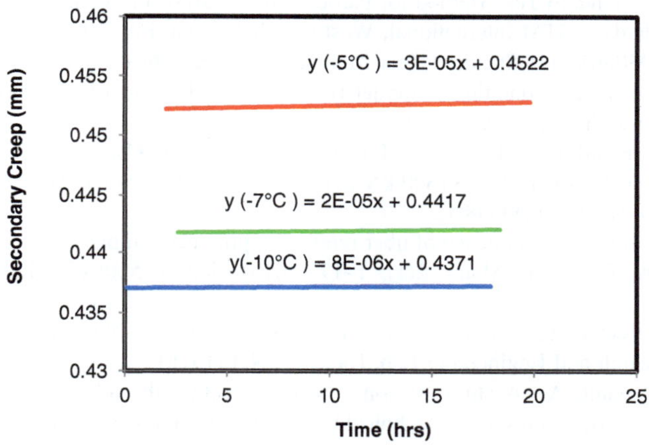

Fig. 5. Variation of pile creep rate with temperature.

6 Conclusions

This study aims at investigating the effect of permafrost degradation on the load carrying capacity and creep behavior of steel piles in ice-poor soils. The adfreeze strength of piles in ice-poor soils is the resultant of the adhesion and frictional resistance at the pile-soil interface. The adhesion showed to have an inverse correlation with temperature below zero degrees Celsius. Piles in ice-poor soils fail in brittle mode when the peak strength exceeded and lose significant capacity after failure. Failure can happen also because of permafrost degradation especially those in warm conditions at relatively high temperatures ranging from −3 °C to 0 °C. Pile creep increases significantly corresponding to temperature increase. The failure may occur due to loss of bearing capacity or excessive settlement following a permafrost warming. The results of this investigation could be used to improve the design approaches available for piles in ice-poor permafrost to better capture piles behavior in warming permafrost. Large scale tests are also in progress to determine the capability of element scale tests in capturing realistic large scale behavior of piles in frozen soil.

Acknowledgments. This study was financially supported by Carleton University and the Natural Sciences and Engineering Research Council of Canada (NSERC). The authors would like to thank both agencies for this support.

References

Aldaeef, A.A., Rayhani, M.T.: Evolution of Adfreeze strength of pile Foundations in Warming Permafrost. In: Proceeding of GeoVancouver 2016. Vancouver, Canada (2016)

ASTM D422-63: Standard Test Method for Particle-Size Analysis of Soils, Annual Book of 484 ASTM Standards. ASTM International, West Conshohocken, PA (2007)

ASTM D3080/D3080M-11: Standard Test Method for Direct Shear Test of Soils Under 494 Consolidated Drained Conditions, Annual Book of ASTM Standards. ASTM International, 495 West Conshohocken, PA (2012)

ASTM D698: Standard Test Methods for Laboratory Compaction Characteristics of Soil Using Standard Effort (12 400 ft-lbf/ft3 (600 kN-m/m3). ASTM International, West Conshohocken, PA (2012). doi:10.1520/D0698-12

Giraldo, J., Rayhani, M.T.: Influence of fiber-reinforced polymers on pile–soil interface strength in clays. Adv. Civ. Eng. Mater. **2**(1), 1–17 (2013). doi:10.1520/ACEM20120043. ISSN 2165-3984

Haynes, F.D.: Effect of temperature on the strength of snow-ice (No. CRREL-78-27). Cold Regions Research and Engineering Lab. Hanover, NH (1978)

Ladanyi, B., Theriault, A.: A study of some factors affecting the adfreeze bond of piles in permafrost. In: Proceedings of Geotechnical Engineering Congress GSP, vol. 27, pp. 213–224 (1990)

Lawrence, D.M., Slater, A.G.: A projection of severe near-surface permafrost degradation during the 21st century. Geophys. Res. Lett. 32(24) (2005)

Linell, K.A., Johnston, G.H.: Engineering design and construction in permafrost regions: a review. Nat. Res. Counc. Can. **13848**, 553–575 (1973)

Linell, K.A., Lobacz, E.F.: Design and construction of foundations in areas of deep seasonal frost and permafrost (No. CRREL-SR-80-34). Cold Regions Research and Engineering Lab., Hanover, NH. Maksimovg, H. 1967. Time of natural freezing around piles driven into permafrost. Osnovaniya Fundamenty I Mekhanika Gruntov, No. 3, pp. 12–14 (1980)

Morgenstern, N.R., Roggensack, W.D., Weaver, J.S.: The behaviour of friction piles in ice and ice-rich soils. Can. Geotech. J. **17**(3), 405–415 (1980)

Nixon, J.F.: Design approaches for foundations in permafrost areas. Can. Geotech. J. **15**, 96–112 (1978)

Nixon, J.F., McRoberts, E.C.: A design approach for pile foundations in permafrost. Can. Geotech. J. **13**(1), 40–57 (1976)

Smith, E.A., McCarter, J. (eds.): Contested Arctic: Indigenous Peoples, Industrial States, and the Circumpolar Environment. University of Washington Press, Seattle (1997)

Weaver, J.S., Morgenstern, N.R.: Pile design in permafrost. Can. Geotech. J. **18**(3), 357–370 (1981)

Effect of Deep Supported Excavation on the Adjacent Deep Foundation

Qassun S. Mohammed Shafiqu(✉) and Ali A. Shawqi Al-Ameri

Al-Nahrain University, Baghdad, Iraq
qassun@yahoo.com, ali.shawqi2000@gmail.com

Abstract. In dense urban environments where land is scarce and buildings are closely spaced, cut and cover excavations are widely used for basement construction and development of underground facilities. One of the main design constraints in these projects is to prevent or minimize damage to adjacent structures.

In this study, a database for soil properties of Baghdad city zones has been obtained using (GIS-technique) which were then used in a 3-D finite element program PLAXIS 3D FOUNDATION to investigate the effect of a common cantilever supported excavation on the behavior of adjacent deep foundation. The supported material is modeled via a linear elastic model while the non-linear stress-strain behavior of cohesive and cohesionless soils is modeled using Mohr-Coulomb model.

In the numerical study, a parametric study is carried out to address the influence of the supported excavation on the lateral displacement of adjacent pile with various diameter, length and distance from the excavation face.

The results showed that the Mohr-Coloumb model parameters c_u and E are the most effective on the behavior of piles near supported excavation. And from the analysis, the maximum lateral deflection of pile depends on the pile-sheet pile wall distance and decreases significantly by increasing the distance. It has been decreases by approximately (76%) when the distance increased from 1 m to 9 m. Also both the diameter and length of the pile dose not play an important role for the lateral displacement when the distance is equal or greater than 0.8H (i.e. 5 m).

1 Introduction

During deep excavation, changes in the state of stress in the ground mass around the excavation and subsequent ground losses inevitably occur. These changes and ground losses affect the surrounding ground in the form of ground movements, which eventually impose direct strains onto nearby structures. The magnitude and distribution of ground movements for a given excavation depend largely on soil properties, excavation geometry including depth, width, and length, and types of wall and support system, and more importantly construction procedures.

Over the years, there have been a number of studies on the subject of wall and ground movements associated with deep excavation. Finno [1] gave an interesting example of this class of problem by reporting on the performance of groups of

M. Abu-Farsakh et al. (eds.), *Advances in Analysis and Design of Deep Foundations*,
Sustainable Civil Infrastructures, DOI 10.1007/978-3-319-61642-1_21

step-tapered piles located adjacent to a 15 m deep tieback excavation. After that Poulos and Chen [2] described a two-stage analysis involving finite element and boundary element methods to study pile responses due to excavation- induced lateral soil movement, focusing on unsupported and braced excavation in clay. Free-field displacements are motions of the soil that occur at a distance from the pile such that the displacements are not affected by the presence of the pile. And Iliadelis [3] studied the behavior of single axially loaded pile located close to a 30 m deep braced excavation in Marine Clay in Singapore. (with Plaxis Foundation 3-D), comparing different pile lengths (17 m, 30 m and 42 m) cross sections (solid concrete sections of 0.4 m and 1.0 m diameter) and proximity to the excavation (2 m–10 m). The results focus on the development of horizontal deformations and bending moments due to the excavation process.

In Iraq, the construction processes are gradually increasing throughout the country especially at the capital Baghdad. The number of projects of different types has become intensive such as multi-story buildings, industrial plants, electrical substations, bridges, stadiums, electrical power plants and others. Therefore, the supported excavation work adjacent to existing structures has become a common construction activity in most cities as utilization of underground space. One main concern in a supported excavation in urban area is the damage to the adjacent structures. To date, much of the research has focused on the lateral movements of the supported wall system and predictions of ground movements. Since buildings are supported on deep and shallow foundations, there is a concern that lateral ground movements resulting from the soil excavation can damage the piles and have effect on shallow footings. Although an excavation will cause both vertical and lateral soil movements, the later component is considered to be more critical, as piles are usually designed to sustain significant vertical loads. In contrast, lateral loads imposed by soil movements induce bending moments and deflections on the pile, which may lead to structural distress and even failure. For this reason, this study pays special attention to the development of lateral pile deformations caused by supported excavations in Baghdad soils.

Al-Adili and Al-Safi [4] defined Baghdad soil properties as of alluvial origin and generally consists of cohesive fill material at the top layer with variable thickness ranging from 0.5 to 4.0 m. This is generally underlain by clay, silty clay or clayey silt, with sand lenses in some places. At depths of more than 8 to 12 m sediments are generally sandy, knowing that the clayey soil may reach to depth of about 15 to 20 m. Hence, most of the projects above will be constructed in the saturated cohesive soils and need to use supported excavation system to carry out the foundation construction works. And in order to ensure the stability of the excavation and reduce the effect on the neighbor buildings and underground utilities caused by excavation in Baghdad city, commonly steel sheet pile wall structures are often used. In these cases, the use of cantilever support system is often desirable in order to reduce movements and to achieve relatively low economical benefits. Thus it is essential to conduct a study in order to investigate the influence of supported excavation on adjacent pile and shallow foundation.

The main objectives of this study is to prepare a database for the properties of Baghdad region different soils using GIS-Technique, and then use it to investigate the nonlinear behavior of pile foundations under the influence of construction of a typical and most common adjacent supported excavation system. This will require getting information about soil properties of different zones in Baghdad city from varying soil investigation reports collecting from engineering consulting bureaus of some Iraqi Universities.

In the analysis, the 3-dimensional analyses are performed using **PLAXIS 3-D FOUNDATION**, which is capable of simulating supported excavation, embedment pile element (friction or end-bearing).

2 Database for the Investigation of Baghdad Soils

The current study based on experimental results for underground conditions and the engineering properties of the various strata from many soil investigation reports for projects in Baghdad, which were collected from engineering consulting bureaus of Baghdad, Al-Nahrain and Technology Universities in Baghdad city. Soil investigation reports were for bridges, water treatment plants, multi-story buildings, medical clinic centers, electrical substations, and other projects in various locations in Baghdad Governorate.

In order to determine a data base for the soil properties that will be used for the purpose of this study, Baghdad area is divided into thirteen zones depending on border of municipalities (using Arc GIS technique), as shown in Fig. 1.

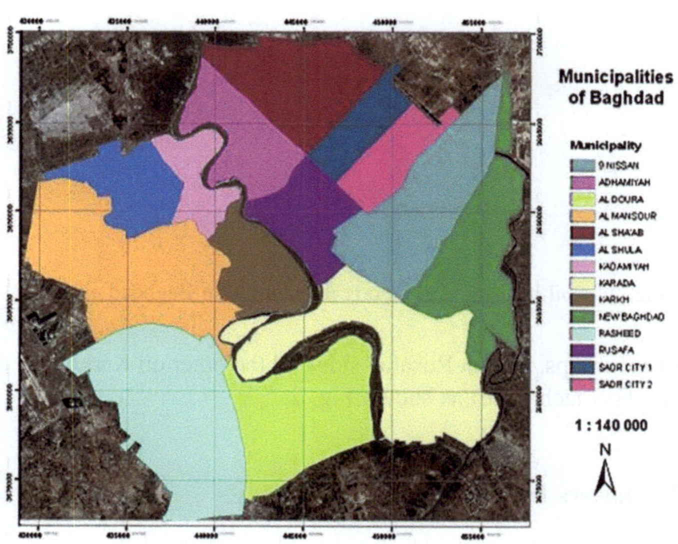

Fig. 1. Baghdad city municipalities

The available soil investigation reports have been collected trying to cover all the zones, 51 investigation reports including 174 boreholes were collected (Table 1)

Table 1. Summary of projects and the number of boreholes

No.	Zone	No. project	No. B.H	Zone symbol
1	Al-Sha'ab	3	9	I
2	Al-Adhimiyiah	4	23	II
3	Al-Sader 1 and 2	2	7	III
4	Rusafah	7	21	IV
5	Al-Karadah	11	27	V
6	Al-Kadhimiyiah	4	20	VI
7	Al-Mansour	11	28	VII
8	Al-Karkh	6	32	VIII
9	Al-Doura	3	7	IX

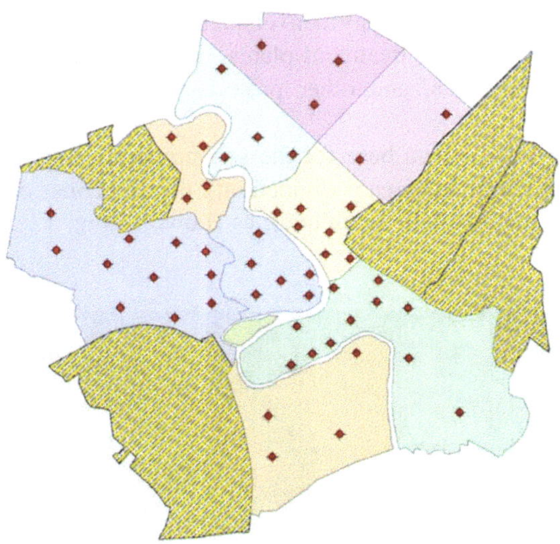

Fig. 2. Soil investigation reports distribution in Baghdad city area

dispensed as two groups, one on Rusafah side and the other on Karkh side on Baghdad map (using arc GIS technique) as shown Fig. 2.

2.1 Soil Parameters Investigated

For this study the parameters, which were used as an input parameter for PLAXIS software were investigated. These parameters are:

c_u = Undrained shear strength.
Ø = Angle of internal friction.
γ_{sat} = Unit weight of saturated soil.
E = Young's modulus
ν = Poisson's ratio

The most soil investigation reports were collected do not involve tests to determine the elastic parameters of soils therefore for this study empirical correlations were used to determine these parameters depending on the undrained shear strength, plasticity index, cone penetration and standard penetration values [5–12].

2.2 Empirical Correlations

For clays, the undrained shear strength is often used in correlations with stiffness parameters. Bowles [5] proposed the correlation between undrained shear strength and the modulus of elasticity for clay as follows:

Normally consolidated clay:

$$E = (200 - 500)\, cu \text{ (kPa)}$$

Over-consolidation clay:

$$E = (750 - 1000)\, cu \text{ (kPa)}$$

The US Army Corps Engineers [6] tries to estimate the modulus of elasticity as,

$$\begin{aligned} E &= kccu \\ kc &= 4200 - 142.54\,PI + 1.73\,PI^2 - 0.007\,PI^3 \end{aligned}$$

where E is Young's soil modulus (MPa), k_c is correlation factor, c_u is undrained shear strength in MPa, and PI is the plasticity index.

Schmertmann [7] measured the modulus of elasticity (E) estimated from the cone resistance from a static cone penetration test as,

$$E = 2q_c$$

The elastic modulus E for sand may be estimated from the standard penetration test N values. D'Appolonia [8] proposed the correlation between the SPT 'N' value and Young's modulus E as follows

For loose sand,

$$E = 106N + 2160\,(t/m)$$

For dense sand,

$$E = 500(N + 15) \text{ (kPa)}$$

According to Japanese National Railways [9], building code for design of foundation, earth retaining structures and underground structures, the correlation between SPT 'N' value and Young's modulus of soils based on extensive study of borehole horizontal loading tests is as follows:

$$E_s = 2500\,N\ (\text{kPa})$$

Also Bowles [5] proposed the correlation between the SPT 'N' value and Young's modulus E

$$E = 250\,(N + 15)\ \textit{for Sand (saturated)}$$

With regard to the Poisson's ration v there are many approximate values for various soils [10], Bowles [5] proposed that the Poisson's ratio for saturated clay is (0.4–0.5), and (0.2–0.3) for sandy soil.

Poulose [11] proposed v as (0.3–0.35) for medium clay and (0.25–0.3) for dense sand.

Cernica [12] proposed v equal to (0.25) for hard clay, and (0.3) for dense sand

For the present study the values for Young's modulus and Poisson's ratio are taken equal to the following:

$$E = 200.\,c_u(\text{kPa}),\ \textit{and}\ v = 0.35\ \ldots\ldots\ldots\ \textit{for clay soil},$$
$$E = 250.\,(15 + N),\ \textit{and}\ v = 0.3\ \ldots\ldots\ldots\ \textit{for sand soil}.$$

2.3 Spatial Data Presentation and Analysis

In this study, the data collected from available soil investigation reports in the study area are presented with respect to their variation with depth. Nine zones are identified; each zone has data (soil properties) analyzed in global by using statistical operation; also the results from this operation are integrated with GIS-Techniques to extract digital maps which contain some missing data in the soil investigation reports.

2.4 Applying GIS Technique

The test results collected from site investigation reports are linked to their geographic locations, through their easting and northing records, in ArcMap9.3 software to build a GIS database for the study region.

Borehole locations are projected on the created map (using ARC GIS) and attribute table of the geotechnical properties recorded in each borehole of the various points in the study region constructed. Figure 2 shows the distribution of borehole locations in the study region. The boundaries of each district (zones) in the study region are also shown in the Fig. 1.

2.5 Spatial Data Analysis

Spatio-analytic capabilities distinguish GIS from other data processing systems. These capabilities use the spatial and non-spatial data in the spatial database to answer questions and solve problems.

The principal advantage of using spatial data analysis in GIS software's (such as ArcMap9.3) is to perform operations that relate values of one location to those at neighboring locations. Since the geotechnical properties change, not only with depth (vertically), but also across the area (horizontally), then using capabilities of GIS to represent the variation in the geotechnical properties across the area, at a particular depth, can be produced as thematic maps that show the variation in values as graduated colors which will help in understanding the distribution in a better way, since a quick glance at such a map will generate the impression of how the property is distributed across the area.

This is done by transferring the information mode from the vector information to the raster information. On the other hand, it is the observed description process in the specific points to change them to thematic maps. Thus, the special data on the kind of soil and its geotechnical properties have been manipulated (using inverse distance weighted method) for borehole points of the study region through its generation and its transformation from its dot shape to the grid shape by submitting it to the counting processing to achieve the purpose of this chapter to produce digital geotechnical maps that show the distribution of any property across the study region as new layer.

So physical and some geotechnical properties can be arranged through the GIS software ArcMap 9.3 by producing the layers which show the distribution of these properties across the study region.

Figures 3, 4, 5, 6, 7, 8, 9, 10, 11 and 12 shows the thematic maps for all Baghdad zones of the soil properties. Mohr-Coloumb model parameters (c_u, *Ø*, γ_{sat}, *and E*) for Baghdad soil appear to be in the following ranges:

Parameters	c_u (kPa)	Ø°	γ_{sat} (kN/m^3)	E (MPa)
Range	28–75	31–41	18.8–20.6	5.6–15

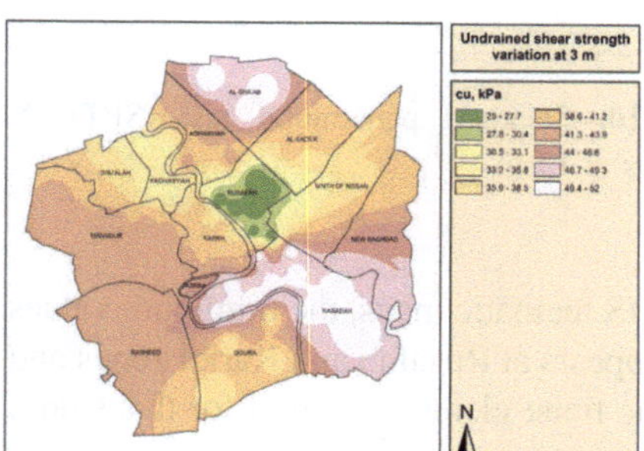

Fig. 3. Undrained shear strength (c_u) at 3 m

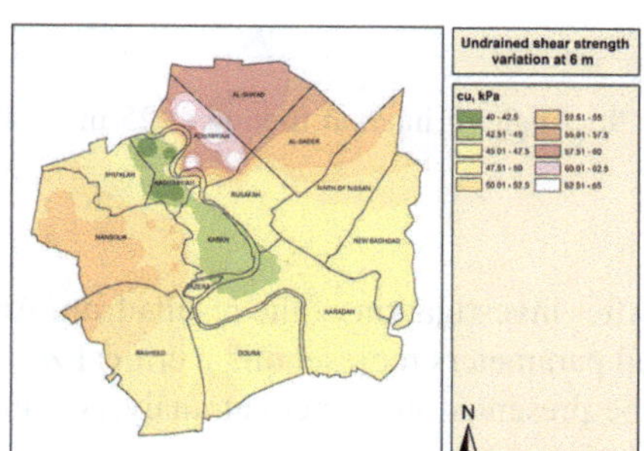

Fig. 4. Undrained shear strength (c_u) at 6 m

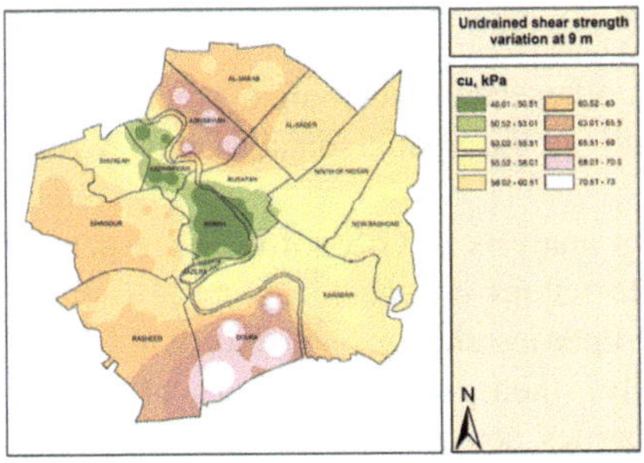

Fig. 5. Undrained shear strength (cu) at 9 m

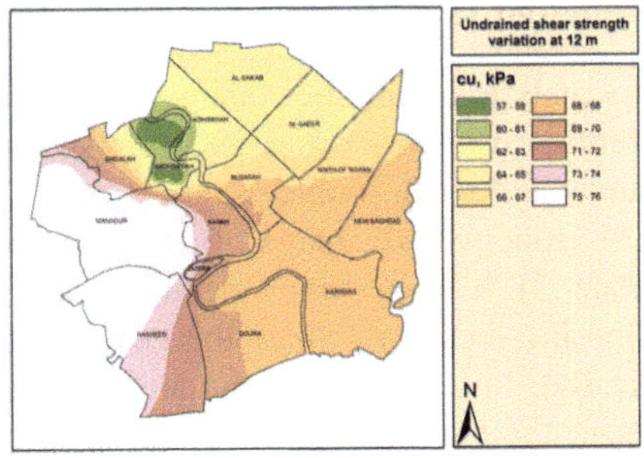

Fig. 6. Undrained shear strength (cu) at 12 m

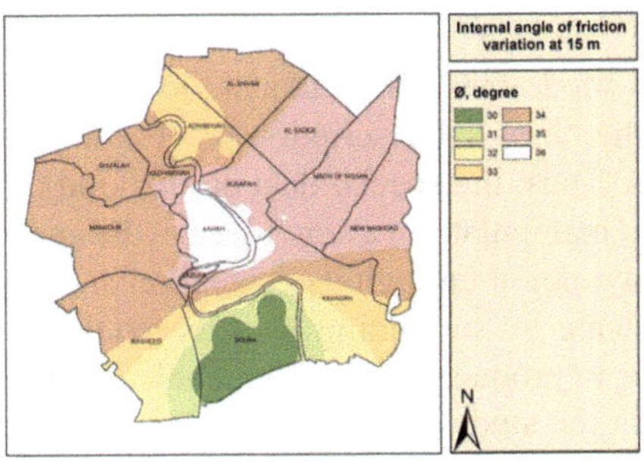

Fig. 7. Angle of internal friction at 15 m

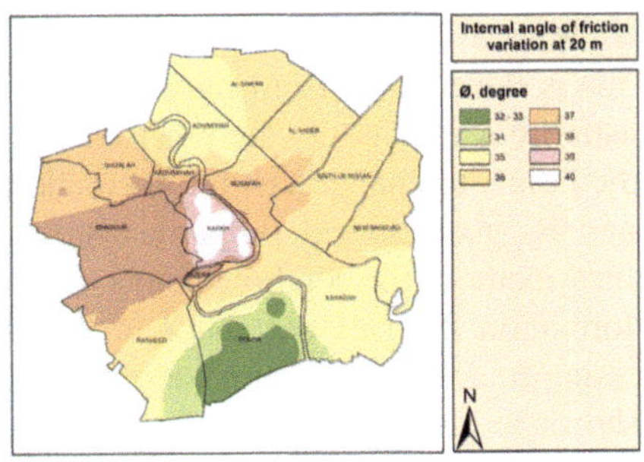

Fig. 8. Angle of internal friction at 20 m

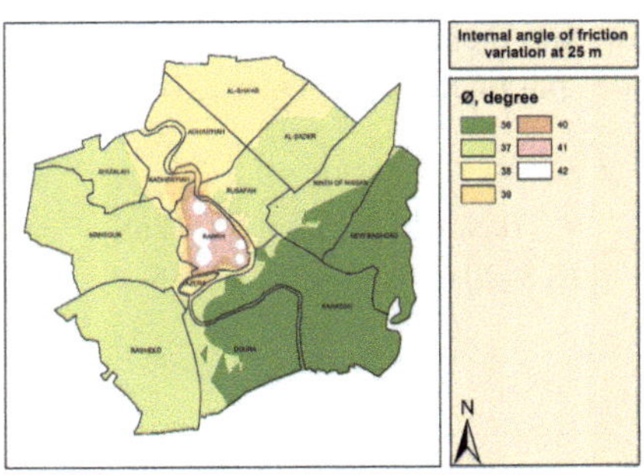

Fig. 9. Angle of internal friction at 25 m

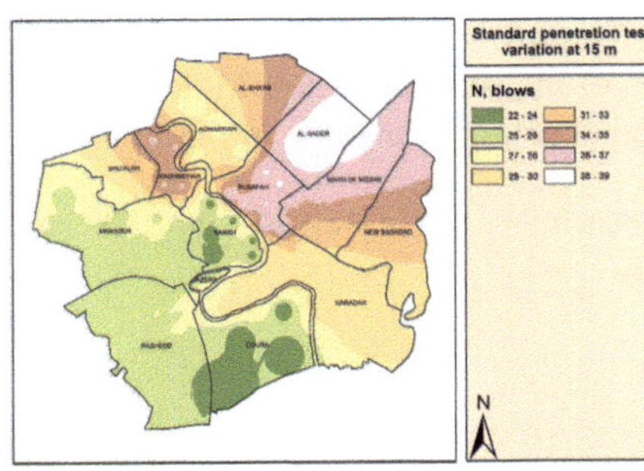

Fig. 10. Standard penetration test (SPT), N values at 15 m

After investigation of the results from Arc GIS thematic maps, the minimum values of soil parameters representing a critical zones appears at Rusafah and Karkh zones and will be presented in numerical analysis using the finite element method for this study.

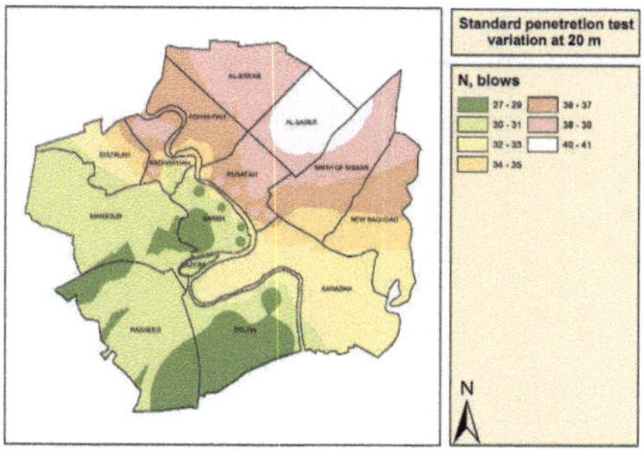

Fig. 11. Standard penetration test (SPT), N values at 20 m

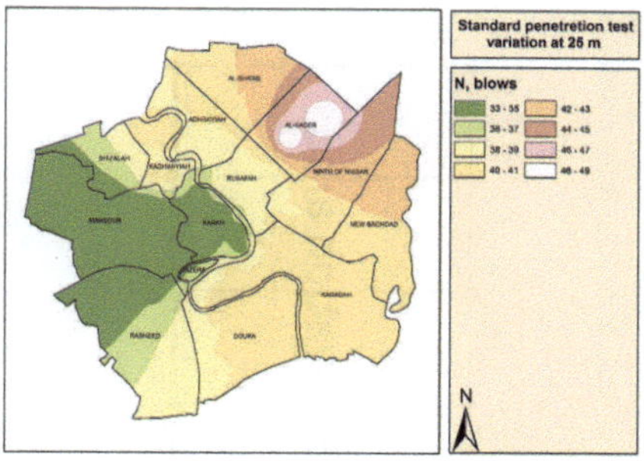

Fig. 12. Standard penetration test (SPT), N values at 25 m

3 Supported Excavation and Substructure Finite Element Modelling

Predicting the behavior of structures adjacent to deep supported excavations is a very complex geotechnical problem, and it cannot be solved explicitly. Therefore, numerical method will be used. The finite element method is a very powerful tool for solving problems in geotechnical engineering in which the domain is divided into sub domains called elements connected with each other at selected points called nodes.

The finite element analysis includes modeling of deep foundations and soil surrounding them with the actual dimensions and properties corresponding to a typical common supported excavation generally used in Baghdad soils.

3.1 Geometry of the 3-D Model

The 3-D finite element model for a typical supported excavation system in Baghdad city is presented in this section assuming symmetry such that, only half of the excavation is simulated. The model considers the effect of a 16 m length of an excavation on the response of adjacent pile. The 3-D finite element model geometry is modelled using a top view approach. Working planes at the top and bottom of the foundations and the sheet pile wall should be introduced. Moreover, working planes corresponding to the excavation stages are needed.

Figure 13 presents the top working plane of the case study corresponding to the ground surface. The dimensions of these models are 16 m x 32 m for excavation adjacent to deep foundation; the vertical dimension is defined by the program to be 3 m below the lowest working plane.

The 3-D finite element meshes used in the analyses are shown in Fig. 14. The mesh represents a sheet pile wall supporting a saturated soil. The total depth of the model is taken to be 25 m, divided to seven layers, the four upper layers consist of cohesive soil and the three lower consist of cohesionless soil.

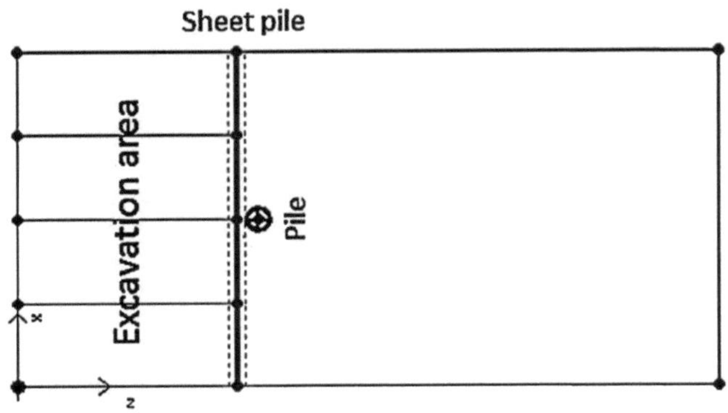

Fig. 13. The top working plane of the model

A sheet pile wall, an AU14 type is used, the embedment depth of the sheet pile wall was calculated, thus 10 m embedment depth of sheet pile wall is used to simulate this common supported excavation system in Baghdad city.

This study assumes a pre-existing pile foundation at a working load corresponding to 50% of the ultimate pile capacity (prior to excavation).

In staged excavation adopted in the simulations, the principle is that one segment of 2 m depth at a time is excavated. The segments are excavated to reach 6 m depth (end of excavation). The finite element mesh consists of about 3600 brick elements each of 15-nodes with a total number of 54000 nodes.

3.2 Material Properties

3.2.1 Soil

According to the thematic maps for all Baghdad zones of the soil properties, the ground condition at each zone used for PLAXIS 3D FOUNDATION were previously appointed in above.

3.2.2 Sheet Pile Wall Properties

A sheet pile wall is a type of retaining wall. Retaining walls are installed, mainly, when a difference in ground elevation with a greater angle than the soil's angle of repose is wanted, where the angle of repose is the maximum angle a soil can withstand without sliding. When the soil is forced into this state the soil wants to transform back to its natural state and lateral pressure towards the wall occurs, called lateral earth pressure. The main purpose of the wall is to resist this pressure, Ryltenius [13] (Fig. 15).

The wall selected in the present study is a sheet pile wall, (AU14). This is a relatively stiff wall, and should be able to withstand the stresses induced by the excavation, meaning that a possible failure in the system will be in the soil and not in the sheet pile wall. And the technical specification for AU14 sheet pile wall and used for PLAXIS 3D FOUNDATION are given in Table 2.

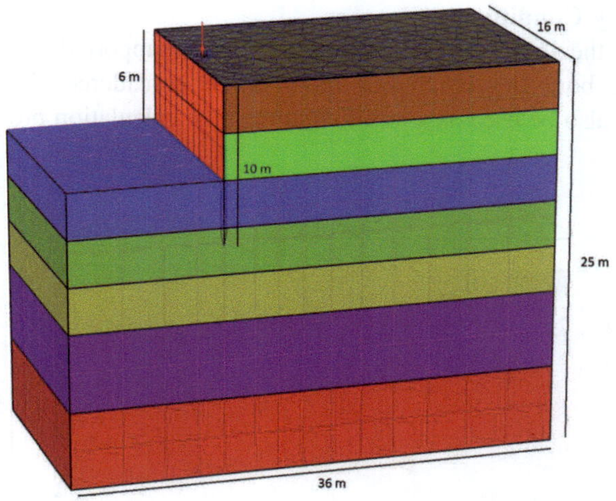

Fig. 14. Finite element mesh of supported excavation adjacent to pile

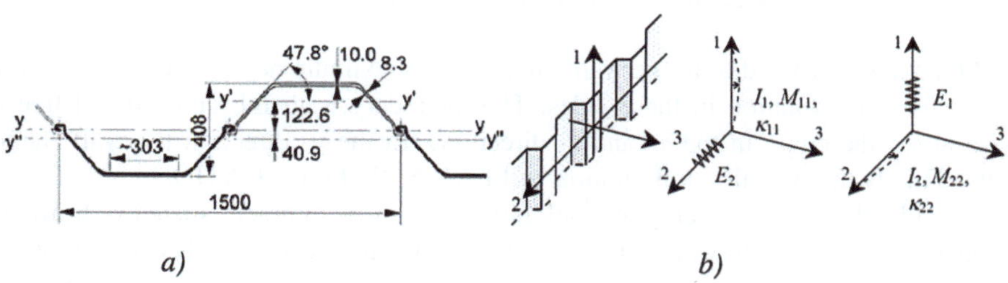

Fig. 15. (a) Sheet pile profile AU14 in plane including dimensions ArcelorMittal, (b) the local system of axes in the wall element and various quantities

Table 2. Technical specifications AU14 (ArcelorMittal, 2008)

AU14	Sectional area cm^2	Mass/m kg/m	Moment of inertia cm^4	Elastic sec. modulus cm^3
Per m wall	132.3	103.8	28710	1410

3.2.3 Pile Properties

The properties of pile foundation used for PLAXIS 3D FOUNDATION are given in Table 3.

Table 3. Pile properties

Pile length, L (m)	Type	Dia., D (m)	γ (kN/m^3)	E_c (MPa)	ν
8, 15 and 20	Concrete	0.5, 0.8 and 1	24	3000	0.2

3.2.4 Boundary Conditions

When modelling the sheet pile wall and foundation in supported excavation problem, the effect of the boundary conditions needs to be considered. This is not only a limitation; it can also be used as a tool to optimize the calculation process as shown in Fig. 16.

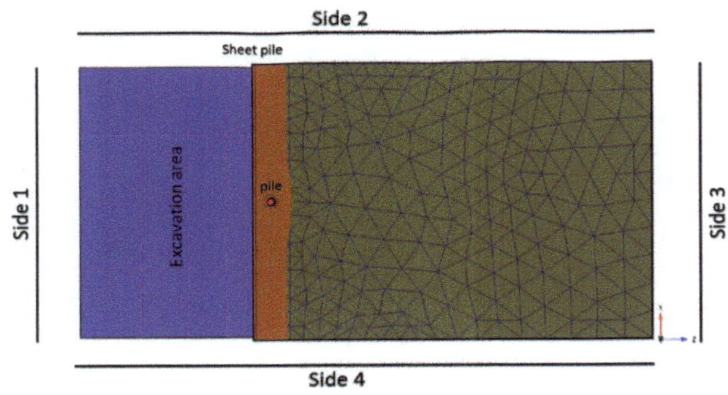

Fig. 16. Boundary condition of the case study

Along sides two and four, the wall will have its maximum deflection. The sheet pile wall is fixed for rotation in these sides. This boundary is thereby considered free to move along the edge, in the x- and z- directions. In the y- direction the wall cannot move. This is the default configuration in PLAXIS 3D FOUNDATION.

On side three, the sheet pile wall must end in a sufficient distance from the boundary. The sheet pile wall must be free to move, unaffected by this boundary. On side one, the sheet pile wall restrained to move in any direction. Since the boundary condition allows free movement along the side without friction, the wall in the model will have a larger displacement than the real case, however, this assumption will be on the safe side, for the bottom face of the mesh is assumed to be fully fixed.

The opposite extreme would instead be to lock the displacements at the boundary, which is possible in PLAXIS 3D FOUNDATION.

3.2.5 Finite Element Analysis in PLAXIS

Excavation is a complicated, three-dimensional soil-structure interaction problem in which the induced ground movements may be affected by a large number of factors. Many studies have demonstrated the reliability of the finite element method (FEM) for the analysis of deep excavations. Therefore, the present study uses the computer software **PLAXIS 3D FOUNDATION** for performing the nonlinear model analysis. The validity and accuracy of the finite element model using **PLAXIS 3D FOUNDATION** for studying effects of deep supported excavation on the adjacent pile is beyond the scope of this paper, and a verification study can be obtained in Al-Ameri [14], were nonlinear F.E.analysis was performed by using **PLAXIS 3D FOUNDATION - 1.6** computer software for predicting the maximum deflection and bending moment of a single pile due to excavation-induced lateral soil movements, with specific attention being focused on braced excavations in clay layers.

4 Effect of Deep Supported Excavation on Adjacent Pile

In this study, a complete set of Mohr-Coulomb model parameters were obtained for Baghdad city soils. Therefore, it is useful to make use of these parameters in studying the effect of common supported excavation system on existing adjacent pile in Baghdad city.

A sheet pile wall, an AU14 type is used, the embedment depth of the sheet pile wall was calculated, thus 10 m embedment depth of sheet pile wall is used to simulate this common supported excavation system in Baghdad city.

Free head single pile will be used in this investigation, as a critical case and in order to simplify the study.

The simulations consider circular solid section concrete piles, with diameters D = 0.5 m, 0.8 m and 1.0 m and lengths L = 8 m, 15 m and 20 m. The bearing capacity of a single pile is the sum of its shaft bearing capacity and its end bearing capacity. The bearing capacity is calculated for each pile for all Baghdad zones. There are many factors that may have effect on the behavior of pile adjacent to supported excavation such as soil properties, diameter of pile, length of pile and distance from face of excavation.

4.1 Influence of Pile Diameter

To study the effect of pile diameter on the lateral displacement, the same model with multi-soil layers which was described before is considered (16 × 32 × 25) m, with three pile diameters (0.5D, 0.8D, 1D), where D (pile diameter) is taken equal to 1 m. The pile is located at 1 m distance from face of supported system.

Figures 17 and 18 show the lateral horizontal displacement of the pile with different lengths and diameters under the influence of cantilever of 6 m deep supported excavation. The figures compare the lateral deflections of these piles located 1 m from the sheet pile wall for the Rusafah and Karkh zones in Baghdad city.

From the results, in general, it can be observed that the soil properties (especially c_u values) are the effective parameters on the pile behavior. Also it can be noted that the pile displacements are the highest in Rusafah zone.

The piles which have the same length with different diameters exhibit a very similar deformation behavior in each zone.

From Figs. 17, 18, 19, 20, 21, 22, 23 and 24, it can be observed that the maximum lateral displacement appears at pile head of 1 m diameter and decreases with depth, but for pile of 0.5 m and 0.8 m diameter the maximum lateral displacement occurs at variation depth from the ground level depending on cu values.

Figure 17 shows the maximum lateral displacement occurs in Rusafah zone for three 8 m pile length with different diameters, for 0.5D the result shows that the top of pile displaced 55 mm, and when the pile diameter increase to 0.8D the top of pile displaced 58 mm. While for pile having 1 diameter the top head displaced 60 mm.

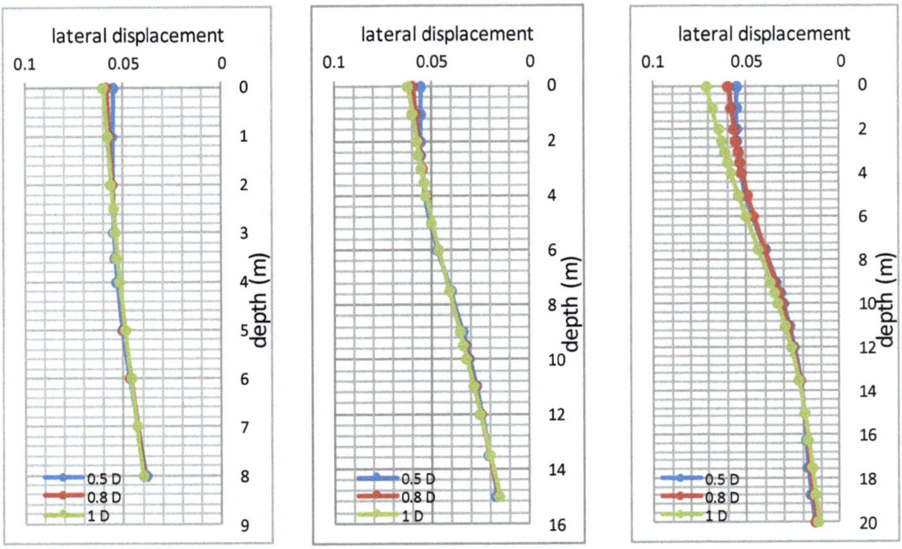

Fig. 17. Lateral displacement of piles having 8, 15 and 20 m lengths with different diameters located 1 m from excavation face for Rusafah zone's

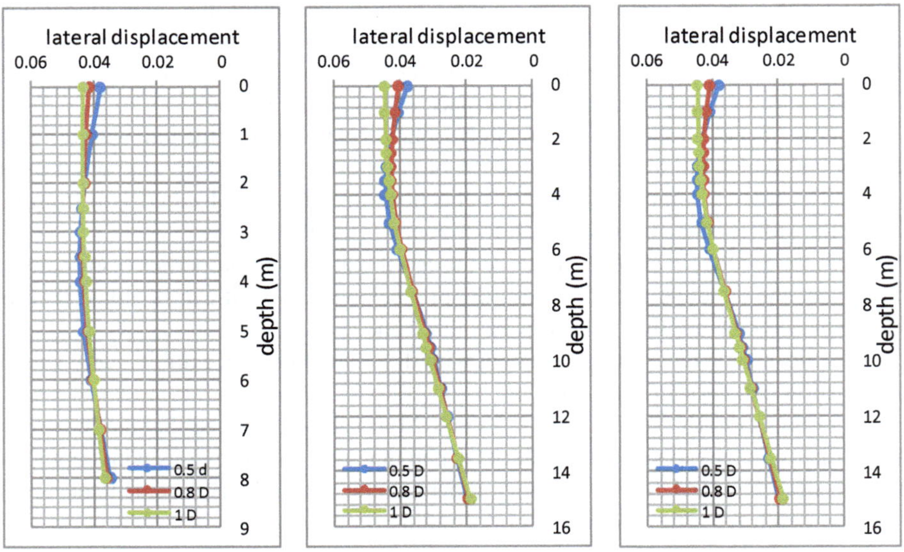

Fig. 18. Lateral displacement of piles having 8, 15 and 20 m lengths with different diameters located 1 m from excavation face for Karkh zone's

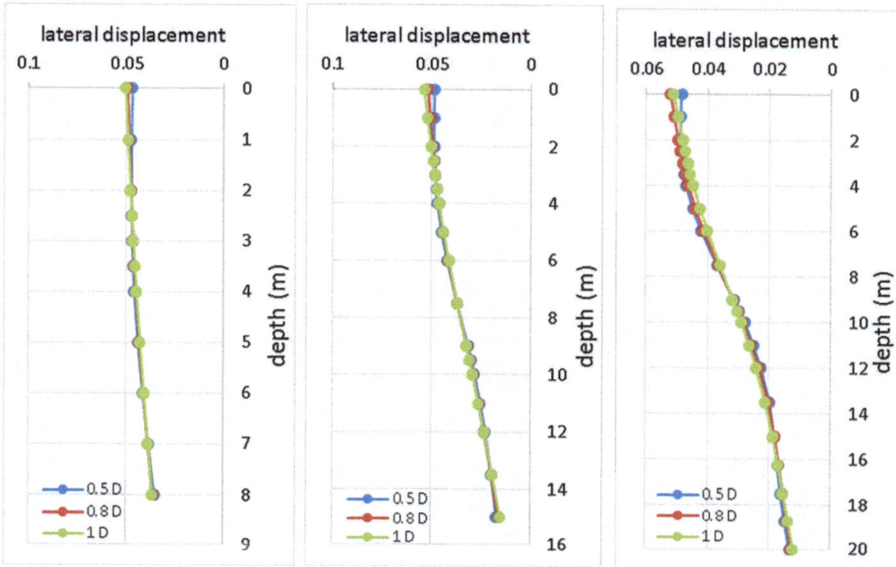

Fig. 19. Lateral displacement of piles having 8, 15 and 20 m lengths with different diameters located 3 m from excavation face for Rusafah zone

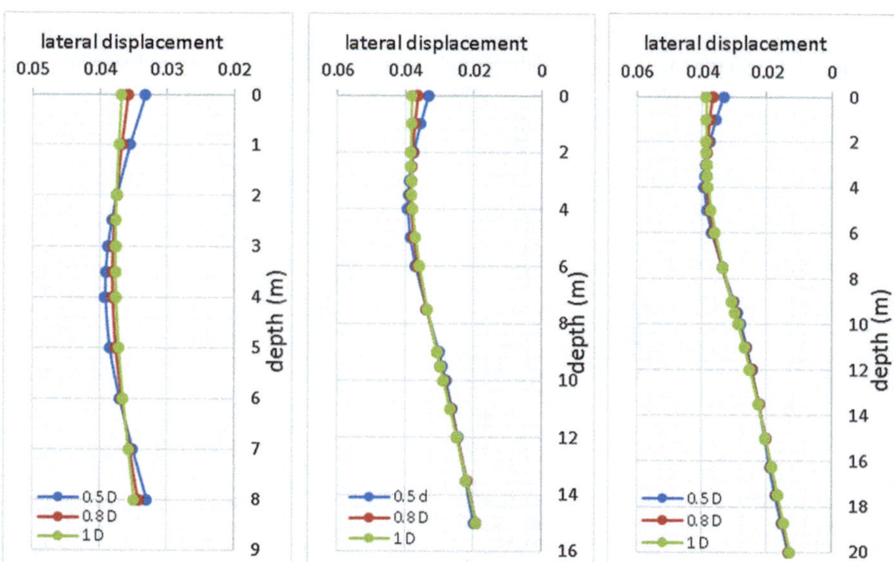

Fig. 20. Lateral displacement of piles having 8, 15 and 20 m lengths with different diameters located 3 m from excavation face for Karkh zone's

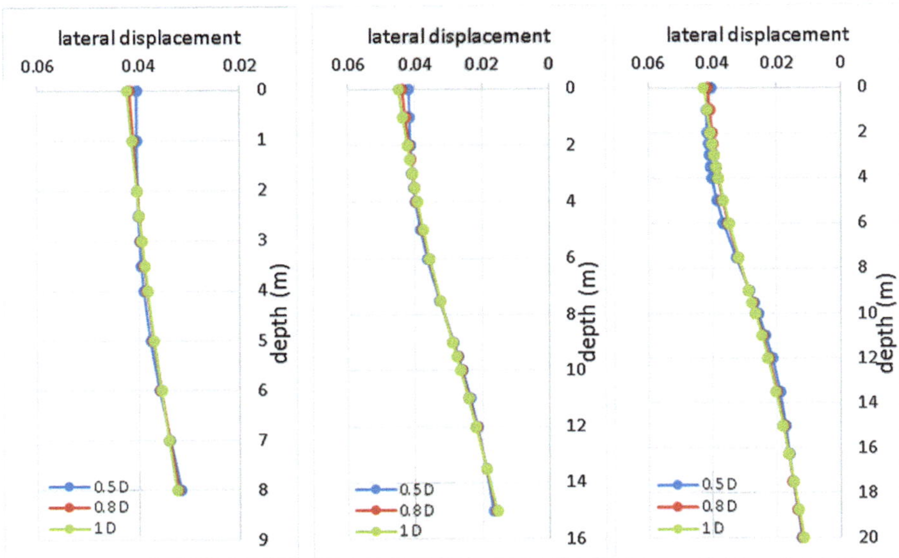

Fig. 21. Lateral displacement of piles having 8, 15 and 20 m lengths with different diameters located 5 m from excavation face for Rusafah zone

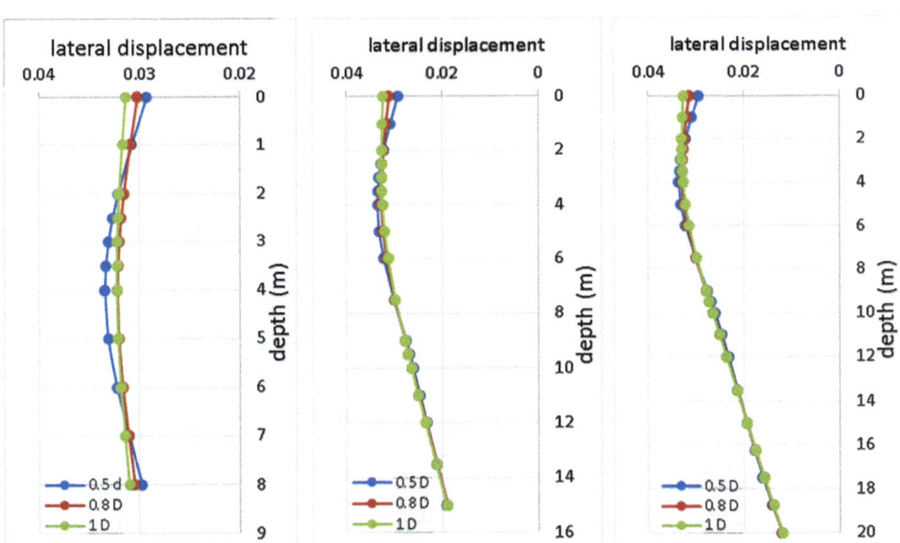

Fig. 22. Lateral displacement of piles having 8, 15 and 20 m lengths with different diameters located 5 m from excavation face for Karkh zone

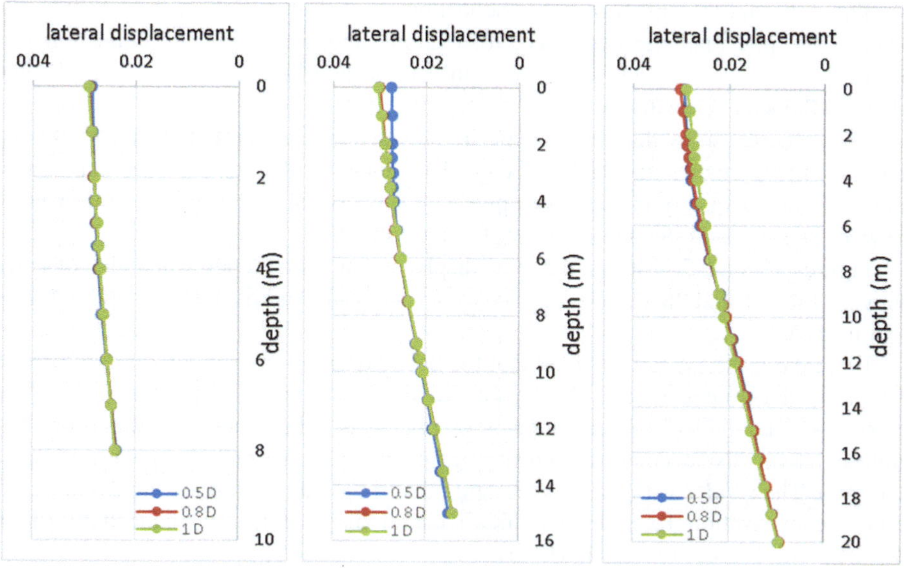

Fig. 23. Lateral displacement of piles having 8, 15 and 20 m length with different diameters located 9 m from excavation face for Rusafah zone

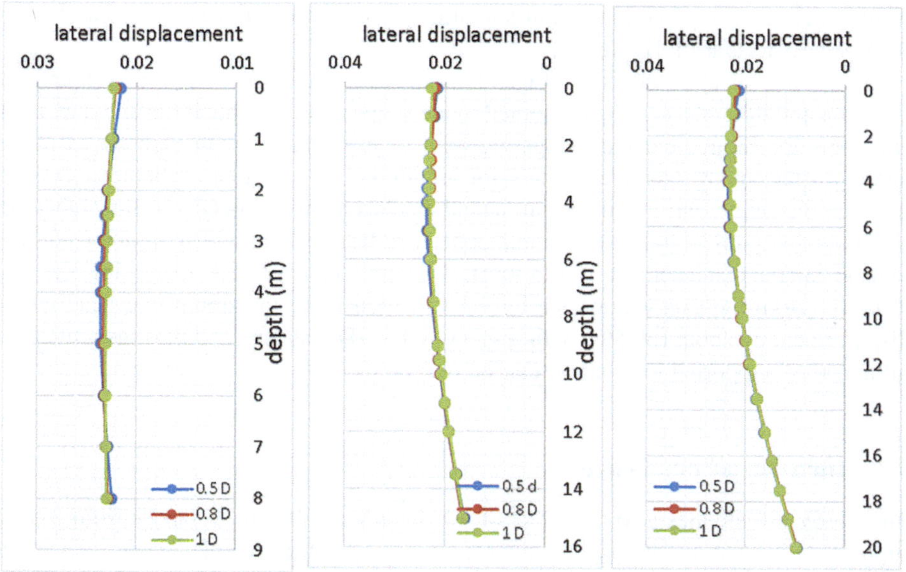

Fig. 24. Lateral displacement of piles having 8, 15 and 20 m length with different diameters located 9 m from excavation face for Karkh zone

The position of the maximum pile deflection was found to move away from the pile head as the pile diameter decreased, for example the maximum pile deflection was 56 mm, 58 mm, and 60 mm at 3.5 m, 3 m, and at the pile head, for 0.5D, 0.8D, and 1D pile diameter respectively.

These figures show that the increase in diameter of pile causes increase in pile top and toe displacements in Baghdad city. This may be due to the fact that higher allowable load capacity is applied to pile with highest diameter, and the soil distance between the pile and the sheet pile wall is the lowest.

Figures 19 and 20 shows the pile displacement with depth when the distance between the pile and the excavation face for Rusafah and karkh zones is increased from 1 m to 3 m.

From the results it can be noted that the behavior of all piles in all zones located at 3 m from sheet pile wall has little change with decease in lateral displacements, and the maximum lateral displacement appears at pile head which has 1 m diameter, The highest lateral displacement obtained is equal to (52 mm) and the lowest value is (31 mm). The results also give an indication that if the pile location is increased by 3D from the excavation face, the maximum lateral displacement is reduced by about (26.7%) with respect to that of 1D distance.

From the Figs. 21 and 22, it can be noted that the behavior of all piles in all zones located at 5 m from sheet pile wall have little change from pile behavior located at 3 m, with deceasing lateral displacement, also it can be observed that the maximum lateral displacement appears in Rusafah zone at pile head of 1 m diameter with value equal to 30 mm. The results also give an indication that if the pile location is increased by 5D from excavation face, the maximum lateral displacement is reduced by about (57.7%) with respect to that of 1D distance.

From the results shows in Figs. 23 and 24, it can be noted that there are insignificant differences in the computed mode shape and the behavior of the pile, in all zones no appearance change comparing with other distances with high deceasing in lateral displacement, The pile of the same length and no variation in diameter exhibits the same behavior, and the maximum lateral displacement still occurs in Rusafah zone, also the maximum lateral displacement appears at head of pile with 1 m diameter. The highest lateral displacement is (28 mm). This will indicate that increasing the pile distances from the excavation face by 9D will reduce the maximum pile head lateral displacement by about (60.5%) with respect to that 1D distance, and thus insignificance influence of supported excavation can be observed.

4.2 Influence of Pile Length

This study investigates also the effect of pile length at different distances from face of supported excavation on the behavior of pile and lateral displacement. Since the maximum displacement occurs for pile of 1 m diameter, the results of lateral displacement with depth for pile having 8, 15, and 20 m length and 1 m diameter located 1, 3, and 9 m distances from sheet pile wall in Rusafah and Karkh zones are presented in Figs. 24 and 25.

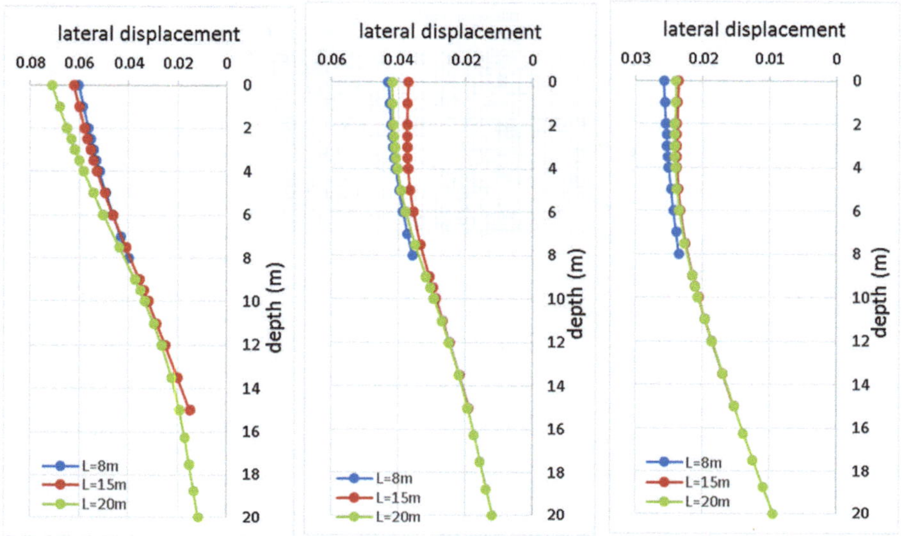

(a) (b) (c)

Fig. 25. Lateral displacement of pile having 8, 15, and 20 m length located (a) 1 m distance, (b) 3 m distance, (c) 9 m distance. For Rusafah zone's

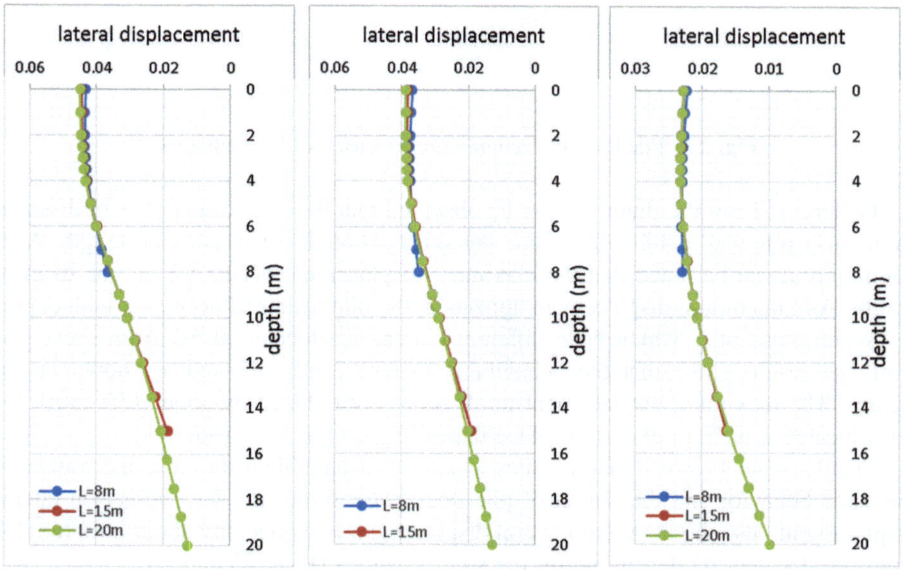

Fig. 26. Lateral displacement of pile having 8, 15, and 20 m length located. (a) 1 m distance, (b) 3 m distance, (c) 9 m distance. For Karkh zone's

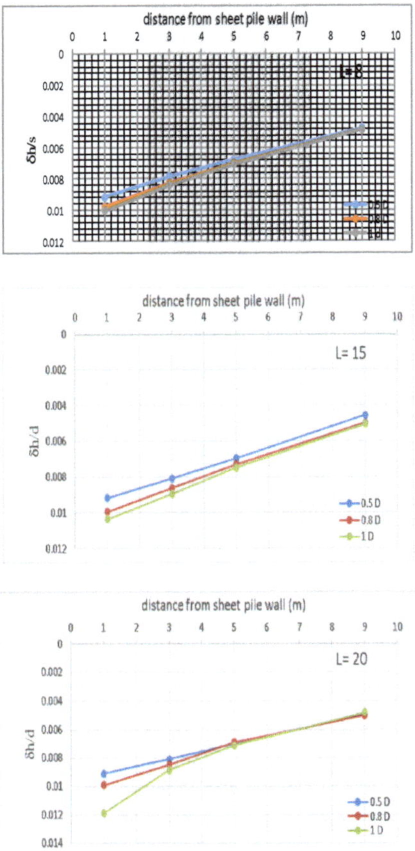

Fig. 27. Pile head deflection with distances for Rusafah zone

Generally, from the figures it can be observed that the pile located at 1 m distance from sheet pile wall exhibits the same behavior although it has different length. Also commonly it can be noted that the maximum displacement occurs for pile of 20 m in length. And maximum displacement appears at the pile head and decreases with depth.

When these piles which have different lengths are located at 5D from sheet pile wall, the results show that the behavior is insignificantly affected as shown in the figures. The maximum displacement in all zones are reduced by about 15%–40% for piles located 1 m from excavation face depending on the soil properties.

Finally, it can be noted that the piles which are located 9D from sheet pile wall have the same behavior and the length of pile does not significantly have effect on lateral displacement, also the maximum lateral displacement is reduced by about 30% to 62% from pile located 3D depending on the soil properties (Figs. 25 and 26).

In order to investigate the effect of distance from supported excavation face on the pile head deflection, Figs. 27 and 28 shows the normalized results of the displacement of pile head of 0.5 m, 0.8 m and 1 m diameter with 8 m, 15 m and 20 m length at various pile-sheet pile distances for Rusafah and Karkh Zones.

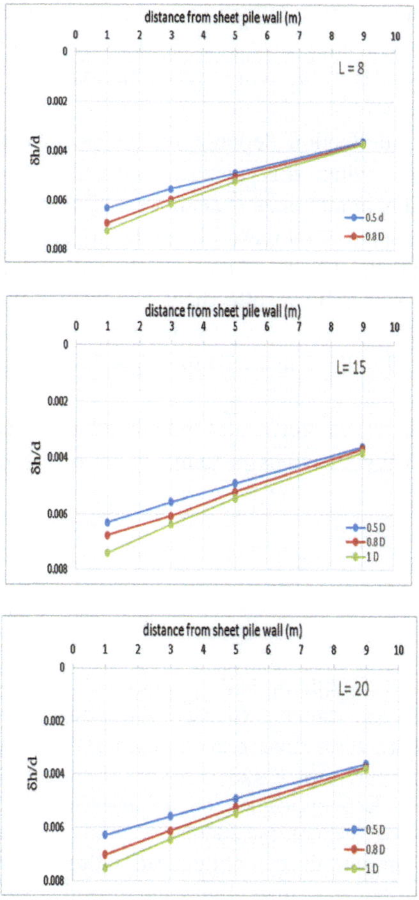

Fig. 28. Pile head deflection with distances for Karkh zone

5 Conclusions

The main conclusion drawn from present study can be summarized as follows:

1. The study shows that, the Mohr- Coloumb model parameters c_u and $\emptyset$ are the most effective on the behavior of deep foundation under the influence of adjacent supported excavation system.
2. It was clear that analyzing the effect of supported excavation on adjacent deep foundations by using non-linear finite element method has many benefits as the problem is modelled in detail and all the important parameters relating to soil parameters and excavation supported system are included. Moreover, the model can be established for the site specific condition.

3. The maximum pile head lateral deflection is obtained equal to (71 mm) in zone IV at 1 D distance from excavation face. It is worth noting that increasing the pile-sheet pile wall distance from 1D to 9D will reduces the maximum lateral displacement in zone IV by about (60.5%).
4. The maximum lateral deflection depends on the pile-sheet pile wall distance and decreases significantly when this distance increases, the maximum horizontal deformation of the piles in all zones is decreased by approximately (76%) when the pile to excavation distance is increased from 1D to 9D.
5. Both the diameter and the length of the pile do not play an important role in the lateral displacement especially at pile-sheet pile wall distance equal to or greater than about 0.8H (i.e. 5 m). And it was found that the increase the pile diameter and length from 0.5D to 1D and 8 m to 20 m respectively slightly increases the horizontal deformation.
6. The ratio of the maximum horizontal pile head displacement to the depth of excavation ($\delta h_{max}/H$) for all zones is found to be in the range of about (1.2% to 0.28%).

References

1. Finno, R.J., Lawrence, S.A., Allawh, N.F., Harahap, I.S.: Analysis of performance of pile groups adjacent to deep excavation. J. Geotech. Eng. ASCE **117**(6), 934–955 (1990)
2. Poulos, H.G., Chen, L.T.: Pile response due to excavation-induced lateral soil movement. J. Geotech. Geoenviron. Eng. ASCE **123**(2), 94–99 (1997)
3. Iliadelis, D.: Effect of deep excavation on an adjacent pile foundation. M.Sc. thesis, Massachusetts Institute of Technology (2006)
4. Al-Adili, A.S.: Geotechnical evaluation of Baghdad soil subsidence and their treatments. Ph. D. thesis, University of Baghdad, Iraq (1998) and Al-Safi: Empirical correlations between physical and engineering properties of clayey soil in Baghdad City. M.Sc. thesis, College of Engineering, Al-Nahrain University (2005)
5. Bowles, J.E.: Foundation Analysis and Design, 4th ed. McGraw-Hill Book Company, New York (1988). Clough, G.W., Smith, E. W., Sweeney, B.P.: Movement control of excavation (1989)
6. Department of the Army U.S. Army Corps of Engineers: Engineering and Design Settlement Analysis (1990)
7. Schmertmann, J.H.: Static cone to compute static settlement over sand. J. Soil Mech. Found. Div. ASCE **96**(3), 1011–1043 (1970)
8. D'Appolonia, E.: Settlement of spread footings on sand. JSMFD, ASCE, **94**(SM 3), 735–760 (1968)
9. Japanese National Railways: Building Code for Design of Foundation, Earth Retaining Structures and Underground Structures (1986)
10. Bardet, J.P.: Experimental Soil Mechanics. Prentice-Hall Inc, Upper Saddle River (1997)

11. Poulose, H.G.: Settlement of isolated foundation. In: Proceeding of the Symposium on Soil Mechanics—Recent Developments, Sydney, Australia, pp. 181–212 (1975)
12. Cernica, J.N.: Geotechnical Engineering, Soil Mechanics, p. 241. Wiley, New York (1995)
13. Ryltenius A.: FEM modelling of piled raft foundation in two and three dimension. M.Sc. thesis, Dept. of Construction Sciences, LTH, Lund (2011)
14. Al-Ameri, A.A.: Effect of deep supported excavation on the adjacent structures. M.Sc. thesis, College of Engineering, Al-Nahrain University (2014)

Side Resistance Assessment of Drilled Shafts Socketed into Rocks: Empirical Versus Artificial Intelligence Approaches

Asmaa M.H. Mahmoud[1(✉)] and Ahmed M. Samieh[2]

[1] ABCAN Engineering Consultancy, El-Demerdash, Cairo, Egypt
abcams98@gmail.com
[2] Department of Civil Engineering, Helwan University, Mataria, Cairo, Egypt
abcan98@yahoo.com

Abstract. Drilled shafts socketed into rock are widely used to transfer heavy structural loads through weak overburden strata to underlying bedrock, which can sustain the load. Numerous studies have been conducted in the recent years to predict the side resistance of rock socketed shafts under vertical loads. The problem is extremely complex owing to the large number of factors that affect the socketed shafts behavior. This study investigates the applicability of the existing empirical equations to predict the side shear resistance of drilled shafts socketed into rock using a compiled shaft load tests database. The compiled database is, also, analyzed to investigate the possibility of establishing an empirical equation for improving the prediction of side shear resistance of the drilled socketed shafts. In addition, an artificial intelligence approach, a fuzzy logic scheme, is established in this study to evaluate the applicability of such approaches to predict the side resistance of drilled shafts socketed into rock formation. The established approaches exhibited a good comparison between the predicted and the monitored values.

1 Introduction

When the structure loads are relatively large or where the soil is of relatively poor quality, drilled shafts are a common foundation selection to support these types of structures. The shafts are often drilled through the weak surface soil to the underlying rock mass, these shafts could be founded or seated on the rock mass surface, or they could be drilled into the rock mass to form a rock socket. The applied butt load/stress is supported by the socket through both tip and side resistances. Design loads might be limited for shafts relying upon end bearing on the rock surface if the rock is weak and weathered at/near the rock formation surface. For such formations, rock sockets are constructed to increase the loading capacity of the shafts.

Up until approximately the mid 1970's, prediction of the side shear resistance of rock sockets was based on crude extrapolation from empirical relations that were originally developed from drilled shafts in clay (Johnston 1992). The adhesion factor of drilled shafts in clay decreases as the clay shear strength increases. As a result, very low adhesion factors were commonly assigned to strong geomaterials such as rock.

M. Abu-Farsakh et al. (eds.), *Advances in Analysis and Design of Deep Foundations*,
Sustainable Civil Infrastructures, DOI 10.1007/978-3-319-61642-1_22

The mechanism of side resistance development in rock sockets is complex and includes both adhesion and frictional effects. Several empirical correlations were proposed for the determination of the average side shear resistance at the socket interface by different researchers (e.g. Rosenberg and Journeaux 1976; Horvath and Kenney 1979; Williams and Pells 1981; Horvath et al. 1983; Rowe and Armitage 1984; Horvath et al. 1989; Kulhawy and Phoon 1993; Kulhawy et al. 2005) whom reported the results of loading tests on full-scale shaft models. Most of these correlations relate the side shear resistance to the uniaxial compressive strength of the weaker of the shaft concrete or the rock core.

Although both empirical and analytical methods have been used over the years to predict the socketed shafts side resistance, the mechanisms are not yet entirely understood because of its complexity. Therefore, there is a need for developing alternative methods that might be capable of resolving the considerable uncertainties involved in predicting the side resistance of the socketed shafts. Recently, artificial intelligence techniques, such as fuzzy logic system have been successfully applied to many applications in geotechnical engineering.

Fuzzy logic, in a broad sense, refers to all of the theories that employ fuzzy sets, which are classes with unsharp boundaries. In simple words, a conventional black-and-white concept is generalized to a matter of degree. In such a way, two goals are accomplished; ease of describing human knowledge involving vague concepts and enhanced ability to establish a cost-effective solution to real-world problems. Traditionally, fuzzy logic has been viewed as a theory for handling uncertainty about complex systems and an approach for approximation theory (Ross 1995; Yen and Langari 1999; Kartalopoulos 2002).

Fuzzy logic systems have been successfully utilized in many geotechnical engineering applications. Juang et al. (1991) applied the concept to estimate the ultimate capacity of single piles driven in sand formations. The fuzzy logic concept was also adopted to predict the friction capacity of driven piles (Samieh 2003) and bored piles (Samieh 2005) in clay formations. Juang et al. (1992) reported a low-cost, qualitative evaluation scheme using the fuzzy logic concept for mapping slope failure potential. In addition, the fuzzy logic concept was extended to investigate the shear strength of soils (Chung 1995), soil identification using piezocone data (Pradhan 1998) and assessment of the relative density of sand formations from cone penetration test data (Juang et al. 1996), liquefaction potiential (Elton et al. 1995; Chen and Chen 1997) and geotechnical risk analysis (Rahman and El-Zahaby 1997).

Fuzzy logic and artificial neural networks (Haykin 1998; Mehrotra et al. 1997) are complementary technologies in the design of intelligent systems. The combination of these two technologies into an integrated system appears to be a promising path toward the development of intelligent hybrid systems. This integrated system will has the advantages of both neural networks (e.g. learning abilities, optimization abilities and connectionist structures) and fuzzy systems (e.g. humanlike IF-THEN rules thinking and ease of incorporating expert knowledge), Brown and Harris (1995). Adaptive Neuro-Fuzzy Inference System (ANFIS) is one of the most successful schemes that combine the benefits of neural networks and fuzzy logic systems into a single capsule (Jang 1993). The ANFIS is, from a topology point of view, an implementation of a representative fuzzy inference system using a back propagation neural network like structure. ANFIS has several features that enable it to achieve great success in a wide range of scientific applications. These features include: ease of implementation, fast and accurate learning, strong generalization abilities, excellent explanation facilities through fuzzy rules, and easy to incorporate both linguistic and numeric knowledge for problem solving (Jang and Sun 1995; Jang et al. 1997).

This paper examines the existing empirical equations to predict the side shear resistance of socketed shafts using a compiled shaft load tests database. The compiled database is, also, analyzed to investigate the possibility of establishing a mathematical expression for improving the prediction of the side shear resistance of socketed shafts. In addition, this study investigates the possibility of establishing a fuzzy logic system that may be capable of predicting the side shear resistance of socketed shafts using the compiled database. Furthermore, the study compares the prediction of the established fuzzy logic system with the prediction of the empirical equations to decide on the adequacy of the established artificial intelligence system and to help in answering the question of which system/empirical equation would be more reliable to predict the side shear resistance of socketed shafts.

2 Load Testing Database of Rock Socketed Shafts

An axial load tests database of rock socketed shafts was compiled in this study from published worldwide records that are primarily concerned with shaft resistance. The database includes a total of 95 load tests. The database includes socketed shafts into different types of rocks; including shale, mudstone, sandstone, siltstone, limestone and marl. The loading tests considered in the current study were those continued until the ultimate load was reached. In addition, all the shafts of the database records were constructed in one type of rock formation, along the shaft socket and underneath its base. The database consisted of 67 tests in compression where the load was applied to the top of the shaft, 9 tests in uplift and 19 Osterberg load cell tests. The friction capacity of the tested shaft was taken directly from the field measurements for instrumented and O-cell piles except a few tests without field measurements of the skin friction. For these tests, the approach of Carter and Kulhawy (1988) was adopted to estimate the shaft friction from the monitored Load-Settlement curve.

3 Assessment of the Rock Socketed Shafts Side Resistance Using Empirical Equations

Several empirical correlations exist for the assessment of the average side shear resistance of socketed shafts. Some of these equations are summarized in Table 1. These equations relate the side shear resistance of the socketed shafts to the rock compressive strength. Table 1 equations are plotted in Fig. 1 together with the records of the current study database. The figure exhibits the wide variability of the different equations and scatter of the testing records with respect to the different empirical equations. Figure 1 exhibits that while Horvath et al. (1983), using a b-value of 0.63, sought a conservative lower bound value of the socketed side resistance while the

Table 1. Summary of the most used empirical equations to predict the sockets side resistance

References	Empirical Correlation
Reynolds and Kaderbek (1980)	$f_s = 0.3q_u$
Gupton and Logan (1984)	$f_s = 0.2q_u$
Williams et al. (1980)	$f_s = 0.44(q_u)^{0.36}$
Horvath et al. (1983) *[b ranges from 0.63 to 0.94]*	$\left(\frac{f_s}{p_a}\right) = b\left(\frac{q_u}{p_a}\right)^{0.5}$
Row and Armitage (1984)	$\left(\frac{f_s}{p_a}\right) = 1.42\left(\frac{q_u}{p_a}\right)^{0.5}$
Kulhawy et al. (2005)	$\left(\frac{f_s}{p_a}\right) = \left(\frac{q_u}{p_a}\right)^{0.5}$

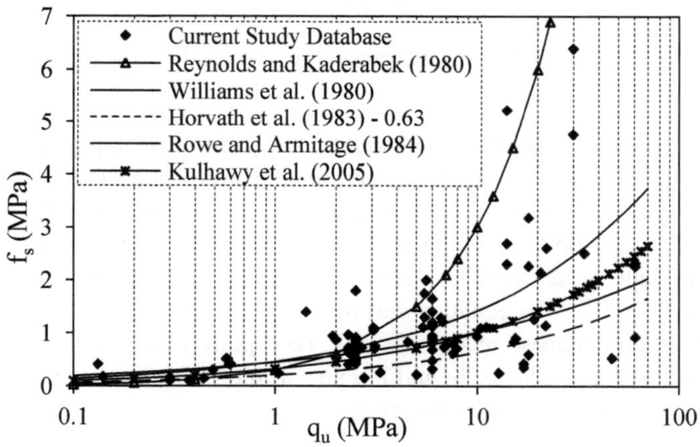

Fig. 1. Socketed shafts average side resistance versus rock compressive strength.

Reynolds and Kaderabek (1980) equation may be over-estimating the skin resistance of socketed shafts.

In a trail to improve the state of practice in regard to assessing the side resistance capacity of socketed shafts, the current study compiled database was analyzed to establish an empirical equation that may be adopted to assess the average side shear resistance of socketed shafts. Figure 2 shows the log-log plot of the database records of the socketed shafts side resistance, f_s, and the unconfined compressive strength of the rock, q_u, normalized to the atmospheric pressure, p_a. Based on the least square fit of the 95 database records, the following relationship between (f_s/p_a) and (q_u/p_a) can be obtained:

$$\left(\frac{f_s}{P_a}\right) = 1.25\sqrt{\left(\frac{q_u}{p_a}\right)} \tag{1}$$

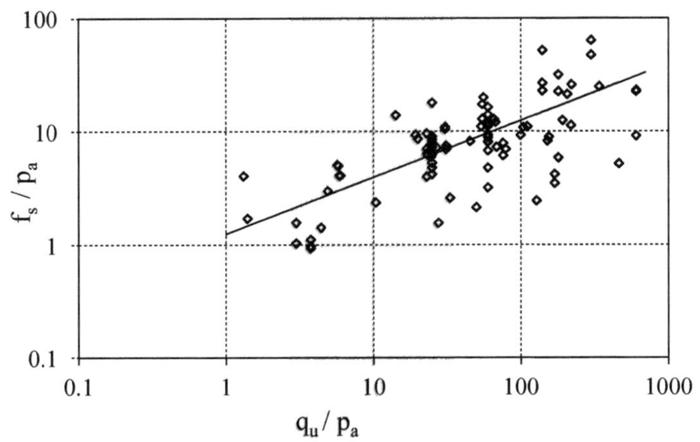

Fig. 2. Normalized average side shear resistance, f_s/p_a, versus the rock normalized unconfined strength, q_u/p_a.

The reasonableness of the proposed equation, Eq. (1), was supported by evaluating the correlation coefficient "R" which is a measure of a perfect linear correlation between two random variables. The proposed equation exhibited a correlation degree of 0.71.

Empirically lower bound value of rock socketed shaft is conventionally estimated using the Horvath et al. (1983) equation, Table 1, utilizing a b-value of 0.63; $(f_s/p_a) = 0.63\ (q_u/p_a)^{0.5}$. This equation was found to represent the lower bound for 90% of the current study database records, Fig. 3. To capture nearly all the records of the database, a recommended lower bound b-value would be about 0.3, Fig. 3. In such a case, the side resistance-unconfined strength correlation is:

$$\left(\frac{f_s}{p_a}\right) = 0.3\left(\frac{q_u}{p_a}\right)^{0.5} \tag{2}$$

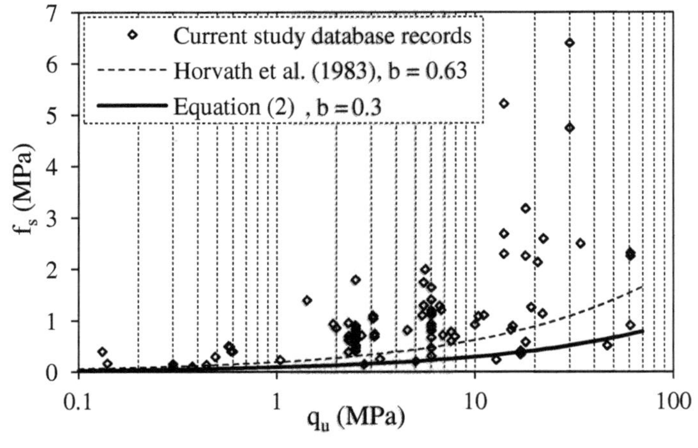

Fig. 3. Empirically lower bound variation of the average side resistance with the rock unconfined strength.

The empirical relationships of Table 1 and that of the current study, Eq. (1), are appealing for utilization by the geotechnical engineers because of their succinct form and dependence on a single conventionally measured parameter, the rock unconfined strength. However, these relationships may be misleading because they try to correlate a single and simple rock property, q_u, with the mechanical performance of a complicated shaft-rock interacting system. In an attempt to extend the f_s-q_u empirical correlations to include some of the characteristics of jointed rock masses, an additional factor has been introduced to the f_s-q_u equation by O'Neill and Reese (1999). The factor was based on the ratio between the rock mass modulus and the modulus of the intact rock material. O'Neill and Reese (1999) modified the Horvath and Kenney (1979) empirical equation by applying an empirical reduction factor α_E to account for the degree of fracturing as follows:

$$\left(\frac{f_s}{p_a}\right) = 0.65\,\alpha_E \left(\frac{q_u}{p_a}\right)^{0.5} \tag{3}$$

where the coefficient α_E is a function of the estimated ratio of the rock mass modulus to the intact rock modulus (E_m/E_r). This ratio was correlated to the rock quality designation, RQD, of the rock formation, as shown by O'Neill and Reese (1999). However, Turner (2006) addressed the issue that application of the α_E factor may be questionable because the RQD and rock mass modulus were not accounted for explicitly in the original correlation analysis by Horvath and Kenney (1979). Turner (2006) pointed out, also, that since the load testing database included sites with RQD less than 100 and modulus ratio values less than one, it would appear that these factors (RQD and E_m/E_r ratio) were indirectly affecting the load test results and therefore they are already incorporated into the resulting empirical equations.

In a trail to include the effect of rock discontinuities within the current practice frame of f_s-q_u relationship, the records of the compiled database of the current study were utilized to reach such an equation. The Matlab software (Math Works 2009) has been utilized using the surface fit toolbox, the available records with RQD-values, 17 records of current study database, together with the corresponding q_u and f_s to reach the following f_s-q_u-RQD relationship:

$$\left(\frac{f_s}{p_a}\right) = 1.8\left(\frac{q_u}{p_a}\right)^{0.35} (\mathrm{RQD})^{0.5} \tag{4}$$

where p_a is the atmospheric pressure and RQD is in percentage. The minimum RQD value used in establishing Eq. (4) was 11% and the maximum value was 100% and it should be used within these limits.

The database records with RQD-values were analyzed using Eqs. (1) and (4), of the current study, and compared with the measured values. Figure 4 exhibits the results of such analysis. The figure shows the equal sloping line, 45° line. The points below this line are within the under-prediction zone while the points above this line are within the zone of over-prediction. Considering a factor of 2 with respect to both sides of the equal slopping line, Fig. 4 shows that most of the data point predicted using Eq. (4) are

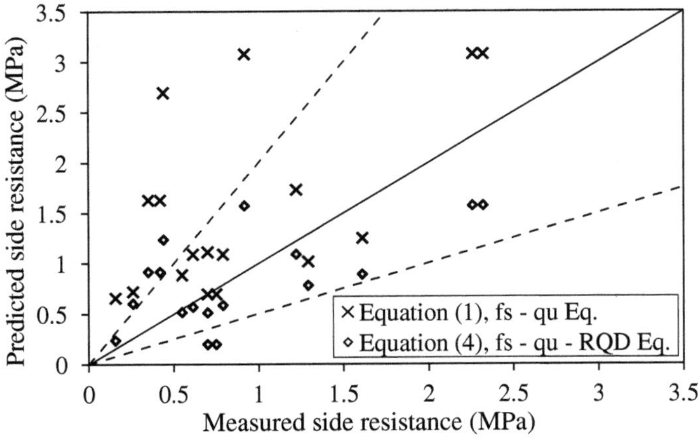

Fig. 4. Measured versus predicted side resistance using Eqs. (1) and (4).

included between these two lines. In other words, the prediction of Eq. (4) is much better than that of Eq. (1). Conventionally, on designing rock sockets, a factor of safety greater than 2.5 is adopted. Based on the trend shown in Fig. 4, it is recommended to adopt a factor of safety greater than 2.5 on using the established empirical equations of the current study to design rock sockets.

The predicted side shear resistance using Eqs. (1) and (4), which are shown in Fig. 4, were compared to exhibit the efficiency of the proposed empirical correlations. The predictive differences are evident in Fig. 5. As shown in this figure, the predicted average side resistance using Eq. (4), f_s-q_u-RQD equation, is on the order of 48% less than that estimated using Eq. (1).

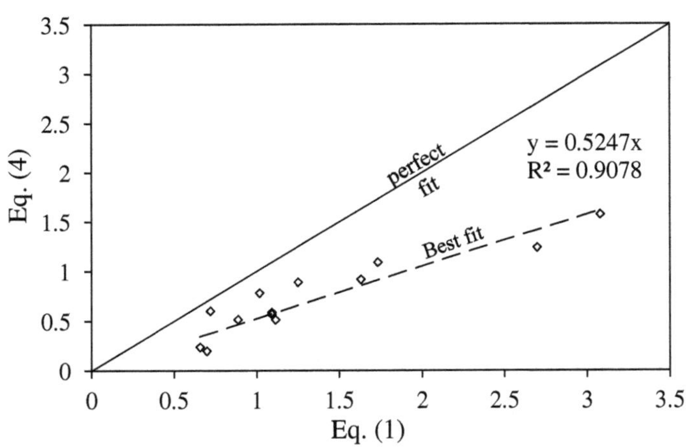

Fig. 5. Comparison of the predicted side resistance using Eqs. (1) and (4).

4 Assessment of the Socketed Shafts Side Resistance Using a Fuzzy Logic System

The parameters that are typically available for estimation of socketed shafts friction capacity at early design stages of projects are an initial assessment of the shaft socketed length (L_s), shaft diameter (D_s), and the average uniaxial compressive strength (q_u) of the rock along the socketed length. The influence of each of these parameters on the shaft capacity is difficult to measure/quantify by field testing or experimentally. However, these parameters are expected to have different impacts on the final skin frictional capacity of the socketed shaft. In other words, there is a hidden/undefined relationship among the properties of the rock formation, socketed shaft diameter, socketed shaft length and the friction capacity of the shaft-rock interface. This undefined and vague nature of the problem gives rise to the potential success of artificial intelligence systems to solve the current problem. This is referred to the flexibility of the artificial intelligence systems and its ability to account for various uncertainty and hidden nonlinear relationship between variables.

In this study, the shaft socketed length, shaft diameter and unconfined compressive strength of the rock formations are used to establish a fuzzy logic system. These parameters are conventionally assessed based on the conventional field investigation program at the project site, laboratory testing, and the available shaft drilling machines. In addition, these parameters can be easily determined and are expected to impose a direct impact on the shaft capacity. In brief, shaft diameter, socketed shaft length, and rock unconfined strength are the only parameters considered in this study for establishing the fuzzy logic system required to infer the frictional capacity of socketed shafts.

4.1 The Fuzzy Logic System: Modeling Scheme and Results

The fuzzy logic system developed in this study utilized the adaptive neuro-fuzzy inference technique, ANFIS. This technique provides a method for the fuzzy system to learn the characteristics of a given data set. Finding the characteristics of the data set leads the system to determine the parameters of the membership functions that allow the fuzzy inference system to capture the variations and unseen properties of the given input/output data. In brief, using a given input and output data set, a fuzzy inference system is constructed with membership function parameters adjusted using a back-propagation approach in combination with a least squares technique to track the characteristics of the problem that is being modeled. In such a way, the fuzzy system learns from the data that is being modeled (Math Works 2009).

The Matlab software (Math Works 2009) has been utilized in the current study to establish the fuzzy logic system. The database records were randomly divided into two parts. The first part, consisting of 73 records - around 75% of the database records, was used as a training set. The rest of the database records were used as a testing/verifying data to see how well the fuzzy system predicts the output of data not used in training the fuzzy system. Gaussian membership functions were utilized in the established fuzzy model for the input data. Weighted average defuzzification algorithm with linear output membership functions and Sugeno's inference mechanism were utilized in the current model to assess the single output value of the system, the frictional capacity of the socketed shaft.

A comparison between the measured average side shear resistance and the fuzzy system prediction for both the training and testing data are shown in Fig. 6. A good comparison is revealed between the measured and the predicted values of the training data, with a degree of correlation of 0.99. The predicted average side shear resistance of the testing data showed more scatter than the training data, with a correlation coefficient of 0.97. The figure shows, also, that most of the testing data are included within the lines of a factor of 2 with respect to the equal sloping line.

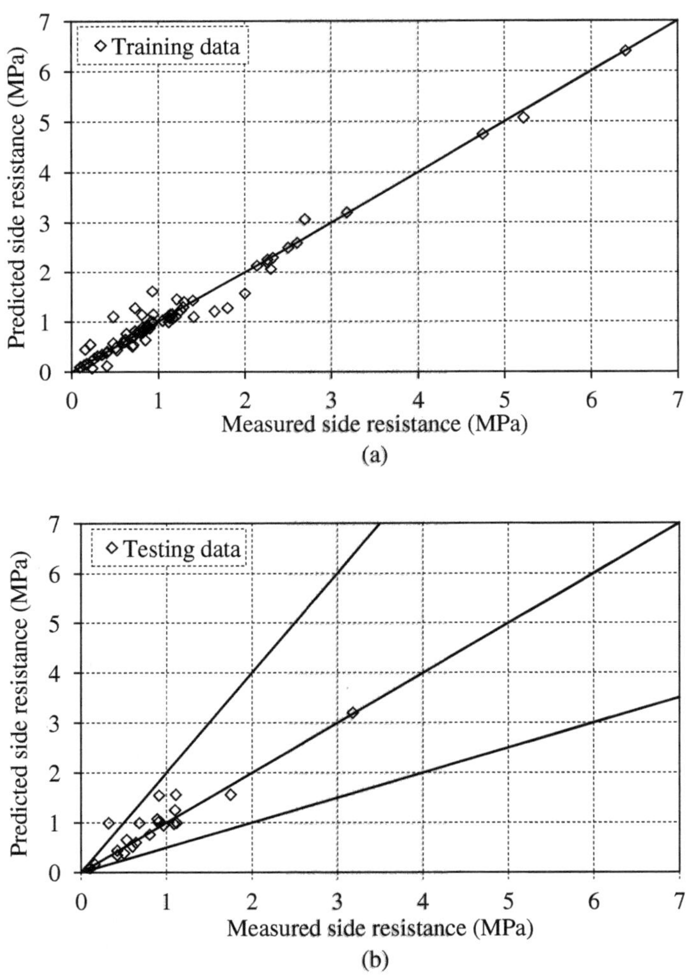

Fig. 6. Comparison of the measured and fuzzy predicted side shear resistance of the socketed shafts database. **a** Comparison for the training database records. **b** Comparison for the testing database records.

5 Emprical Equations Versus the Artificial Intelligence System

A second axial load testing database of rock socketed shafts was compiled, consisting of nine load tests that have neither used in establishing the empirical equations nor the training or testing phases of the fuzzy logic system of this study. This second database is used in this section to further verify the proposed empirical equation, Eq. (1): $(f_s/p_a) = 1.25\ (q_u/p_a)^{0.5}$, and the fuzzy logic system of this study. Figure 7 compares the measured and predicted sockets side resistance using the ANFIS system in addition to the empirical Eq. (1) considering the records of the verifying/second database. It can be inferred from the figure that the established fuzzy logic system exhibits remarkably less scatter than that of the empirical equation. The correlation degree and root mean-squared error (RMSE) of the fuzzy logic system and Eq. (1) are summarized in Table 2. As shown in Table 2, the prediction of the fuzzy logic system gave higher degree of correlation and lower root mean squared error when compared with that of the established empirical equation, Eq. (1), of this study. It is clear that the ANFIS prediction is more consistent than Eq. (1).

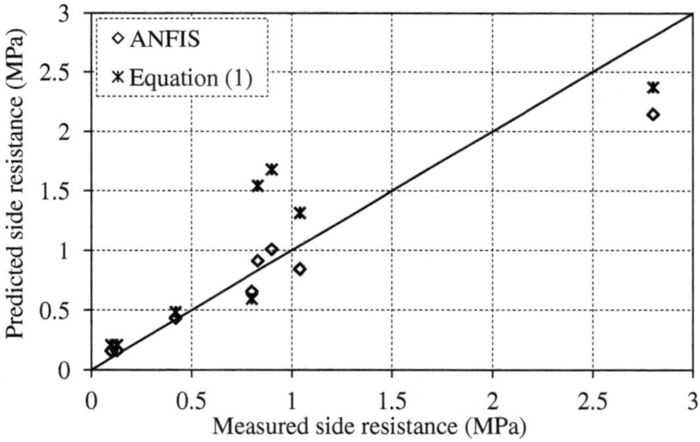

Fig. 7. Measured versus predicted sockets side resistance using the established fuzzy logic system and Eq. (1).

Table 2. Summary of degree of correlation and RMSE for the records of the verification database.

Type	Degree of correlation	RMSE
ANFIS	0.99	24.96
Equation (1)	0.94	60.03

To further analyze the current study model and equation, the instrumented pile load test described by Panozzo et al. (1993), was analyzed in this study using the empirical equations, Eqs. (1) and (4), and the developed fuzzy logic system of the current study.

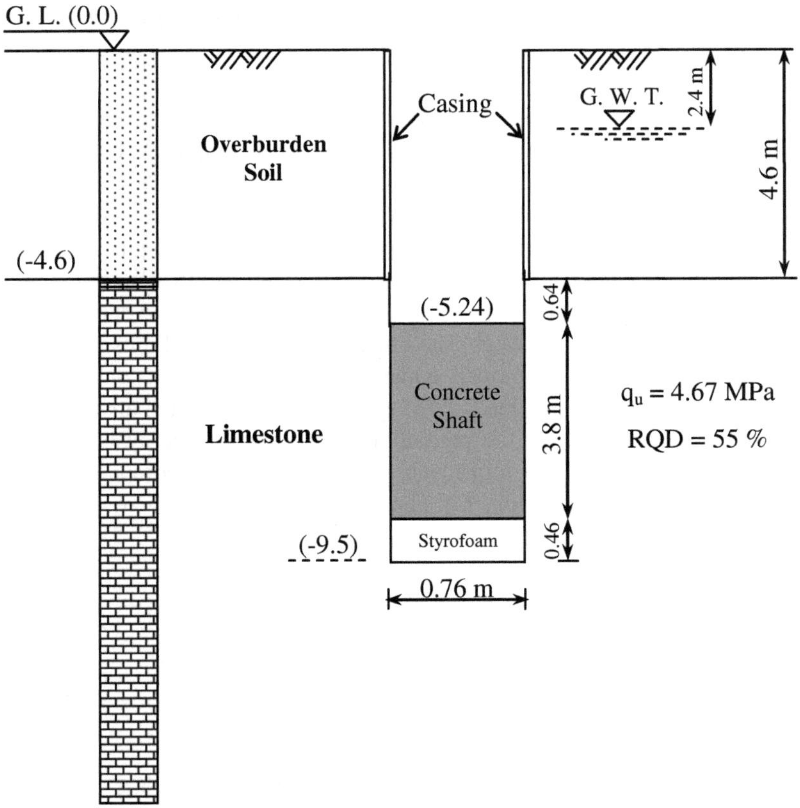

Fig. 8. Configuration of the instrumented pile load test socketed in a limestone formation (Panozzo et al. 1993).

This test results were not used before in establishing any of the empirical equations or fuzzy logic system of this study. This shaft was constructed at the South Parking Garage at the International Airport in Tampa, Florida, USA. The stratification at the shaft site consisted of approximately 4.6 m of overburden soil that is underlain by Limestone bedrock. The groundwater is at a depth of about 2.4 m below the ground surface, Fig. 8. The limestone formation had an average rock quality designation, RQD, of 55%, average unconfined strength of about 4.67 MPa and an average elastic modulus of 2.0 GPa. The shaft was 0.76 m in diameter and was socketed for a length of 3.8 m. The shaft was drilled to the required depth, 9.5 m, using a wet hole construction technique. A biopolymer drilling fluid additive and steel casing along the surface soil layer were used to stabilize the shaft excavations. The rock socket was excavated with auger bits, and a clean-out bucket was used to remove drill cuttings and sidewall slough. The shaft was constructed with a false bottom to eliminate shaft tip resistance, using 0.46 m of Styrofoam plug. Instrumented reinforcing steel cage was lowered into the shaft excavation and a high slump concrete was placed using tremie pipe. The drilled shaft was instrumented with vibrating wire strain gauges and telltales, the

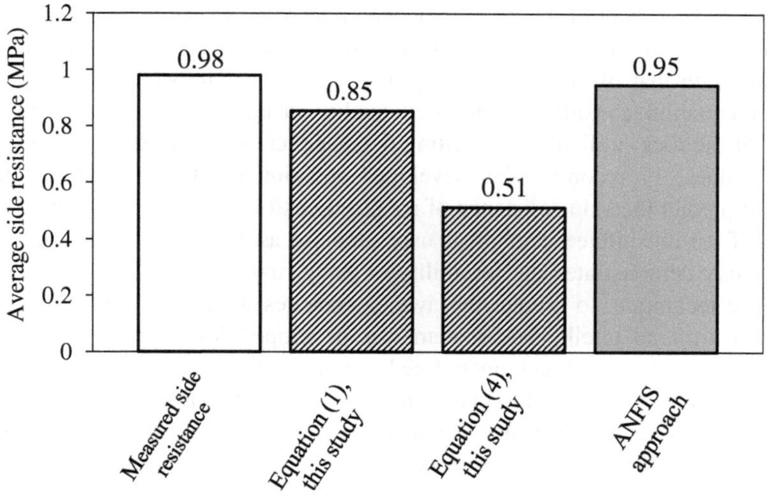

Fig. 9. Comparison of the assessed average side resistance using the empirical and artificial intelligence approach with the monitored side resistance of the socketed shaft of Fig. 8.

telltales displacements were measured with dial gauges. The shaft was loaded using a hydraulic jack with a calibrated load cell. Figure 9 compares the predicted shaft resistance using the empirical equations and fuzzy logic system of this study with the measured shaft average side resistance. The figure exhibits a good comparison between the measured side resistance and that predicted using the fuzzy logic and empirical equation, Eq. (1). However, the empirical equation Eq. (4) underestimated the side resistance by about 47% for the case under consideration.

Figure 9 together with the comparisons and discussions presented in this study delineate that it is recommended for the geotechnical designer to use different approaches to assess the sockets side resistance, specially the fuzzy logic approach of this study. A decision about the appropriate design value is reached through comparison of the different analysis values, as shown in Fig. 9, and experience with the project area. The geotechnical designer choice should be confirmed at pre-construction/construction time by in-situ shaft loading tests.

6 Conclusions

Drilled shafts are often socketed into rock to increase their capacity. Several empirical correlations exist for the determination of the average side resistance as function of the rock unconfined compressive strength. Generally, the data used to derive these empirical correlations are highly scattered and consequently the estimated average side resistance are widely different. This study presented, first, a review of the empirical equations currently/mostly in use by the geotechnical designers to assess the side resistance of socketed shafts. The study compiled a database of socketed shafts side resistance. This database was used in the current study to establish an empirical equation for assessment of the side resistance of socketed shafts as function of the rock unconfined strength. The prediction of the proposed equation is generally acceptable comparing with that of the existing empirical equations. The study proposed another empirical equation considering the rock quality designation value and unconfined strength of the rock, utilizing the existing database records with RQD and unconfined strength values, 17 records. This developed equation exhibited some scatter when utilized to predict the skin resistance of socketed shafts. However, the scatter was less than that if the unconfined strength is used alone to assess the skin friction.

The study demonstrated the feasibility of using fuzzy logic system, as an artificial intelligence technique, to predict the average side resistance of socketed shafts. The established artificial intelligence system was developed for a wide range of drilled shafts diameter, length and rock unconfined compressive strength. The prediction of the fuzzy logic system exhibited good comparison with the field measurements. It is suggested that the established fuzzy logic system is a decision support tool/assistant to help geotechnical designers make a good estimate of the average side resistance of the socketed shafts.

In the quest for continuous development, it should be delineated that any new method is better than/has an advantage over the other methods in use at the time of development. To point out the superiority of the established fuzzy logic system, the compiled database was analyzed by the empirical equations and the established artificial intelligence system and the results were compared all together with the measured frictional resistance utilizing a separate database records that were not used in establishing or testing the current study systems/equations. The results of the fuzzy logic system showed remarkably less scatter than the empirical equations results when compared with the measured capacity.

References

Brown, M., Harris, C.J.: Neurofuzzy Adoptive Modeling and Control, 1st edn. Prentice Hall, Hemel Hempstead (1995)

Carter, J.P., Kulhawy, F.H.: Analysis and design of drilled shaft foundations socketed into rock. Report EL-5918. Electric Power Research Institute, Palo Alto (1988)

Chen, J.W., Chen, C.Y.: A fuzzy methodology for evaluation of the liquefaction potential. Microcomput. Civil Eng. **12**, 193–204 (1997)

Chung, P.H.: Use of fuzzy-sets for evaluating shear strength of soil. Comput. Geotech. **17**, 425–446 (1995)

Elton, D.J., Juang, C.H., Sukmaran, B.: Liquefaction susceptibility evaluation using fuzzy sets. Soil Found. **35**(2), 40–60 (1995)

Gupton, C., Logan, T.: Design guidelines for drilled shafts in weak rocks of south florida. In: Proceedings of South Florida Annual ASCE Meeting, ASCE (1984)

Haykin, S.: Neural Networks. A Comprehensive Foundation, 2nd edn. Prentice Hall, Upper Saddle River (1998)

Horvath, R.G., Kenney, T.C.: Shaft resistance of rock socketed drilled piers. In: Proceedings of Symposium On Deep Foundations, ASCE, New York, pp. 182–214 (1979)

Horvath, R.G., Kenney, T.C., Kozicki, P.: Methods of improving the performance of drilled piers in weak rock. Can. Geotech. J. **20**(4), 758–772 (1983)

Horvath, R.G., Schebesch, D., Anderson, M.: Load-displacement behavior of socketed piers-hamilton general hospital. Can. Geotech. J. **26**(2), 260–268 (1989)

Jang, J.S.R.: ANFIS: adaptive-network-based fuzzy inference system. IEEE Trans. Syst. Man Cybern. **23**(3), 665–685 (1993)

Jang, J.S.R., Sun, C.T.: Neuro-fuzzy modeling and control. Proc. IEEE **83**, 378–406 (1995)

Jang, J.R., Sun, C., Mizutani, E.: Neuro-fuzzy and Soft Computing. Prentice hall, Upper Saddle River (1997)

Johnston, I.W.: New developments in the predictions of side resistance of piles in soft rock. Civil Engineering Department, The University of Monash, Clayton, Victoria, Australia (1992)

Juang, C.H., Wey, L.J., Elton, D.J.: Model for capacity of single piles in sand using fuzzy sets. J. Geotech. Eng. Div. ASCE **118**(GT3), 475–494 (1991)

Juang, C.H., Lee, D.H., Sheu, C.: Mapping slope failure potential using fuzzy sets. J. Geotech. Eng. Div. ASCE **118**(GT3), 475–494 (1992)

Juang, C.H., Huang, X.H., Holtz, R.D., Chen, J.W.: Determining relative density of sands from CPT using fuzzy sets. J. Geotech. Eng. Div. ASCE **122**(GT1), 1–6 (1996)

Kartalopoulos, S.V.: Understanding Neural Networks and Fuzzy Logic. Prentice Hall, New Delhi (2002)

Kulhawy, F.H., Phoon, K.K.: Drilled shaft side resistance in clay soil to rock. In: Proceedings of Conference on Design and Performance of Deep Foundation: Piles and Piers in Soil and Soft Rock, Geotechnical Specification, ASCE, New York, vol. 38, pp. 172–183 (1993)

Kulhawy, F.H., Prakoso, W.A., Akbas, S.O.: Evaluation of capacity of rock foundation sockets. In: Chen G. et al (eds.) Proceedings of forty US Symposium on Rock Mechanics. Anchorage, pp. 05–767 (2005)

Math Works Inc.: Matlab 9.0 User's Guide. MATLAB, MA, USA (2009)

Mehrotra, K., Mohan, C.K., Ranka, S.: Elements of Artificial Neural Networks. The MIT Press, Cambridge (1997)

O'Neill, M.W., Reese, L.C.: Drilled shafts: construction procedures and design methods. Department of Transportation, Fedral Highway Administration (FHWA), Office Implementation, McLean, Va. FHWA Pbl. No. FHWA-IF-99-025 (1999)

Panozzo, G.L., Bauhof, F.C., Kulhawy, F.H., O'Brien, A.J.: Testing of drilled shafts socketed into limestone. In: Proceedings of 3rd International Conference on Case Histories in Geotechnical Engineering, St. Louis, MO, paper no. 1.47 (1993)

Pradhan, T.B.S.: Soil identification using piezocone data by fuzzy method. Soils Found. **38**(1), 255–262 (1998)

Rahman, M.S., El-Zahaby, K.E.: Inclusion of fuzzy variables in geotechnical risk analysis. In: Yuan J.-X. (ed.) Computer Methods and Advances in Geomechanics: Proceedings of the Ninth International Conference of the Association for Computer Methods and Advances in Geomechnics: Wuhan, China, 2–7 November, pp. 567–572 (1997)

Reynolds, R.T., Kaderbek, T.J.: Miami limestone foundation design and construction. J. Geotechn. Eng. Div. ASCE **107**(7), 859–872 (1980)

Rosenberg, P., Journeaux, N.: Friction and end bearing test on bedrock for high capacity socket design. Can. Geotech. J. **13**(3), 324–333 (1976)

Ross, J.T.: Fuzzy Logic with Engineering Applications. McGraw Hill, New York (1995)

Row, R.K., Armitage, H.H.: The design of piles socketed into weak rock. Research Report. GEOT-11-84, Faculty of Engineering Science, the University of Western Ontario, London (1984)

Samieh, A.M.: Prediction of friction capacity of driven piles in clay: a fuzzy logic approach. In: Proceeding of the tenth International Colloquium on Structural and Geotechnical Engineering, Cairo, pp. 22–24 (2003)

Samieh, A.M.: Assessment of the friction capacity of bored piles in clay formations: artificial intelligence approaches. Eng. Res. J. **100**, C123–C136 (2005)

Turner, J.: Rock-socketed shafts for highway structure foundation. In: A Synthesis of Highway Practice, NCHRP Synthesis 360, Transportation Research Board of the National Academies, Washington (2006)

Williams, A.F., Johnston, I.W., Donald, I.B.: The design of socketed piles in weak rock. In: Proceedings of International Conference on Structural Foundations on Rock, Sydney, vol. 1, pp. 327–347 (1980)

Williams, A.F., Pells, P.J.N.: Side resistance rock sockets in sandstone, mudstone, and shale. Can. Geotech. J. **18**(4), 502–513 (1981)

Yen, J., Langari, R.: Fuzzy Logic: Intelligence, Control and Information. Prentice Hall, Upper Saddle River (1999)

Author Index

M. Abu-Farsakh et al. (eds.), *Advances in Analysis and Design of Deep Foundations*,
Sustainable Civil Infrastructures, DOI 10.1007/978-3-319-61642-1

MIX
Papier aus verantwortungsvollen Quellen
Paper from responsible sources
FSC® C105338

If you have any concerns about our products,
you can contact us on
ProductSafety@springernature.com

In case Publisher is established outside the EU,
the EU authorized representative is:
Springer Nature Customer Service Center GmbH
Europaplatz 3, 69115 Heidelberg, Germany

Printed by Libri Plureos GmbH
in Hamburg, Germany